DATE DUE

OCT 2 2 2009			
4/20/10			

Demco No 62-0549

Physics of Classical Electromagnetism

Minoru Fujimoto

Physics of Classical Electromagnetism

Springer

Minoru Fujimoto
Department of Physics
University of Guelph
Guelph, Ontario
Canada, N1G 2W1

Library of Congress Control Number: 2007921094

ISBN: 978-0-387-68015-6 e-ISBN: 978-0-387-68018-7

Printed on acid-free paper.

9 8 7 6 5 4 3 2 1

springer.com

Contents

Preface

The Maxwell theory of electromagnetism was well established in the latter nineteenth century, when H. R. Hertz demonstrated the electromagnetic wave. The theory laid the foundation for physical optics, from which the quantum concept emerged for microscopic physics. Einstein realized that the speed of electromagnetic propagation is a universal constant, and thereby recognized the Maxwell equations to compose a fundamental law in all inertial systems of reference. On the other hand, the pressing demand for efficient radar systems during WWII accelerated studies on guided waves, resulting in today's advanced telecommunication technology, in addition to a new radio- and microwave spectroscopy. The studies were further extended to optical frequencies, and laser electronics and sophisticated semi-conducting devices are now familiar in daily life. Owing to these advances, our knowledge of electromagnetic radiation has been significantly upgraded beyond plane waves in free space. Nevertheless, in the learning process the basic theory remains founded upon early empirical rules, and the traditional teaching should therefore be modernized according to priorities in the modern era.

In spite of the fact that there are many books available on this well-established theme, I was motivated to write this book, reviewing the laws in terms of contemporary knowledge in order to deal with modern applications. Here I followed two basic guidelines. First, I considered electronic charge and spin as empirical in the description of electromagnetism. This is unlike the view of early physicists, who considered these ideas hypothetical. Today we know they are factual, although still unexplained from first principle. Second, the concept of "fields" should be in the forefront of discussion, as introduced by Faraday. In these regards I benefited from Professor Pohl's textbook, *Elektrizitätslehre*, where I found a very stimulating approach. Owing a great deal to him, I was able to write my introductory chapters in a rather untraditional way, an approach I have found very useful in my classes. In addition, in this book I discussed microwave and laser electronics in some depth, areas where coherent radiation plays a significant role for modern telecommunication.

I wrote this book primarily for students at upper undergraduate levels, hoping it would serve as a useful reference as well. I emphasized the physics of electromagnetism, leaving mathematical details to writers of books on "mathematical physics." Thus, I did not include sections for "mathematical exercise," but I hope that readers will go through the mathematical details in the text to enhance their understanding of the physical content.

In Chapter 21 quantum transitions are discussed to an extent that aims to make it understandable intuitively, although here I deviated from classical theories. Although this topic is necessary for a reader to deal with optical transitions, my intent was to discuss the limits of Maxwell's classical theory that arise from phase coherency in electromagnetic radiation.

It is a great pleasure to thank my students and colleagues, who assisted me by taking part in numerous discussions and criticisms. I have benefited especially by comments from S. Jerzak of York University, who took time to read the first draft. I am also grateful to J. Nauheimer who helped me find literature in the German language. My appreciation goes also to Springer-Verlag for permission to use some figures from R. W. Pohl's book *Elektrizitätslehre*.

Finally, I thank my wife Haruko for her encouragement during my writing, without which this book could not have been completed.

M. Fujimoto
September 2006

1
Steady Electric Currents

1.1. Introduction

The macroscopic *electric charge* on a body is determined from the quantity of electricity carried by particles constituting the material. Although some electric phenomena were familiar before discoveries of these particles, such an origin of electricity came to our knowledge after numerous investigations of the structure of matter. Unlike the mass that represents mechanical properties, two kinds of electric charges different in sign were discovered in nature, signified by attractive and repulsive interactions between charged bodies. While electric charges can be combined as in algebraic addition, carrier particles tend to form neutral species in equilibrium states of matter, corresponding to *zero* of the charge in macroscopic scale.

Frictional electricity, for example, represents properties of rubbed bodies arising from a structural change on the surfaces, which is unrelated to their masses. Also, after Oersted's discovery it was known that the magnetic field is related to moving charges. It is well established that electricity and magnetism are not independent phenomena, although they were believed to be so in early physics. Today, such particles as electrons and atomic nuclei are known as *basic elements* composed of masses and charges of materials, as substantiated in modern chemistry. The electromagnetic nature of matter can therefore be attributed to these particles within accuracies of modern measurements. In this context we can express the law of electromagnetism more appropriately than following in the footsteps in early physics.

Today, we are familiar with various sources of electricity besides by friction. Batteries, for example, a modern version of Voltà's pile[1], are widely used in daily life as sources of *steady electric currents*. Also familiar is *alternating current* (AC) that can be produced when a mechanical work is converted to an *induction current* in a magnetic field. Supported by contemporary chemistry, in all processes where

[1] Voltà's pile was constructed in multi-layers of Zn and Cu plates sandwiched alternately with wet rags. Using such a battery, he was able to produce a relatively high emf for a weak current by today's standard.

electricity is generated, we note that positive and negative charges are separated from neutral matter at the expense of internal and external energies. Positive and negative charges $\pm Q$ are thus produced simultaneously in equal quantities, where the rate dQ/dt, called the *electric current*, is a significant quantity for electrical energies to be used for external work. Traditionally, the current source is characterized by an *electromotive force*, or "emf," to signify the driving force, which is now expressed by the *potential energy* of the source.

Historically, Voltà's invention of an electric pile (ca. 1799) played an important role in producing steady currents, which led Oersted and Ampère to discoveries of the fundamental relation between current and magnetic fields (1820). In such early experiments with primitive batteries, these pioneers found the law of the magnetic field by using small compass needles placed near the current. Today, the magnetic field can be measured with an "ammeter" that indicates an induced current. In addition to the *electron spin* discovered much later, it is now well established that charges in motion are responsible for magnetic fields, and all electrical quantities can be expressed in practical units of emf and current. As they are derived from precision measurements, these units are most suitable for formulating the laws of electromagnetism.

On the other hand, *energy* is a universal concept in physics, and the unit "joule," expressed by J in the MKS system[2], is conveniently related to practical electrical units. In contrast, traditional CGS units basically contradict the modern view of electricity, namely, that it is independent from mechanical properties of matter.

In classical physics, the electrical charge is a macroscopic quantity. Needless to say, it is essential that such basic quantities be measurable with practical instruments such as the aforementioned ammeter and the voltmeter; although the detailed construction of meters is not our primary concern in formulating the laws, these instruments allow us to set the standard for currents and emf's.

1.2. Standards for Electric Voltages and Current

Electric phenomena normally observed in the laboratory scale originate from the gross behavior of electrons and ions in metallic conductors and electrolytic solutions. It is important to realize that the *electronic charge* is the minimum quantity of electricity in nature under normal circumstances[3]. The charge on an electron has been measured to great accuracy: $e = -(1.6021892 \pm 0.000029) \times 10^{-19}$ C, where C is the practical MKS unit "coul."

---

[2] MKS stands for meter-kilogram-second, representing basic units of length, mass, and time, respectively. Electric units can be defined in any system in terms of these units for mechanical quantities, however, in the MKSA system, the unit *ampere* for an electric current is added as independent of mass and space-time. CGS units are centimeter, gram, and second, representing mass and space-time similar to MKS system.

[3] According to high-energy physics, charged particles bearing a fraction of the electronic charge, such as $e/2$ and $e/3$, called "quarks" have been identified. However, these particles are short-lived, and hence considered as insignificant for classical physics.

In a battery, charges $+Q$ and $-Q$ are produced by internal chemical reactions and accumulate at positive and negative electrodes separately. Placed between and in contact with these electrodes, certain *materials* exhibit a variety of conducting behaviors. Many materials can be classified into two categories: *conductors* and *insulators*, although some exhibit a character between these two categories. Metals, e.g., copper and silver, are typical conductors, whereas mica and various ceramics are good insulators. Microscopically, these categories can be characterized by the presence or absence of mobile electrons in materials, where mobile particles are considered to be moved by charges $\pm Q$ on the electrodes. As mentioned, traditionally, such a driving force for mobile charges was called an electromotive force and described as a *force* F proportional to Q, although differentiated from a voltage difference defined for a battery. Mobile electrons in metals are by no means "free," but moved by F, drifting against an internal frictional force F_d. For *drift motion at a steady rate*, the condition $F + F_d = 0$ should be met, giving rise to what is called a *steady current*.

For ionic conduction in electrolytic solutions, Faraday discovered the law of electrolysis (1833), presenting his view of the ionic current. Figure 1.1 shows a steady electrolysis in a dilute $AgNO_3$ solution, where a mass M_{Ag} of deposited silver on the negative electrode is proportional to the amount of charge q transported during a time t, that is,

$$M_{Ag} \propto q. \tag{1.1}$$

The mass M_{Ag} can be measured in precision in terms of *molar number* N of Ag^+, hence the transported charge q can be expressed by the number N that is proportional to the time t for the electrolysis.

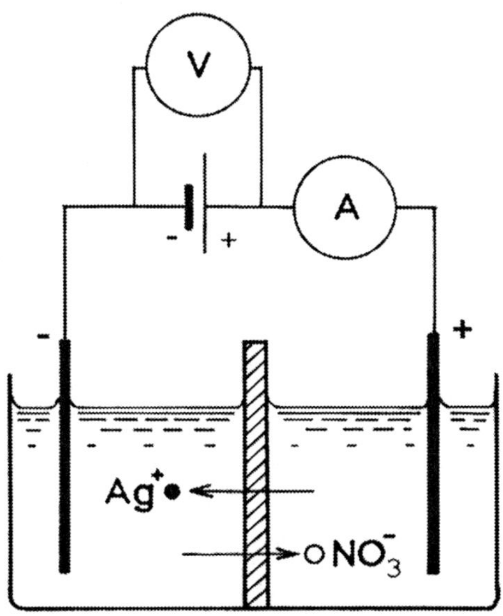

FIGURE 1.1. Electrolysis of $AgNO_3$ solution.

With accurately measured M_{Ag} and t, the current $I = q/t$ can be determined in great precision. Using an electrolytic cell, a current that deposits 1.1180 mg of silver per second is defined as one *ampere* (1A). Referring to the current of 1A, the unit for a charge is called 1 *coulomb*, i.e., 1C = 1A-sec.

To move a charge q through a connecting wire, the battery must supply an energy W that should be proportional to q, and therefore we define the quantity $V = W/q$, called the *potential* or *electric potential*. The MKS unit for W is J, so that the unit of V can be specified by J-C^{-1}, which is called a "volt" and abbreviated as V.

In the MKS unit system the ampere (A) for a current is a fundamental unit, whereas "volt" is a derived unit from ampere. Including A as an additional basic unit, the unit system is referred to as the "MKSA system." Nevertheless, a cadmium cell, for example, provides an excellent voltage standard: $V_{\mathrm{emf}} = 1.9186 \times 0.0010$ V with excellent stability 1μV/yr under ambient conditions.

A practical passage of a current is called a *circuit*, connecting a battery and another device with conducting wires. For a steady current that is time-independent, we consider that each point along a circuit can be uniquely specified by an electric potential, and a potential difference between two points is called *voltage*. The potential V is a function of a point x along the circuit, and the potential difference $V(+) - V(-)$ between terminals + and − of a battery is equal to the emf voltage, V_{emf}. If batteries are removed from a circuit, there is, naturally, no current; the potentials are equal at all points in the circuit—that is, $V(x) = \mathrm{const}$ in the absence of currents.

1.3. Ohm Law's and Heat Energy

Electric conduction takes place through conducting materials, constituting a major subject for discussion in solid state physics. In the classical description we consider only idealized conductors, either metallic or electrolytic. In the former electrons are charge carriers, whereas in the latter both positive and negative ions are mobile, contributing to the electrolytic current. These carriers can drift in two opposite directions; however, the current is defined for expressing the amount of charges $|Q|$ transported per unit time, which is basically a *scalar* quantity, as will be explained in the following discussion. In this context how to specify the current direction is a matter of convenience. Normally, the current is considered to flow in the direction for decreasing voltage, namely from a higher to a lower voltage.

Because it is invisible the current is "seen" by three major effects, of which magnetic and chemical effects have already been discussed. The third effect is *heat* produced by currents in a conducting passage, for which Ohm (1826) discovered the basic law of *electrical resistance*. Joule (ca. 1845) showed later that heat produced by a current is nothing but dissipated energy in a conductor.

Consider a long conducting wire of a uniform cross-sectional area S. Figure 1.2 illustrates a steady flow of electrons through a cylindrical passage, where we consider a short cylindrical volume $S\Delta x$ between $x - \Delta x$ and x along the wire.

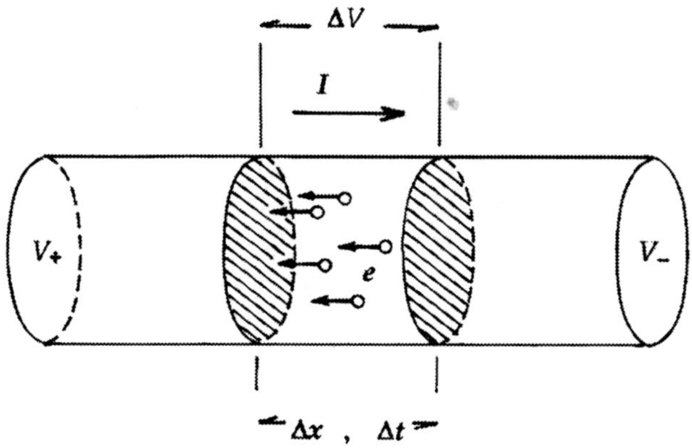

FIGURE 1.2. A simplified model for electronic conduction.

For a steady flow we can assume that the carrier density n is a constant of time between these points, at which the corresponding times are $t - \Delta t$ and t, respectively. If the potential is assumed to be higher to the left and lower to the right, the current should flow from the left to the right, as indicated in the figure. Assuming also that all electrons move in the direction parallel to the wire, the amount of charge $\Delta Q = neS\Delta x$ can be considered to move in and then out of the volume $S\Delta x$ during the time interval Δt, where $e < 0$ is the negative electronic charge. Therefore, the current is expressed as

$$I = \frac{\Delta Q}{\Delta t} = neS\frac{\Delta x}{\Delta t} = nev_{\mathrm{d}}S \tag{1.2}$$

where $v_{\mathrm{d}} = \Delta x/\Delta t$ is the drift velocity of electrons. We can therefore write

$$I = jS, \quad \text{where} \quad j = nev_{\mathrm{d}}, \tag{1.3}$$

called the *current density* in the area S. It is noted from (1.3) that the sign of j depends on the direction of v_{d}. For negative electrons, $e < 0$, the current is positive, $I > 0$, if $v_{\mathrm{d}} < 0$. For an ionic conduction of positive and negative carriers whose charges are $e_1 > 0$ and $e_2 < 0$, respectively, the equation (1.3) can be generalized as

$$j = n_1 e_1 v_1 + n_2 e_2 v_2, \tag{1.4}$$

where both terms on the right give positive contributions to j, despite the fact that ions 1 and 2 drift in opposite directions.

In the above simple model, the current density for a steady flow may not be significant, but for a distributed current it is an important measure, as will be discussed later for a general case. The MKSA unit of a current density is given by $[j] = \text{A-m}^{-2} = \text{C-m}^{-2}\text{-s}^{-1}$.

Next, we consider the potential difference between $x - \Delta x$ and x, which is responsible for the current through these points. For a small Δx, the potential difference ΔV can be calculated as

$$\Delta V = V(x - \Delta x) - V(x) = -F\Delta x,$$

where the quantity $F = -\Delta V/\Delta x$ represents the driving force for a *hypothetical unit charge* 1C. Nevertheless, there is inevitably a frictional force in the conducting material, as can be described by *Hooke's law* in mechanics, expressed by $F_d = -kv_d$, where k is an elastic constant.

For a steady current, we have $F = -F_d$, and therefore we can write

$$\Delta V = F_d \Delta x = kv_d \Delta x.$$

Eliminating v_d from this expression and (1.2), we obtain

$$\Delta V = \Delta RI, \quad \text{where} \quad \Delta R = \left(\frac{k}{neS}\right)\Delta x,$$

the *electrical resistance* between $x - \Delta x$ and x. For a uniform wire, this result can be integrated to obtain the total resistance from the relation, that is,

$$-\int_A^B dV = V(A) - V(B) = V_{emf} = \frac{Ik}{neS}\int_0^L dx = RI,$$

where

$$R = \frac{kL}{neS} = \rho\frac{L}{S} \tag{1.5}$$

is the resistance formula for a uniform conductor that can be calculated as the length L and cross-sectional area S are specified. The constant $\rho = \dfrac{k}{ne}$ is called the *resistivity*, and its reciprocal $1/\rho = \sigma$ is the *conductivity* of the material. Writing a potential difference and resistance between two arbitrary points as ΔV and R, Ohm's law can be expressed as

$$\Delta V = RI. \tag{1.6}$$

Obviously, the current occurs if there is a potential difference in a circuit. That is, if no current is present the conductor is static and characterized by a constant potential at all points. Values of ρ listed in Tables 1.1 and 1.2 are for representative industrial materials useful for calculating resistances.

The current loses its energy when flowing through a conducting material, as evidenced by the produced heat. Therefore, to maintain a steady current, the connected battery should keep producing charges at the expense of stored energy in the battery. Although obvious by what we know today, the energy relation for heat generation was first verified by Joule (ca. 1845), who demonstrated equivalence of heat and energy.

Figure 1.3 shows Joule's experimental setup. To flow a current I through the resistor R, a battery of V_{emf} performs work. The work to drive a charge ΔQ out of the battery is expressed by $\Delta W = V_{emf}\Delta Q$, and hence we can write $\Delta W = (RI)\Delta Q$, using Ohm's law.

TABLE 1.1. Electrical resistivities of metals

Metals	Temp, °C	$\rho \times 10^{-8}$ Ωm	$\alpha \times 10^{-3}$	Metals	Temp, °C	$\rho \times 10^{-8}$ Ωm	$\alpha \times 10^{-3}$
Ag	20	1.6	4.1	Mg	20	4.5	4.0
Al	20	2.75	4.2	Mo	20	5.6	4.4
As	20	35	3.9	Na	20	4.6	5.5
Au	20	2.4	4.0	Ni	20	7.24	6.7
Be	20	6.4			−78	3.9	
Bi	20	120	4.5	Os	20	9.5	4.2
Ca	20	4.6	3.3	Pb	20	21	4.2
Cd	20	7.4	4.2	Pd	20	10.8	3.7
Co	0	6.37	6.58	Pt	20	10.6	3.9
Cr	20	17			1000	43	
Cs	20	21	4.8		−78	6.7	
Cu	20	1.72	4.3	Rb	20	12.5	5.5
	100	2.28		Rd	20	5.1	4.4
	−78	1.03		Sb	0	38.7	5.4
	−183	0.30		Sr	0	30.3	3.5
Fe (pure)	20	9.8	6.6	Th	20	18	2.4
	−78	4.9		Tl	20	19	5
Hg	0	94.08	0.99	W	20	5.5	5.3
	20	95.8			1000	35	
In	0	8.2	5.1		3000	123	
Ir	20	6.5	3.9		−78	3.2	
K	20	6.9	5.1	Zn	20	11.4	4.5
Li	20	9.4	4.6	Zr	20	49	4.0

The temperature coefficient α of a resistivity is defined as $\alpha = (\rho_{100} - \rho_0)/100\rho_0$ between 0 and 100°C. From *Rika Nenpyo* (Maruzen, Tokyo 1962).

Because of the relation $\Delta Q = I\Delta t$, we obtain $\Delta W = (RI^2)\Delta t$, and hence the *power* delivered by the battery is expressed by

$$P = RI^2 = \frac{V_{\text{emf}}^2}{R} = V_{\text{emf}} I. \tag{1.7}$$

In Joule's experiment, the resistor R was immersed in a water bath that was thermally insulated from outside, where the heat produced in R was detected

TABLE 1.2. Electrical resistivities of alloys*

Alloys (°C)	$\rho \times 10^{-8}\Omega\mu$	$\alpha \times 10^{-3}$
Alumel (0)	33	1.2
Invar (0)	75	2
Chromel	70–110	0.11–0.54
Brass	5–7	1.4–2.0
Bronze	13–18	0.5
Nichrome (20)	109	0.10
German silver	17–41	0.04–0.38
Phos. bronze	2–6	

*From *Rika Nenpyo*, (Maruzen, Tokyo, 1962).

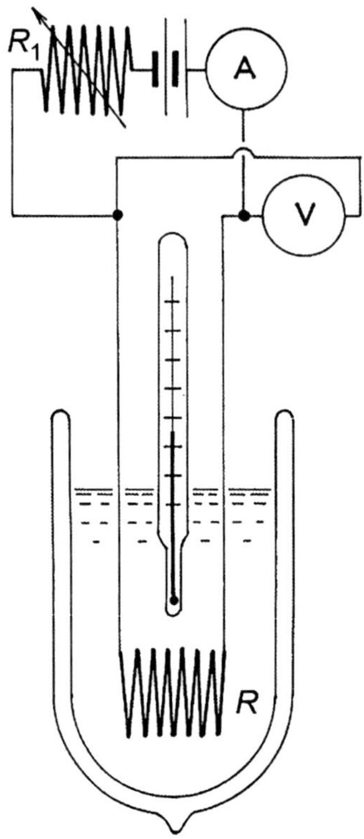

FIGURE 1.3. A set-up for Joule's experiment.

by the thermometer immersed in the water. Joule showed that the energy delivered by the battery, i.e., $\Delta W = P\Delta t$, was equal to the product $C\Delta\theta$, where C is the heat capacity of water bath, and $\Delta\theta$ the observed temperature rise. From (1.7), the unit of power—the watt (W)—is given by the electrical units of volt $\times$ ampere, making the MKSA system more practical than the traditional CGS system.

Owing to Joule's experiment, a traditional unit for heat—"calorie" (cal)—is no longer necessary, replaced by "joule"(J) although "calorie" and "kilocalorie" are still familiar units in biological and chemical sciences. The conversion between J and cal can be made by the relation 1 cal = 4.184 J. With the Ohm law, the MKSA unit of a resistance R is V-A^{-1}, which is called "ohm", and expressed as Ω

1.4. The Kirchhoff Theorem

In an electric circuit, a battery is a current source in conducting resistors, providing distributed potentials $V(x)$ along the circuit. In a simple circuit where a battery of V_{emf} is connected with one resistor R, the Ohm law $V_{\text{emf}} = RI$ can be interpreted to mean that the current I circulates in the direction from the positive terminal to the negative terminal along a *closed loop* circuit of V_{emf} and R. In this description the current direction inside the battery is opposite to the current direction in the resistance; however such a contradiction can be disregarded if we consider that the mechanism inside the battery is not our primary concern. Kirchhoff (1849) extended such an interpretation to more general circuits consisting of many emf's and resistors, and formulated the general theorem to deal with currents through these resistors.

In a steady-current condition, we consider that the potential $V(x)$ is signified by a *unique* value at any point x along the current passage, whereas at specific points x_{i} where batteries are installed, we consider discontinuous voltages $V_{\mathrm{emf}}(x_{\mathrm{i}})$. A steady current I_{j} can flow through each resistor R_{j} that are indexed $\mathrm{j} = 1, 2, \ldots,$ but the current direction and the potential difference across R_{j} can be left as primarily unknown. Also in a circuit are *junctions* in the current passage, indexed by $\mathrm{k} = 1, 2, \ldots,$ where more than two currents are mixed.

For a general circuit, we can consider several *closed loops*, $\alpha, \beta, \ldots,$ which can be either part of the circuit or can make up the whole circuit. It is significant that $V(x)$ takes a *unique* value at any given point x, so that the potential value should be the same, no matter how the current circulates along a closed loop in either direction. Mathematically, this can be expressed by $\oint_C \mathrm{d}V = 0$ along any of these loops C of a given circuit, for which we can write

$$\sum_{\mathrm{i}} V_{\mathrm{emf}}(x_{\mathrm{i}}) = \sum_{\mathrm{j}} R_{\mathrm{j}} I_{\mathrm{j}}. \tag{1.8}$$

Known as the *Kirchhoff theorem*, (1.8) can be applied to any closed loop in the circuit. Next, at any junction k of steady currents, we note that total incoming currents must be equal to total outgoing currents, namely, there is no accumulated charge at the point k; that is the law of *current continuity*. Assuming all directions of currents exist arbitrarily at the junction, we must have the relation

$$\sum_{\mathrm{k}} I_{\mathrm{k}} = 0 \tag{1.9}$$

at all junctions, which is known as *Kirchhoff's junction theorem.* As these currents are primarily unknown, we can usually assume their directions for solving equations (1.8) and (1.9); the assumed current directions are then corrected by signs of the solutions.

We solve (1.8) and (1.9) as a set of algebraic equations for unknown currents $I_{\mathrm{j}}, \mathrm{j} = 1, 2, \ldots.$ However, all of these equations are not required, and hence we can select equations to solve sufficiently for the given number j of R_{j}, as shown by the following examples.

Example 1. Two Resistances in Series Connection.

Figure 1.4(a) shows a circuit, where two resistors R_1 and R_2 are connected in series with a battery of V_{emf}. In a steady current condition, the points A, B, and C are signified by unique potentials V_{A}, V_{B}, and V_{C}, respectively.

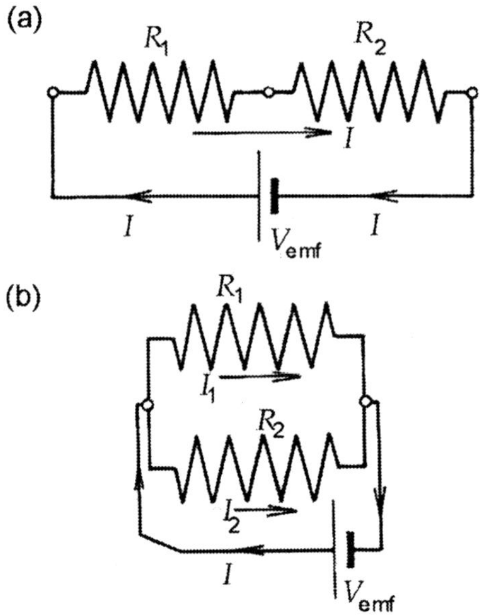

FIGURE 1.4. (a) Resistances R_1 and R_2 in series, and (b) in parallel.

Referring to the junction theorem (1.9), we see that for these currents in the figure we have $I_1 = I_2 = I$, representing the "total" current. Accordingly, using the Ohm law, we can write

$$V_A - V_B = R_1 I \quad \text{and} \quad V_B - V_C = R_2 I,$$

hence

$$V_A - V_C = (V_A - V_B) + (V_B - V_C) = (R_1 + R_2)I = V_{emf},$$

which expresses *Ohm's law for total resistance* $R = R_1 + R_2$.

Example 2. Two Resistances in Parallel Connection.

Figure 1.4b shows a parallel connection of two resistances R_1 and R_2. The junctions A and B are characterized by potentials V_A and V_B under a steady current condition. In this case, I_1 and I_2 are generally different, but related to the current I from the battery by the junction theorem (1.9). That is,

$$I = I_1 + I_2$$

at both A and B.

For loops α and β that include (V_{emf}, R_1) and (V_{emf}, R_2), respectively, we obtain from (1.8) that

$$V_A - V_B = V_{emf} = R_1 I_1 = R_2 I_2.$$

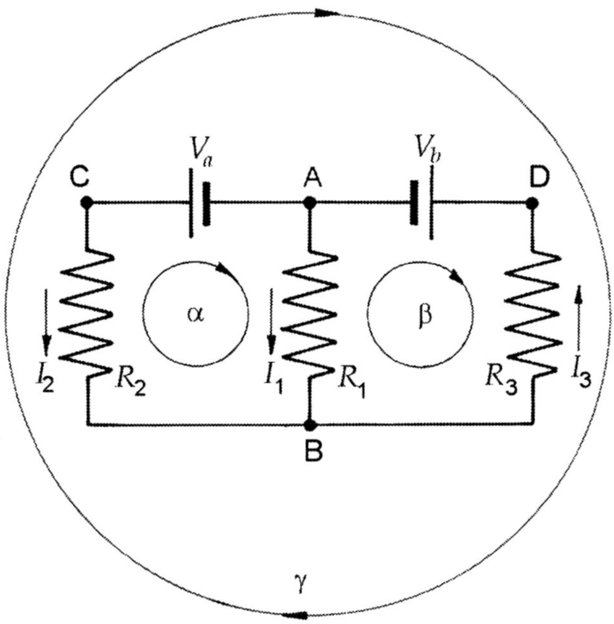

FIGURE 1.5. Kirchhoff's theorems. Junctions A and B, loops ABC, ADB, and CDB are named α, β, and γ, respectively.

Solving these equations for I_1 and I_2, we obtain

$$I = \left(\frac{1}{R_1} + \frac{1}{R_2}\right) V_{\text{emf}},$$

and the total resistance R in parallel connections can be determined from

$$\frac{1}{R} = \frac{1}{R_1} + \frac{1}{R_2}.$$

Example 3. Calculation of currents in a circuit shown in Figure 1.5. In this circuit, three loops α, β, and γ and two junctions A and B can be considered for Kirchhoff's theorem, as shown in the figure; however, we only need three equations to solve (1.8) and (1.9) for I_1, I_2, and I_3. Also, to set up three equations, the directions of currents can be assumed arbitrarily, as indicated in the figure.

Here, we apply (1.8) to the loops α, and β, and (1.9) to the junction A. Referring to assigned current directions, we can write three equations, respectively, for α, β, and A, which are

$$I_1 R_1 + (-I_2)R_2 + V_a = 0,$$
$$(-I_1)R_1 + (-I_3)R_3 + (-V_b) = 0,$$

and

$$I_1 + I_2 + (-I_3) = 0.$$

Solving these equations, we obtain

$$I_1 = -\frac{V_b R_2 - V_a R_3}{R_1 R_2 + R_2 R_3 + R_3 R_1},$$
$$I_2 = -\frac{(V_a - V_b) R_1 + V_b R_3}{R_1 R_2 + R_2 R_3 + R_3 R_1}$$

and

$$I_3 = -\frac{(V_a - V_b) R_1 + V_b R_2}{R_1 R_2 + R_2 R_3 + R_3 R_1}.$$

Part 1
Electrostatics

2
Electrostatic Fields

2.1. Static Charges and Their Interactions

In early physics, static electricity was studied as a subject independent from magnetism; it was after Oersted's experiment that the relation between an electric current and the magnetic field was recognized. Today, static and dynamic electricity are viewed as clearly exclusive phenomena; however, many early findings on static phenomena significantly contributed to establishing present day knowledge of electromagnetism.

First recognized as frictional electricity, static charges were also observed to be produced chemically from batteries, although static effects were only primitively measured in early experiments. Also recognized early were mobile charges in conducting materials, which were noticed as distributed on surfaces of a pair of metal plates separated by a narrow air gap, in particular, when the plates were connected with battery terminals. In early physics, such a device, called a *capacitor*, was used for studying the nature of static charges. On these plates, charges $+Q$ and $-Q$ appeared to condense from the battery, and hence such a device was often referred to as a *condenser*. Empirically, it was found that these opposite charges attract each other across the narrow gap, whereas charges of the same kind on each plate repel each other, so that such a "condensation" process ceases eventually at finite $\pm Q$. Such static interactions between charges constitute the empirical rule for electric quantities, although the origin is still not properly explained by the first principle. In this context, we simply accept the empirical rule as a fact in nature on which the present electromagnetic theory is founded.

It is notable that static charges reside at stationary sites on conducting surfaces in equilibrium with their surroundings. In the static condition, the absence of moving charges signifies the state of a conductor that is characterized by a constant electric potential, namely $V(x) = \text{const}$ at all points x.

The amount of charge Q that accumulates on condenser plates is proportional to the voltage V_{emf} of the battery, depending on the geometrical structure of the device. Therefore, we write the relation

$$Q = CV_{\text{emf}}, \tag{2.1}$$

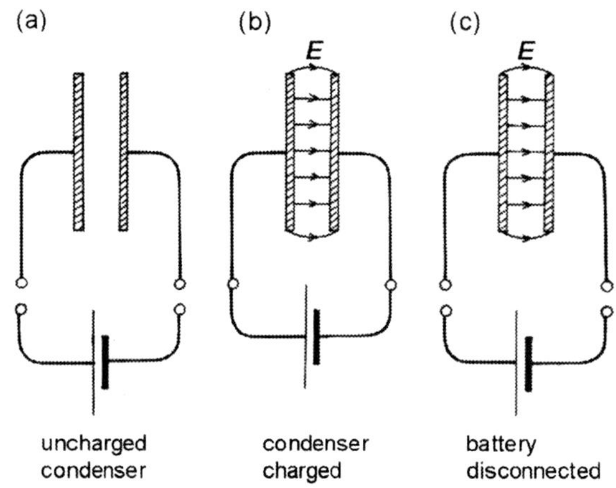

FIGURE 2.1. A capacitor C and a battery V_{emf} (a) before connection, (b) connected, (c) disconnected after charged.

where the constant C, referred to as *capacity*, is determined by the device geometry. In equation (2.1), Q represents the maximum charge that is transferred to the capacitor. The MKSA unit for the capacity is C-V^{-1}, which is called *farad* (F).

Figure 2.1 illustrates a capacitor of two parallel metal plates facing across a narrow gap, which is connected with a battery as shown. Such a capacitor is a standard device for electrostatics and discussed as a theoretical model as well. Figures 2.1(a) and (b) show a parallel-plate capacitor and a battery before and after connection, respectively. The empty capacitor (Figure 2.1(a)) can be charged by the battery (Figure 2.1(b)). Figure 2.1(c) shows the charged capacitor disconnected from the battery. Figures 2.1(b) and (c) show the space between the plates is signified by an *electric field* $\mathbf{E}$ that is related to charges on the plates. It is significant that the charges and field remain in the capacitor even after the battery is removed, as seen in the attractive force that exists between the plates across the gap.

Interactions among carrier particles are responsible for a charge distribution in the capacitor, resulting in a significant attraction between plates of opposite charge, whereas like charges are distributed at uniform densities on surface areas that face each other. Hence, the charged capacitor should be characterized by a potential energy due to an attractive force between the plates on which the battery performed work to transport $\pm Q$.

2.2. A Transient Current and Static Charges

When a capacitor is connected with a battery, charges are transported in the circuit until the circuit reaches a steady state. Accordingly, the charging rate $\mathrm{d}Q/\mathrm{d}t$, called a *transient current*, flows during the charging process. Such a current depends on the resistance R of wire connecting the capacitor with the battery, as shown in Figure 2.2.

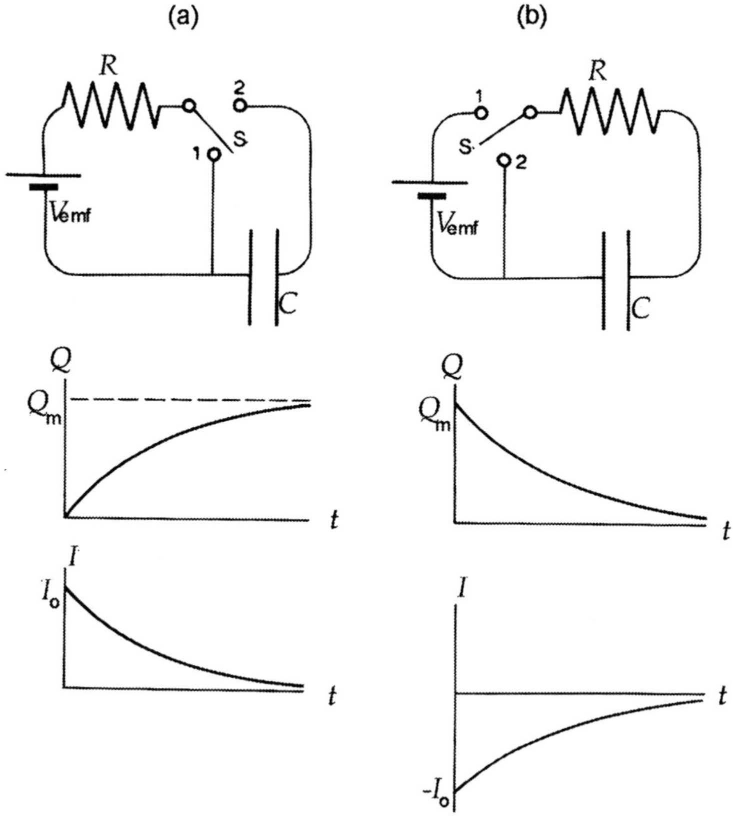

FIGURE 2.2. (a) Charging a capacitor C through a resistance R. (b) Discharge through R.

During the process, the charges $\pm Q$ on the plates change as a function of time t, and hence there should be a varying potential difference $\Delta V_C(t)$ between the two plates. Therefore, similar to the static relation (2.1), we can write

$$\Delta V_C(t) = \frac{Q(t)}{C}. \tag{2.2}$$

In Figure 2.2(a), if the switch S is turned onto the position 2 at $t = 0$, the capacitor C starts to be charged through the resister R, where the transient current is given by $I(t) = \mathrm{d}Q(t)/\mathrm{d}t$, and a potential difference $\Delta V_R(t) = RI(t)$ appears across R. Also, a potential difference ΔV_C, as given by equation (2.2), appears on the capacitor across the gap. These voltages should be related to V_{emf} of the battery by the conservation law; that is,

$$V_{emf} = \Delta V_R + \Delta V_C = RI(t) + \frac{Q(t)}{C}.$$

Multiplying both sides by $I(t)$, the power delivered by the battery can be calculated as

$$P(t) = V_{\mathrm{emf}} I(t) = RI(t)^2 + \frac{Q(t)}{C}\frac{\mathrm{d}Q(t)}{\mathrm{d}t}.$$

Hence, we can obtain the energy $W(t)$ spent by the battery between $t = 0$ and a later time t by integrating $P(t)$; i.e.,

$$W(t) = R\int_0^t I(t)^2\,\mathrm{d}t + \frac{1}{2C}Q(t)^2.$$

The first term on the right represents heat dissipated in R, and the second one is the energy stored in the capacitor during the time t.

To obtain the transient current, we solve the following differential equation, assuming that V_{emf} is a constant of time. Namely,

$$R\frac{\mathrm{d}Q(t)}{\mathrm{d}t} + \frac{Q(t)}{C} = \text{const}, \tag{2.2a}$$

or

$$R\frac{\mathrm{d}I(t)}{\mathrm{d}t} + \frac{I(t)}{C} = 0 \tag{2.2b}$$

with a given initial condition. In the circuit shown in Figure 2.2(a), when the switch S is turned to position 1, a steady current I_o determined by $V_{\mathrm{emf}} = RI_o$ is flowing initially through R. In this case, when S is turned on to 2 at $t = 0$, charging begins to take place in the capacitor C. On the other hand, equation 2.2(b) shows a case where a charged capacitor C starts discharging at $t = 0$, after the switch S is turned to the position 2.

In the latter case, equation (2.2b) can be integrated with the initial condition $I = I_o$ at $t = 0$, resulting in the transient current expressed by

$$I(t) = I_o \exp(-\frac{t}{RC}), \tag{2.3a}$$

showing an exponential decay. In this case, the charge $Q(t)$ on the capacitor plate can be expressed by

$$Q(t) = \int_0^t I(t)\,\mathrm{d}t = I_0 RC \exp\left(-\frac{t}{RC}\right)\Big|_0^t = Q_{\max}\left[1 - \exp\left(-\frac{t}{RC}\right)\right], \tag{2.3b}$$

where $Q_{\max} = I_o RC$ is the maximum amount of charge reached as $t \to \infty$. We can also solve equation (2.2b) with the initial condition $Q = Q_{\max}$ at $t = 0$ to obtain similar expressions to discharging the capacitor C. In Figure 2.3, the transient charge and current given by equations (2.3a) and (2.3b) and similar transients for discharging are shown below circuit diagrams. For these transient currents, it is noted that the parameter defined by $\tau = 1/RC$ represents a characteristic time of the transient current, and hence is called the *time constant*. For $R = 100\,\mathrm{k}\Omega$ and $C = 0.1\,\mu\mathrm{F}$, for example, the time constant is $\tau = 10^{-2}$s, giving the "timescale" of the transient phenomena in the RC circuit. Figure 2.3 illustrates a modernized

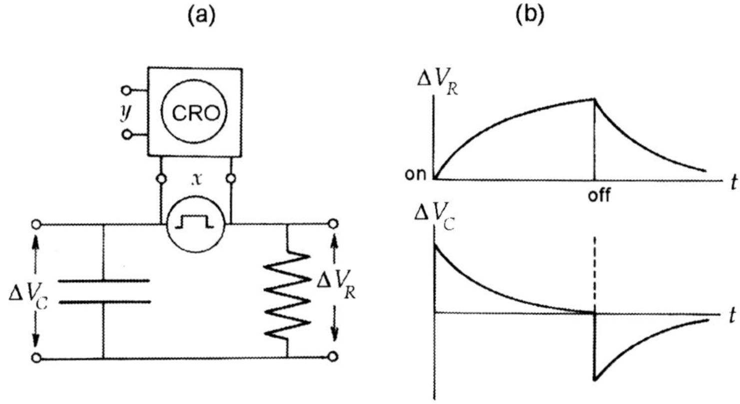

FIGURE 2.3. Transient voltages ΔV_R and ΔV_C observed by a cathode-ray oscilloscope (CRO).

circuit using a cathode-ray oscilloscope (CRO) for demonstrating such transients on the screen.

In the above argument, when a capacitor is charged to $\pm Q$, a potential difference ΔV appears across the plates, between which an energy $Q^2/2C = \frac{1}{2}C(\Delta V)^2$ should be stored. A capacity signified by the variables Q and ΔV indicates that there is an electric field in the space between the plates.

2.3. Uniform Electric Field in a Parallel-Plate Condenser

In a charged condenser, the plates attract each other across the gap, originating from electrostatic forces between $+Q$ and $-Q$, as implied by the Coulomb law (1785) for point charges. It was postulated in early physics that such a force propagates through the empty space, called the *field*; this idea was, however, not easily accepted unless it was duly verified. Nevertheless, Faraday (ca. 1850) described such a field with line drawings, which led Maxwell (1873) to establish his mathematical theory of electromagnetic fields. Today, supported by much evidence, the electromagnetic field is a well-established concept, one that represents a real physical object. Therefore, we can define a static electric field as implemented by Faraday and Maxwell.

2.3.1. The Electric Field Vector

In a charged parallel-plate capacitor signified by the potential difference ΔV across the empty gap, we can consider a potential function $V(z)$ for the field in the empty space, where z is a position on a straight line perpendicular to the plates. The conducting plate surfaces of a large area are signified by constant voltages V_A and V_B, so that the potential in the uniform field can be expressed as

$$V(r, z) = -Ez + \text{const}, \tag{2.4a}$$

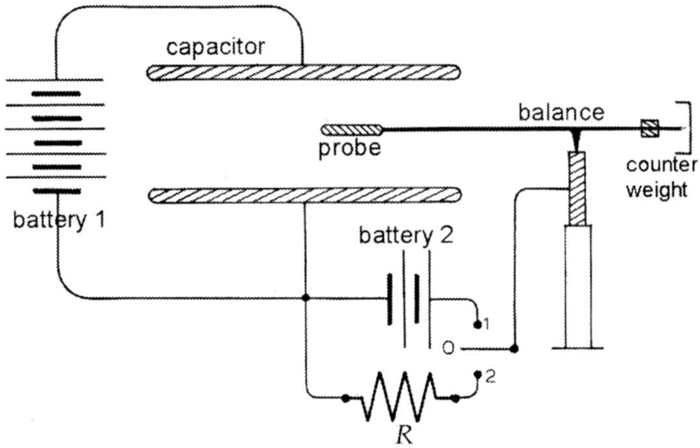

FIGURE 2.4. Measuring a static force. A small metal probe placed in the charged capacitor is charged by the battery 2, resulting in a force that is detected by the balance.

where r is the coordinate in a direction perpendicular to z. Applying this equation to arbitrary points $(r_A, 0)$ and (r_B, d) on the plates A and B, we have $V_A = \text{const}$ and $V_B = Ed +$ const on these plates separated by the gap d, and hence

$$E = \frac{V_A - V_B}{d} \quad \text{or} \quad \Delta V = Ed, \tag{2.4b}$$

a relationship that permits us to define a vector $\boldsymbol{E} = (0, 0, E)$ from $E = -\mathrm{d}V/\mathrm{d}z$. Both V_A and V_B are independent of positions A and B, respectively, on these plates, and hence the magnitude E should be constant as determined by equation (2.4b), and such a vector $\boldsymbol{E}$ can be regarded as representing a *uniform field* that is parallel to the z direction. It is noted that $|\boldsymbol{E}|$ has a unit V-m^{-1} = N-C^{-1}, where N is the MKSA unit of a mechanical force. Therefore, writing $E = 1 \times E$ in V-m^{-1}, E as such can be interpreted as a force on a hypothetical charge of 1C.

In fact, when a small charge q_o is placed in a charged condenser, q_o is moved by a force $F = q_o E$, as verified by such a primitive experiment as shown in Figure 2.4, in which the static field can be defined as given by the limit of F/q_o as $q_o \to 0$.

In this case, unless q_o is sufficiently small in practical scale, E defined from the relation $F = q_o E$ may not be fully acceptable because E is no longer uniform as modified by q_o. However, in modern experiments, we know, for example, electrons injected into the static field E are deflected from their path, providing solid evidence for the static force on a small charge q_o to be expressed as

$$F = q_o E. \tag{2.5}$$

Therefore, the electric field defined by $\Delta V/d$ can be justified by equation (2.5) on a test charge of q_o =1C, assuming it (the field) is sufficiently small.

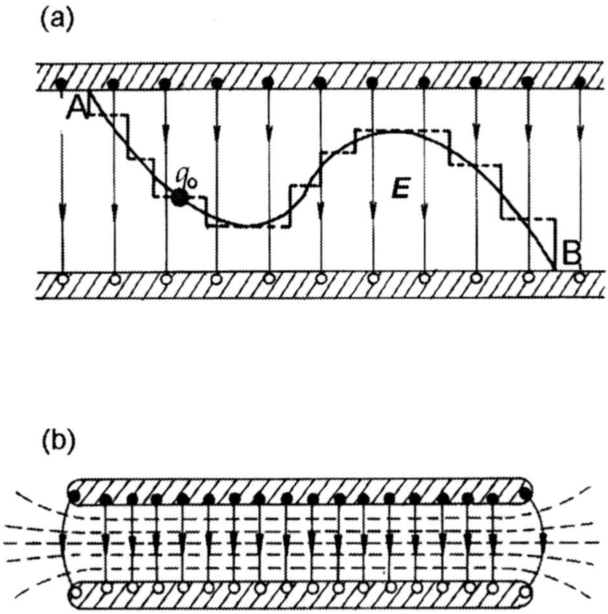

FIGURE 2.5. A uniform electric field. (a) Quasi-static work for moving a test charge q_o in a uniform electric field. (b) Electric field-lines and equipotential surfaces are indicated by solid and broken lines, respectively.

The static electric field $\boldsymbol{E}$ is *conservative*, as can be verified by calculating the work performed by such a force $\boldsymbol{F} = 1\text{C} \times \boldsymbol{E}$ to move 1C in a uniform field $\boldsymbol{E}$ along an arbitrary path Γ, shown in Figure 2.5(a).

Replacing the path Γ by zigzag steps of small displacements $\mathrm{d}\boldsymbol{s}_\perp$ and $\mathrm{d}\boldsymbol{s}_{||}$, the work can be expressed approximately by

$$\int_\Gamma \boldsymbol{E}.\mathrm{d}\boldsymbol{s} = \sum_{\text{horizontal}} \boldsymbol{E}.\mathrm{d}\boldsymbol{s}_\perp + \sum_{\text{vertical}} \boldsymbol{E}.\mathrm{d}\boldsymbol{s}_{||},$$

where $\boldsymbol{E}.\mathrm{d}\boldsymbol{s}_\perp = 0$. Therefore,

$$\int_\Gamma \boldsymbol{E}.\mathrm{d}\boldsymbol{s} = \int_A^B E\mathrm{d}s_{||} = V(\mathrm{B}) - V(\mathrm{A}),$$

and $Ed = -\Delta V$, giving a conservative feature of the field vector $\boldsymbol{E}$.

2.3.2. The Flux Density Vector

We defined the field vector $\boldsymbol{E}$ in Subsection 2.3.1. However, this is not sufficient for describing the uniform field in a parallel-plate capacitor. As illustrated by Faraday, the field is represented by a number of parallel lines, showing a pattern

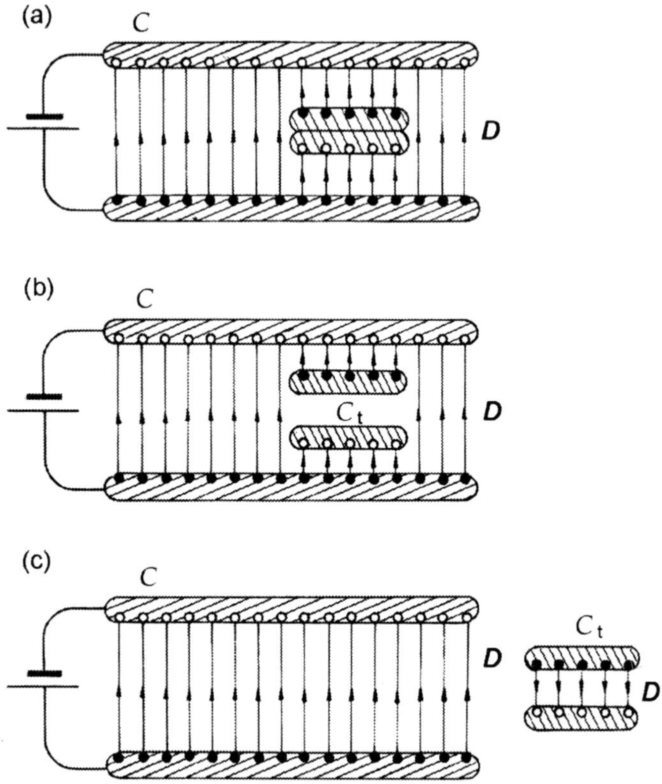

FIGURE 2.6. Flux of field-lines $\boldsymbol{D}$ in a uniform electric field. (a) A test capacitor C_t is inserted in the field with gap closed. (b) The gap is open in the field. (c) C_t is moved to outside of C with the gap open.

of uniformly distributed intensities of the field. We realize that such field-lines are significantly related with distributed charges on conducting surfaces, although no static field can exist inside conducting plates. Faraday sketched such field-lines emerging from positive charges and ending at negative charges.

To express such distributed field-lines mathematically, we can utilize an induction effect on a conductor by an applied electric field, known as *electrostatic induction* or *polarization*. Due essentially to mobile charges, a material placed in a condenser can be *polarized* by an applied electric field, exhibiting positive and negative charges on surfaces close to the negative and positive capacitor plates, respectively. Such a polarization can also occur in an insulator, for which movable molecules are responsible, as discussed in Section 2.5.

As shown in Figure 2.6, we indicate locations of displaced charges schematically by small filled and open circles for positive and negative charges, respectively. In a parallel-plate condenser, these charges attract each other across the empty space, hence residing at the closest distances. Accordingly, charges appear only on inner surfaces of the capacitor plates, whereas no charge shows up on outer surfaces. Also significant is that these charges are *uniformly* distributed on the parallel surfaces, due to repulsive interactions among like charges. Consequently, the field in the condenser can be illustrated by parallel lines that are drawn evenly spaced to indicate the uniformity.

The uniform field in the condenser space can also be illustrated by using potential values given by equation (2.4a), shown in Figure 2.5(b) by broken lines, which represent a group of parallel planes at constant potentials in three dimensions, called *equipotential* surfaces. In fact, the field is almost uniform in the gap of the parallel-plate capacitor, except near the edge where lines are diverted as shown in the figure. Except for such diverted portions of the field, the capacitor can be idealized by uniformly distributed charge densities $\pm\sigma = \pm Q/A$, where A is the effective area on the plates. In this context, it is convenient to set a rule for drawing Q lines theoretically in such a way that each line is terminated at ± 1C at the ends. Accordingly, the uniform field can be expressed by a constant density of field-lines that can be written as $D = Q/A$. Here, it is noted that $\pm Q$ are charges on the plates, whereas the field-line density $D$is a field quantity. There is no charge in the empty space, but by writing $D = \sigma$ we emphasize the fact that the uniform line density in the field is determined by the constant charge density on the plates. With such *flux density* D, we can combine the direction from the positive to negative plates to define the quantity D as the vector $\boldsymbol{D}$.

To evaluate the flux density vector $\boldsymbol{D}$ in a uniform field, we perform a "thought experiment" as illustrated in Figure 2.6.

Figure 2.6(a) shows a small adjustable *test capacitor* C_t is inserted, with the gap closed, into the uniform electric field of C. In this case, C_t is equivalent to a single conductor, and hence polarized by the charged capacitor C, as illustrated. Here, the distributed field-lines in C are not modified from the way before inserting C_t, whereas $E = 0$ inside C_t. It is noted that the induced charge density on C_t is equal to σ i.e., it is the same density on C. Figure 2.6(b) shows that C_t is now open to make a gap, while the plates are kept in parallel with C. In this case, nothing significant occurs, and no field appears between the separated plates of C_t. Next, we take C_t out of C, as shown in Figure 2.6(c), where a field appears inside C_t because polarized charges move from outside to inside surfaces. Thus, we have $D = \sigma$ inside C_t on moving it to outside of C, and the field pattern in C returns to the original one without C_t. In this thought experiment, we obtain $D = \sigma$ in C_t, regardless of where it was initially placed in the uniform field of C. The magnitude $|\boldsymbol{D}|$ at any point in the field is thus confirmed as σ on the plates. Accordingly,

$$D = \sigma \tag{2.6}$$

expresses the flux density at all points in the uniform electric field. The MKSA unit of D is given by the same unit of surface charge density σ, namely C-m^{-2}.

Considering a static force and flux density, we have defined two vector quantities $\boldsymbol{E}$ and $\boldsymbol{D}$ to represent a uniform electric field. Nonetheless, these two vectors should represent the same electric field, and therefore we write the relation

$$D = \varepsilon_o E, \tag{2.7}$$

where the proportionality constant ε_o is a constant of the empty space, and whose value can be left for experimental evaluation. For a uniform field in a parallel-plate capacitor, we have the relations $D = Q/A$ and $E = \Delta V/d$, as expressed from equations (2.6) and (2.4b), respectively, and hence $D = \varepsilon_o \Delta V/d$. Accordingly, by definition, the capacity can be expressed as

$$C_o = \varepsilon_o \frac{A}{d} \tag{2.8}$$

for an empty parallel-plate capacitor. Using measured values of C, A, and d, the constant ε_o can be calculated as

$$\varepsilon_o = 8.654 \times 10^{-12}\ \mathrm{F/m}.$$

Although the physical significance is not immediately clear, the empirical value of the constant ε_o must be used for numerical calculations in static problems, using MKSA units. In the CGS system, ε_o can be set equal to the dimensionless constant 1, for which $\boldsymbol{E}$ and $\boldsymbol{D}$ are essentially the same quantity in a vacuum space. Using CGS units we see that all electromagnetic phenomena are be part of the hierarchy of mechanical space-time, although theoretical discussions are simplified by setting $\varepsilon_o = 1$.

In Section 2.2, it was shown that a charged capacitor can store energy inside. When charging up to certain amounts $\pm Q$, the capacitor plates are characterized by potentials V_A and V_B. Therefore, to transfer additional charges $\pm \mathrm{d}Q$ to the plates, the battery should perform additional work $\mathrm{d}W = V_A \mathrm{d}Q + V_B(-\mathrm{d}Q) = \Delta V_C \mathrm{d}Q$. Since $\Delta V_C = Q/C$, the work W for increasing charges from 0 to $\pm Q$ can be calculated as

$$W = \frac{1}{C}\int_0^Q Q\,\mathrm{d}Q = \frac{Q^2}{2C} = \frac{1}{2}C(\Delta V)^2.$$

Using (2.8) and (2.4b), W can be re-expressed by the field quantities as

$$W = \frac{1}{2}\varepsilon_o E^2(Ad) = \frac{1}{2}DE(Ad), \tag{2.9}$$

where Ad is the volume of the space between the plates. Hence, the quantity

$$u_E = \frac{1}{2}\varepsilon_o E^2 = \frac{1}{2}DE = \frac{1}{2}\frac{D^2}{\varepsilon_o} \tag{2.10}$$

expresses the energy density in the uniform field. Although derived for a parallel-plate capacitor, (2.9) and (2.10) can be generalized to a non-uniform static field, where the energy density u_E is a function of the point $\boldsymbol{r}$ in the field. In general, the stored energy W can be calculated by $\int_v u_E \mathrm{d}v$ integrated over the field volume v.

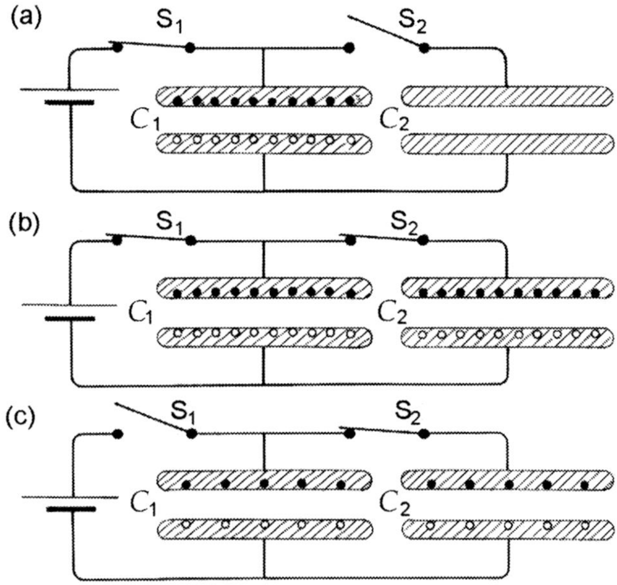

FIGURE 2.7. Distributing charges. (a) C_1 is charged, C_2 is empty, the switch S_1 closed and S_2 open. (b) S_1, S_2 are closed, and C_1, C_2 are both charged. (c) In the case of (a), S_1 is open and then S_2 is closed. The initial charge on C_1 flows partly into C_2 and is distributed between C_1 and C_2.

2.4. Parallel and Series Connections of Capacitors

Combining capacitors is an elementary problem where the charging process is thermodynamically *irreversible* because of an energy loss during the charge distribution process. In this section, we discuss parallel and series connections of two capacitors as examples, to show that an energy loss is inevitably involved.

In Figure 2.7 are shown two capacitors C_1 and C_2 connected in parallel with a battery of V_{emf}. In Figure 2.7(a), C_1 is first charged by switching S_1 on, while S_2 is kept open, in which case charges are only on the plates of C_1. Then, S_2 is closed with S_1 on, as shown in Figure 2.7(b), thereby both capacitors C_1 and C_2 are charged. In contrast, Figure 2.7(c) shows a case where the charged C_1 is isolated from the battery by opening S_1, and C_2 is uncharged if S_2 is open. Closing S_2, however, the charges on C_1 are redistributed between the two capacitors C_1 and C_2.

During the process in Figure 2.7(c), it is noted that the total charge $Q_1 + Q_2$ is constant, and the capacity of the two capacitors connected in parallel is effectively equivalent to a single capacity C given by

$$C = \frac{Q_1 + Q_2}{V_{\text{emf}}} = \frac{Q_1}{V_{\text{emf}}} + \frac{Q_2}{V_{\text{emf}}} = C_1 + C_2. \tag{2.11}$$

Equation (2.11) can be extended to parallel connections of more than two capacitors, for which we obtain the formula, $C = \sum_{\rm i} C_{\rm i}$, where i $= 1, 2, \ldots$.

In the process of Figure 2.7(c), the initial charges Q_1 and $Q_2 = 0$ are redistributed as Q'_1 and Q'_2 in C_1 and C_2, when the total charge is unchanged, i.e., $Q_1 = Q'_1 + Q'_2$. On the other hand, it is noted that the stored energies before and after redistribution are not identical, as seen from the following calculation: The initial energy stored in C_1 is given by $U_1 = \frac{1}{2}Q_1^2/C_1$, whereas after the redistribution, the final energy stored in both capacitors is $U' = \frac{1}{2}Q_1'^2/C_1 + \frac{1}{2}Q_2'^2/C_2$. And, we can see easily that $U_1 > U'$. The difference $U_1 - U'$ can be attributed to heat loss of the transient current during the process.

For two capacitors C_1 and C_2 connected in series, charges on all four plates should be in equal magnitudes but in alternate signs, i.e., $+Q, -Q, +Q, -Q$. The negative plate of C_1 and the positive plate of C_2 are connected together, behaving like a single conductor, in which, obviously, $E = 0$. Hence, we can assign potentials $V_{+,} - V', +V', -V_{-,}$ consistent with $E = 0$ in the connection. Therefore, we can write

$$V_+ - V_- = (V_+ - V') + (V' - V_-),$$

where

$$V_+ - V' = \frac{Q}{C_1} \quad \text{and} \quad V' - V_- = \frac{Q}{C_2}.$$

Thus, we obtain the effective capacity C for series connection, i.e.,

$$\frac{1}{C} = \frac{1}{C_1} + \frac{1}{C_2}, \tag{2.12}$$

and the formula $1/C = \Sigma_{\rm i}(1/C_{\rm i})$ is for a series connection of more than two capacitors.

2.5. Insulating Materials

In most practical capacitors, conducting plates are supported by insulating materials to maintain a narrow gap. Therefore, the filling material should significantly modify properties of such a capacitor. Although signified by absence of electric currents, most insulators exhibit polarization as induced by an electric field arising from charges on the plates, and hence an insulator is often called a *dielectric material*. Due to displaceable molecular charges, dielectric properties of filling insulators are essential in practical capacitors.

The electric polarization is generally a complex phenomenon, depending on the shape of material in some cases. Here, we discuss only a simple case of a dielectric slab inserted in parallel with capacitor plates of area A with a gap d, as shown in Figure 2.8.

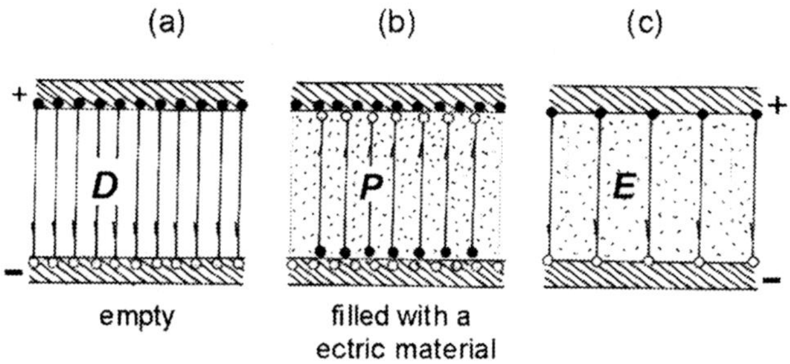

FIGURE 2.8. (a) A charged empty capacitor, where the field-lines are given by ***D***. (b) The gap is filled with dielectric material, where there are flux ***D*** and lines of polarization ***P***. (c) Field-lines of the vector ***E*** determined by (***D*** $-$ ***P***)/ε.

To simplify the argument further, we assume that the entire space of a parallel-plate capacitor is completely filled by a slab. With a battery connected, the capacitor plates are charged not only to certain amount $\pm Q$, but also induced charges $\mp Q_\mathrm{P}$ appear on the surfaces of a material, being distributed at uniform densities $m\sigma_\mathrm{P} = Q_\mathrm{P} = Q_\mathrm{P}/A$. Hence, the surface charge densities are contributed by $\pm\sigma$ as well as $\mp\sigma_\mathrm{P}$ at the boundaries, where $\sigma > \sigma_\mathrm{P}$, because of $Q > Q_\mathrm{P}$. In this case, flux density is given by $\sigma = D$, but in the insulating material there are also fluxes of lines due to induced charges as determined by $\sigma_\mathrm{P} = -P$. At the boundary, the free charges on the platesg cannot mix with bound charges of the insulator, but the total flux $D - P$ inside the insulator appears as if it were originating from *apparent charges* $\pm(\sigma - \sigma_\mathrm{P}) = \pm\sigma'$ on the surfaces. We can therefore write that

$$D - P = \varepsilon_\mathrm{o} E, \tag{2.12}$$

where E is the electric field in the material of the capacitor, for which apparent charges $\pm\sigma'$ are considered to be responsible.

The induced flux P in the insulator is called the *electric polarization*, and considered as the response of the material to the applied field. A linear relation

$$P = \chi\varepsilon_\mathrm{o} E \tag{2.13}$$

can be applied to *normal dielectrics* when E is relatively weak, where the proportionality constant χ is referred to as the *electric susceptibility*. In nature, *spontaneously polarized materials* also exist, originating from an internal polarizing mechanism. The relation (2.13) does not hold for such materials as *ferroelectric* crystals; moreover, even in normal dielectrics the relation between P and E becomes nonlinear in a strong E field. Attributed to internal mechanisms, the solutions to these specific dielectric problems are beyond the scope of this book.

TABLE 2.1. Dielectric permittivities κ of insulators*

Insulators	Permittivities κ
Vacuum	1.00000
Air	1.00054
Water	78
Paper	3.5
Mica	5.4
Amber	2.7
Porcelain	6.5
Fused quartz	3.8
Pyrex glass	4.5
Polyethylene	2.3
Polystyrene	2.6
Neoprene	6.9
Teflon	2.1
Titanium oxide	100

*From D. Halliday & R. Resnick, *Fundamentals of Physics*, Wiley, New York, 1974.

In normal dielectrics to which (2.13) is applicable, the general relation (2.12) can be expressed as

$$D = (1 + \chi)\varepsilon_o E,$$

which can be written as

$$D = \varepsilon E, \tag{2.14a}$$

where

$$\varepsilon = \kappa\varepsilon_o \quad \text{and} \quad \kappa = 1 + \chi. \tag{2.14b}$$

The constants ε and κ are called the *dielectric constant* and electric *permittivity*, respectively. Values of κ for representative materials are listed in Table 2.1.

The capacity of a parallel-plate capacitor filled by a dielectric material is expressed as

$$C = \frac{\varepsilon A}{d} = \kappa C_o, \tag{2.15}$$

where $C_o = \varepsilon_o A/d$ is the capacity of an empty capacitor. Equation (2.15) indicates that the capacity value is increased by a factor κ when filled by a dielectric material of permittivity κ. When capacitors C and C_o are connected to a battery of V_{emf}, we have by definition $C = Q/V_{emf}$ and $C_o = Q_o/V_{emf}$, for which we have $Q = \kappa Q_o$. On the other hand, if the empty capacitor C_o is charged to Q and then isolated from the battery of V_{emf}, the potential difference between the plates is determined by $V_{emf} = (\Delta V)_o = Q/C_o$. If a dielectric slab is inserted into such a charged capacitor, the difference $(\Delta V)_o$ drops to $\Delta V = (\Delta V)_o/\kappa$.

Exercises

1. Calculate a change in the stored energy when a dielectric slab of permittivity κ is inserted in a parallel-plate capacitor $C_o = \varepsilon_o A/d$ for two cases: (a) a battery is kept connected in the circuit, and (b) the slab is inserted after C_o is charged and then isolated from the battery.
2. Obtain the expression for capacity of a parallel-plate capacitor (area A and gap d) when the space is partially filled with a dielectric slab of thickness δ $(< d)$ in parallel to the plates.

3
The Gauss Theorem

3.1. A Spherical Capacitor

Spherical and cylindrical capacitors are important devices in practical applications as well as for simple analysis of a static electric field. In these capacitors, field-lines can be visualized precisely as radial, thereby the model of a point and line charge can be established.

Figure 3.1 shows a sectional view of two concentric conducting spherical shells connected to a battery of V_{emf}. In this case, charges $\pm Q$ are distributed uniformly over the surfaces facing each other, as illustrated in the figure. Denoting the radii of these surfaces by a and b where $a < b$, the surface charge densities are given by $\sigma_a = Q/4\pi a^2$ and $-\sigma_b = -Q/4\pi b^2$, respectively. Assuming $Q > 0$, field-lines start from positive charges on the outer spherical surface of radius a, and end at negative charges on the inner spherical surface of radius b. Inside conducting materials, the field $\boldsymbol{E}$ should be zero in a static condition, and in the outside space of the outer conductor there is no field, since no charges appear on the outermost surface. These conducting spheres can be signified by unique electrical potentials V_a and V_b, where $V_a > V_b$ if $Q > 0$, and all concentric spheres of radius r are equipotential in the range of $a \le r \le b$. According to the drawing rule for field-lines given in Section 2.3.2, the number of lines should, theoretically, be equal to Q, and expressed by $Q = \sigma_a(4\pi a^2) = \sigma_b(4\pi b^2)$. For an equipotential surface with an arbitrary radius r, the flux density D_r can be calculated from the integral

$$\oint_S D_r \mathrm{d}S_r = D_r(4\pi r^2) = Q,$$

where $\mathrm{d}S_r$ is the surface element on a sphere of radius r. Because of the symmetry, field-lines are all radial, and hence D_r represents the radial component of the flux density vector $\boldsymbol{D}$. Therefore, for a spherical capacitor, we can write

$$D_r = \frac{Q}{4\pi r^2} \quad \text{and} \quad E_r = \frac{Q}{4\pi\varepsilon_o r^2}. \tag{3.1}$$

FIGURE 3.1. A spherical capacitor.

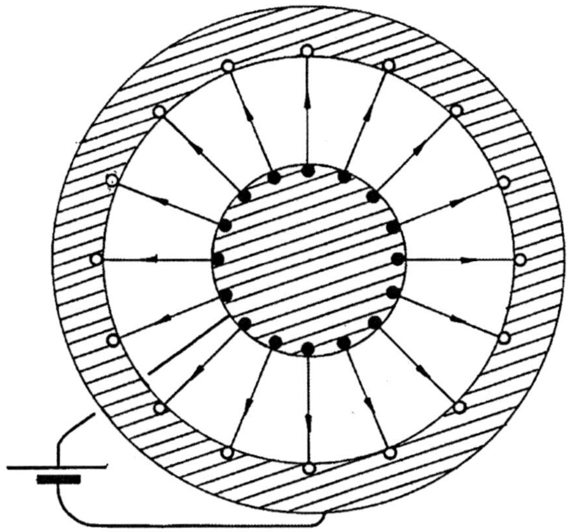

The potential function $V(\boldsymbol{r})$ can be defined, as we discussed for a parallel-plate capacitor. As illustrated in Figure 3.2, placing a test charge 1C in a spherical field, we can calculate the work for moving it along an arbitrary path Γ from a point A on the inner surface at $r = a$ to a point B on the surface at $r = b$. In this case, the arbitrary path Γ can be replaced by small zigzag steps in radial and angular directions in succession approximately.

The electric field vector $\boldsymbol{E}$ has a radial component E_r only, and hence the angular component E_θ is 0 at all θ, so that we can write

$$V_A - V_B = -\int_\Gamma \boldsymbol{E}.\mathrm{d}\boldsymbol{r} = -\int_a^b E_r \mathrm{d}r = -\frac{Q}{4\pi\varepsilon_o}\int_a^b \frac{\mathrm{d}r}{r^2} = \frac{Q}{4\pi\varepsilon_o}\left(\frac{1}{a} - \frac{1}{b}\right), \tag{3.2a}$$

and

$$V(r) = -\frac{Q}{4\pi\varepsilon_o r} + \text{const.} \tag{3.2b}$$

Here, the additional constant in $V(r)$ is physically meaningless and can normally be disregarded by assuming that the potential vanishes in the limit of $r \to \infty$, i.e., $V(\infty) = 0$.

It is noted that the potential $V(r)$ and the corresponding field vector $\boldsymbol{E}(r)$ are *singular* at $r = 0$, where their values cannot be determined. In fact, (3.2b) cannot be considered for a *point charge*, because, for such a finite charge, Q must be postulated as residing inside such a singular point at $r = 0$. However, by assuming that Q is confined to a small sphere of radius $r = a$, the constant in (3.2b) can be set equal to $Q/4\pi\varepsilon_o a$, which may be regarded for the potential value at $r \approx 0$.

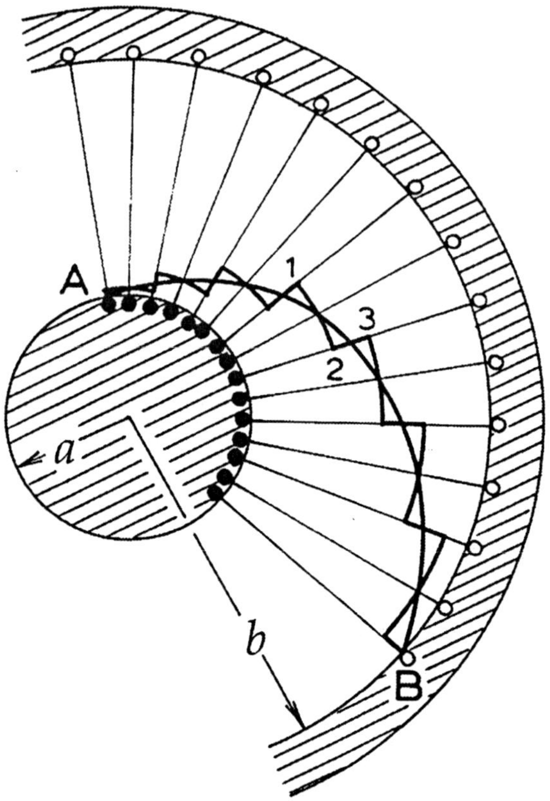

FIGURE 3.2. Quasi-static work for moving a test charge q_o in a spherical electric field.

Thus, for a small charge q, (3.2a) can be written as

$$V(r) = \frac{q}{4\pi\varepsilon_o r}, \tag{3.3}$$

to express the *Coulomb potential* of a point charge q.

Using (3.2b), the capacity of a spherical capacitor is expressed as

$$C_o = \frac{Q}{V_A - V_B} = 4\pi\varepsilon_o \frac{ab}{b-a} \tag{3.4a}$$

if the capacitor is empty inside. On the other hand, if it is filled with a dielectric material of permittivity κ, the capacity is expressed by

$$C = \kappa C_o = 4\pi\varepsilon \frac{ab}{b-a}. \tag{3.4b}$$

3.2. A Cylindrical Capacitor

Referring Figure 3.1, we can discuss a cylindrical capacitor where the electric field-lines are radial between cylindrical surfaces of radii a and b $(a < b)$ and length l. Assuming for a long capacitor that $l \gg (a, b)$, we can ignore both ends, considering the field-lines to be predominantly two-dimensional, perpendicular to the axis. Nevertheless, for a sufficiently long capacitor with a fixed length, charge densities on the facing cylindrical surfaces can be expressed as $\sigma_a = Q/2\pi al$ and $-\sigma_b = -Q/2\pi bl$, and we can draw all Q field-lines in radial directions and evenly distributed in angular directions. In this case, obviously $Q = \sigma_a(2\pi al) = \sigma_b(2\pi bl)$, but $\sigma_a > \sigma_b$, and hence $D_a = Q/2\pi al$, and $D_b = Q/2\pi bl$ at $r = a$ and b, respectively. On an arbitrary surface of radius r between a and b, we can write

$$D_r = \frac{Q}{2\pi rl} \quad \text{and} \quad E_r = \frac{Q}{2\pi\varepsilon_o rl}.$$

Defining the charge density on the cylindrical conductor per length by $\lambda = Q/l$ in unit C-m^{-1}, the radial field can be expressed as

$$D_r = \frac{\lambda}{2\pi r} \quad \text{and} \quad E_r = \frac{\lambda}{2\pi\varepsilon_o r} \tag{3.5a}$$

in addition to the z components $D_z = 0$ and $E_z = 0$.

In such a two-dimensional field, the potential function can be obtained in a way similar to the spherical case. That is,

$$V_A - V_B = -\int_\Gamma \mathbf{E}.d\mathbf{r} = -\int_a^b E_r dr = -\frac{\lambda}{2\pi\varepsilon_o}\int_a^b \frac{dr}{r} = \frac{\lambda}{2\pi\varepsilon_o}\ln\frac{a}{b}, \tag{3.5b}$$

where V_A and V_B are unique values of the potential $V(r)$ at $r = a$ and $r = b$, respectively. Using (3.10b), the capacity of a long cylindrical capacitor can be expressed as

$$C_o = (2\pi\varepsilon_o l)/\ln\frac{b}{a}. \tag{3.6}$$

The potential function can be written as

$$V(r) = \frac{\lambda}{2\pi\varepsilon_o}\ln r + \text{const}, \tag{3.7}$$

which is called the *logarithmic potential*. Notice, however, that for the constant in (3.12) we cannot take the reference point at $r = \infty$. Neither can we define the potential at $r = 0$. Nevertheless, the additional constant of a potential function is physically insignificant, and such an ambiguity in (3.12) can be avoided by taking reference points at $r = a$ and b.

If the space is filled by a dielectric material of permittivity κ, (3.7) should be modified as

$$V(r) = \frac{\lambda}{2\pi\varepsilon} \ln r$$

by simply replacing ε_o by $\varepsilon = \varepsilon_o \kappa$.

3.3. The Gauss Theorem

The electric field can be represented by a static force on a hypothetical charge of 1C and also by field-lines originating from charges on boundary surfaces. The Gauss theorem is a mathematical statement for the latter, expressing the relation precisely. The boundary is composed of continuous conducting surfaces on which charges can be distributed, providing singular sites in the field.

We consider differential area dS at a point $\boldsymbol{r}$, thereby counting the number of field-lines passing through dS. As illustrated in Figure 3.3, such an area can be expressed as $\mathrm{d}S = (r^2 \mathrm{d}\omega) \cos\theta$ with respect to an arbitrary point O, where $\mathrm{d}\omega$ is the *solid angle* subtended by dS at the point $\boldsymbol{r}$, and θ. the angle between the field-line and the *normal* to dS. Here, $r^2 \mathrm{d}\omega = \mathrm{d}S_\perp$ represents the projected area of dS on a sphere of radius r. Therefore, it is convenient to consider the differential area as a vector d$\boldsymbol{S}$ of magnitude *dS* and direction $\boldsymbol{n}$ of the *normal*, i.e., $\mathrm{d}\boldsymbol{S} = \mathrm{d}S\,\boldsymbol{n}$. Flux of the density vector $\boldsymbol{D}$ through d$\boldsymbol{S}$ can then be written as a scalar product $\boldsymbol{D}.\mathrm{d}\boldsymbol{S}$.

With respect to a closed surface S in the field, it is necessary to distinguish the field vector $\boldsymbol{D}$ as to its direction toward either outside or inside of S, for which $\boldsymbol{n}$

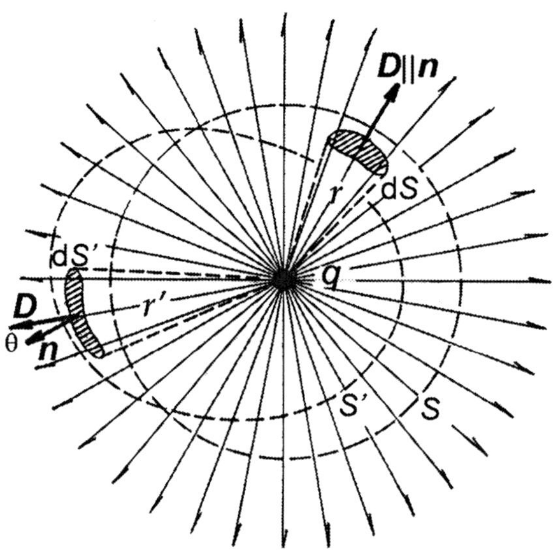

FIGURE 3.3. Flux of field-lines from a point charge q, as counted on a sphere S and on an arbitrary surface S' which are nevertheless identical.

is usually defined as the *outward normal*. With this definition of the vector area we can see that the total flux of field-lines Φ etermined by a closed surface S, i.e., $\Phi = \oint_S \boldsymbol{D}.\mathrm{d}\boldsymbol{S}$, is independent of the choice of S.

Considering a point charge q at the origin O as shown in Figure 3.3, the electric field is radial as expressed by the flux density $D_\mathrm{r} = q/4\pi r^2$, therefore

$$\Phi = \oint_S \boldsymbol{D}.\mathrm{d}\boldsymbol{S} = \oint_{S_\perp} D_r \mathrm{d}S_\perp = \frac{q}{4\pi} \oint_{\text{sphere}} \frac{r^2 \mathrm{d}\omega}{r^2} = \frac{q}{4\pi} \oint_{\text{sphere}} \mathrm{d}\omega = q,$$

which is consistent with the rule for the theoretical number of field-lines to be drawn from q.

If q is located inside or outside S is significant in this argument, and we notice that such a surface can be arbitrary in shape for counting field-lines. Therefore, the above relation $\Phi = q$ holds for any arbitrary S, provided that q is inside S, and this result can be extended to a system of many charges q_i ($\mathrm{i} = 1, 2, \ldots$) as

$$\Phi = \oint_S \boldsymbol{D}.\mathrm{d}\boldsymbol{S} = \sum_\mathrm{i} q_\mathrm{i} = Q_\text{inside}, \tag{3.8}$$

where Q_inside is the total charge enclosed in S.

We can show that a charge q_e located outside S has no contribution to the flux Φ etermined by S. For a charge q_e as shown in Figure 3.4(a), we evaluate fluxes $\mathrm{d}\Phi_1$ and $\mathrm{d}\Phi_2$ determined by differential areas $\mathrm{d}\boldsymbol{S}_1 = \boldsymbol{n}_1(r_1^2 \mathrm{d}\omega_1)$ and $\mathrm{d}\boldsymbol{S}_2 = \boldsymbol{n}_2(r_2^2 \mathrm{d}\omega_2$ at points 1 and 2 in the radial line drawn from q_e.

In this figure, note that the angle θ_1 between $\boldsymbol{D}_1$ and $\boldsymbol{n}_1$ is an obtuse angle, whereas θ_2 between $\boldsymbol{D}_2$ and $\boldsymbol{n}_2$ is an acute one; hence we obtain

$$\mathrm{d}\Phi_1 = \frac{q_\mathrm{e}}{4\pi r_1^2} r_1^2 \mathrm{d}\omega_1 = \frac{q_\mathrm{e}}{4\pi} \mathrm{d}\omega_1 \quad \text{and} \quad \mathrm{d}\Phi_2 = -\frac{q_\mathrm{e}}{4\pi} \mathrm{d}\omega_2,$$

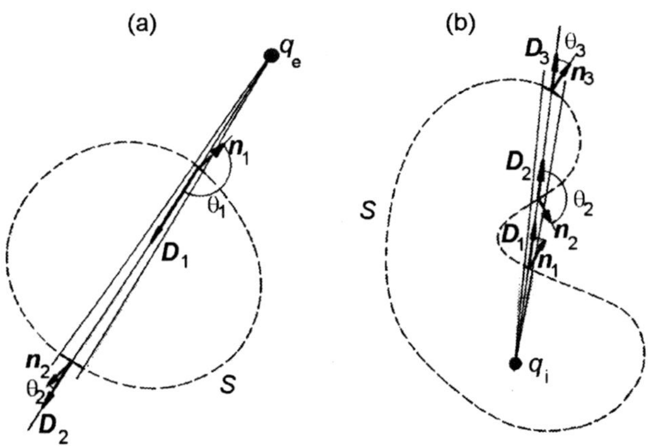

FIGURE 3.4. Flux of field-lines counted on a closed surface in general shape. (a) A charge q_e outside S. (b) A charge q_i inside S.

so that

$$\Phi = \oint_S (d\Phi_1 + d\Phi_2) = \frac{q_e}{4\pi} \oint_S (d\omega_1 - d\omega_2) = 0.$$

We can thus exclude all charges q_e outside S from the calculation of Φ.

To count field-lines the surface S should be continuous and arbitrary in shape. Nevertheless, some indented surface can be permitted in part on S, as shown in Figure 3.4(b). In this case a radial field-line from inside q_i is crosses such an indented S three times (or an odd number of times in general), giving such a total contribution that $d\Phi_1 + d\Phi_2 + d\Phi_3 = d\Phi_1$, because $d\Phi_2 = -d\Phi_3$.

For an arbitrarily chosen surface S in general, some charges can be inside of S but some are outside. However, clearly only those enclosed in S can contribute to the integral $\oint_S \boldsymbol{D}.d\boldsymbol{S}$. Therefore, (3.8) can represent a general case, where Q_{inside} is the total charge enclosed inside S, and the relation is called the Gauss theorem. Such a surface S, as chosen for mathematical purposes, is referred to as a *Gaussian surface*.

In the above we considered point charges q_i and q_e for simplicity; however charges can also be distributed continuously on surfaces and in a volume as well. Such distributed charges are represented by a differential charge element that can be considered as a point charge. For a continuous distribution of charges we can write

$$Q_{\text{inside}} = \int_v \rho(\boldsymbol{r})\, d^3\boldsymbol{r} \quad \text{or} \quad \int_S \sigma(\boldsymbol{r})\, d^2\boldsymbol{r}, \tag{3.9}$$

where $\rho(\boldsymbol{r})$ and $\sigma(\boldsymbol{r})$ are volume and surface densities of charges at $\boldsymbol{r}$, respectively, and $d^3\boldsymbol{r}$ and $d^2\boldsymbol{r}$ are commonly used notations for volume and surface elements dv and dS.

Example 1. A Conducting Plate.

Figure 3.5 shows a conducting plate of a large area on which a flat cylindrical surface with a short height is set up as a Gaussian surface, where the top and bottom surfaces of area A are parallel to the plate.

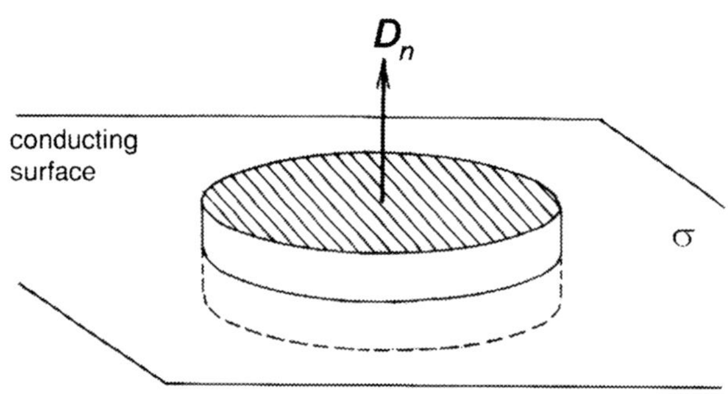

FIGURE 3.5. A Gaussian surface in the shape of a pillbox on a conducting boundary.

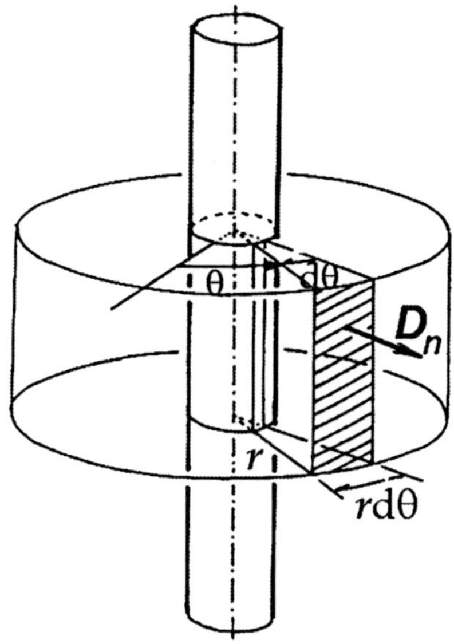

FIGURE 3.6. A Gaussian surface in cylindrical shape for counting field-lines from a long charged conducting rod.

In fact, this problem is hypothetical, unless we consider another conductor with the given plate, which is necessary in principle for forming a capacitor. For such a single charged plate we should consider another one like a spherical conductor with infinite radius. In such a single plate, charges should reside on both surfaces of the plate, whereas if there is another plate nearby as in a parallel-plate capacitor, all charges appear on the facing surface only. Thus, if the surface charge density is $Q/A = \sigma$ on parallel plates, it should be reduced to $Q/2A = \frac{1}{2}\sigma$ on the single isolated plate.

Ignoring the effect near the edges, the field-lines are all parallel to one another and perpendicular to the plate of infinite area. Applied to the surface of a pillbox shape, the Gauss theorem can be expressed as

$$\oint_{\text{box}} \boldsymbol{D}.\mathrm{d}\boldsymbol{S} = (D_n A)_{\text{top}} + (D_n A)_{\text{bottom}} + (D_n A')_{\text{side}} = 2 \times \frac{1}{2}\sigma A.$$

Since $A = A'$, we obtain $D_n = \frac{1}{2}\sigma$, signifying that the field is separated into the top and bottom of the plate due to separated charge densities $\frac{1}{2}\sigma$.

Example 2. A Long a Conducting Rod.

Figure 3.6 shows a long conducting rod, where a concentric cylindrical Gaussian surface S of radius r and height l is indicated. When we charge it with a battery, however, we have to assume that the rod is a part of a cylindrical uniaxial capacitor that is composed with another cylinder of radius $r = \infty$. Further assuming the length is sufficiently long, the charge Q on the rod can be distributed evenly on the surface, which is conveniently expressed by the density per unit length, i.e., $\lambda = Q/l$. Such a concentric Gaussian surface is an obvious choice for a simple analysis of this case, where the flux-density vector $\boldsymbol{D}$ can be radial and perpendicular to the rod, so that the flux $\boldsymbol{D}.\mathrm{d}\boldsymbol{S}$ vanishes on the top and bottom surfaces. Referring the rod axis to the z-axis, we use cylindrical coordinates (r, θ, z) for a point in the field. The flux Φ of field-lines determined by S can be expressed by

$$\int_S \boldsymbol{D}.\mathrm{d}\boldsymbol{S} = \int_0^{2\pi} D_r r l \mathrm{d}\theta = D_r r l \times 2\pi = \lambda l,$$

from which we obtain the expression $D_r = \lambda/2\pi r$, that is identical to (3.5a).

Example 3. A Uniformly Charged Sphere.

J. J. Thomson (1903) considered a uniformly charged sphere to be a model of an atom, one in which an electron can vibrate in harmonic motion to emit light. Assuming it to be electrically neutral, such an atom consists of an electron and a positive charge $+e$. He considered that positive charge uniformly distributed inside a spherical volume of radius R.

Using the Gauss theorem, we can calculate the flux density $\boldsymbol{D}$ at a distance r greater than R. By symmetry, the vector $\boldsymbol{D}$ should be radial with respect to the center and given only by its radial component D_r. Therefore, the Gauss theorem applied to a spherical surface of radius $r > R$ can be expressed as

$$\oint_S \boldsymbol{D}.\mathrm{d}\boldsymbol{S} = \oint_S D_r r^2 \mathrm{d}\omega = D_r r^2 \oint_S \mathrm{d}\omega = D_r \left(4\pi r^2\right) = +e,$$

and hence, at a point $r > R$, we have

$$D_r = \frac{e}{4\pi r^2} \quad \text{and} \quad E_r = \frac{e}{4\pi\varepsilon_o r^2},$$

implying that such a charged sphere looks as if a point charge located at $r = 0$ when observed at a distant point $r > R$.

On the other hand, at a point $r < R$ inside the sphere, the electric field can be calculated as arising from the amount of charge enclosed by a sphere S of radius r, whose flux density is determined by $e(r/R)^3$. Therefore, for such a sphere we can write

$$e\left(\frac{r}{R}\right)^3 = D_r 4\pi r^2,$$

and hence

$$E_r = \frac{e}{4\pi\varepsilon_o R^3} r \quad \text{for} \quad r < R.$$

Placing the electron at an inside point r, the force $F_r = -eE_r = -\dfrac{e^2}{4\pi\varepsilon_o R^3} r$ is a restoring force for the electron to be bonded at $r = 0$, being proportional to r. Considering the kinetic energy, the electron can vibrate in the vicinity of the center, hence emitting its energy as a radiating dipole.

3.4. Boundary Conditions

We generally consider the field confined to a space surrounded by boundaries that are either conductive or dielectric in character. Properties of the field can essentially be determined by surrounding boundaries.

The field vector $\boldsymbol{E}$ is basically a static force acting on a charge 1C, and *conservative* in character, as expressed by a vanishing integral

$$\oint_{\Gamma} \boldsymbol{E}.\mathrm{d}\boldsymbol{s} = 0, \tag{3.10}$$

which can be performed along any closed curve Γ, where $\boldsymbol{ds}$ is the line element. On the other hand, the flux density $\boldsymbol{D}$ is related by charges in the field, as described by the Gauss theorem

$$\oint_{S} \boldsymbol{D}.\mathrm{d}\boldsymbol{S} = \sum_{\mathrm{i}} q_{\mathrm{i}}, \tag{3.8}$$

where these q_{i} are *free* charges, distinctively different from *bound* charges on dielectric surfaces. Equations (3.15) and (3.13) represent the basic laws of electrostatics in integral form, which can be applied to any static problems. These laws can be applied to boundaries to obtain conditions for calculating the field at a given point $\boldsymbol{r}$.

3.4.1. A Conducting Boundary

Figure 3.7 shows a small portion of a conducting boundary, where the shaded side represents the conducting medium. For the integral in (3.10), we consider the closed rectangular path 1-2-3-4-1 as Γ, where the section 1-2 is in the field space, whereas 3-4 is inside the conductor. Denoting edge lengths of Γ as d_{12}, d_{23} and so forth, we notice that $d_{12} = -d_{34}$, and $d_{23} = -d_{41}$, and these lengths are considered short.

Therefore, normal components of $\boldsymbol{E}$ along d_{23} and d_{41} are practically identical, whereas tangential components along d_{12} and d_{34} are E_t and 0, respectively. Therefore, (3.15) can be written as

$$\oint_{12341} \boldsymbol{E}.\mathrm{d}\boldsymbol{s} = E_{\mathrm{t}} d_{12} + (0 \times d_{34}) = 0,$$

where the condition $E_{\mathrm{t}} = 0$ implies that only the normal component E_n is significant on the conducting surface. As shown in Figure 3.5, for a pillbox Gaussian

FIGURE 3.7. A closed rectangular path Γ: $1 - 2 - 3 - 4 - 1$ for the integral $\int_{\Gamma} \boldsymbol{E}.\mathrm{d}s$ across a conducting boundary plane.

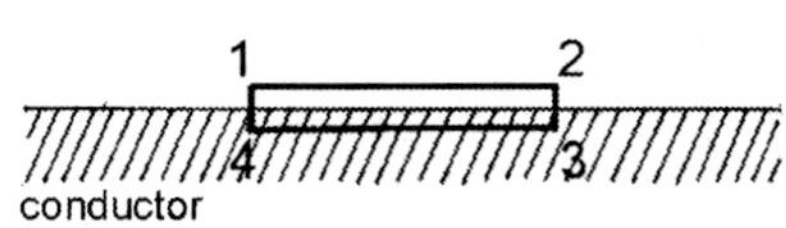

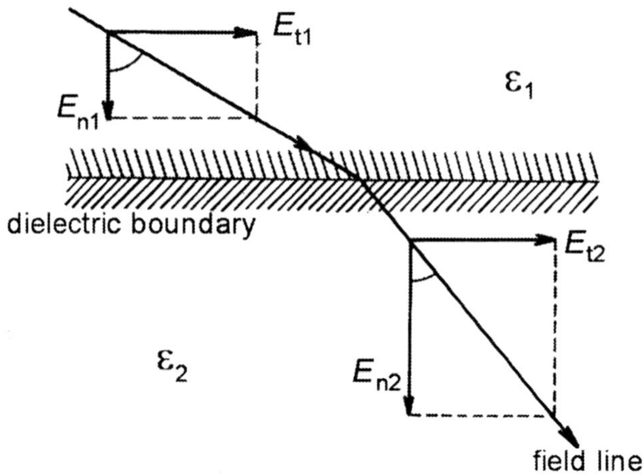

FIGURE 3.8. An E-line bent across a dielectric boundary.

surface the normal component D_n is determined by the surface charge density σ as $D_n = \sigma = \varepsilon_0 E_n$.

Thus, the boundary conditions on a conducting surface are given by

$$E_t = 0 \quad \text{and} \quad D_n = \sigma. \tag{3.11}$$

3.4.2. A Dielectric Boundary

A boundary between different dielectrics is generally complicated because of the presence of induced charges that can neither move nor mix on the surface. Further, induced charges may not be uniformly distributed, except for a planar boundary perpendicular to the applied field. Figure 3.8 illustrates a boundary between different media of dielectric constants ε_1 and ε_2, and $\boldsymbol{E}_1$ and $\boldsymbol{E}_2$ are electric field vectors in these media.

First, we apply (3.10) to a rectangular path shown in Figure 3.7 considering the short sections 1-2 and 3-4 in the medium 1 and medium 2, respectively. The sections 2-3 and 4-1 crossing the boundary may be assumed as shorter and negligible. For the integrand in (3.15), we therefore need only tangential components of $\boldsymbol{E}$, so that

$$\oint_{12341} \boldsymbol{E}.\mathrm{d}\boldsymbol{s} = E_{t1} d_{12} + E_{t2} d_{34} = 0 \quad \text{where} \quad d_{12} = -d_{34}.$$

Therefore,

$$E_{t1} = E_{t2}, \tag{3.12}$$

stating that tangential components of $\boldsymbol{E}$ should be continuous across the boundary.

Next, we apply the Gauss theorem (3.13) to a pillbox surface that is set perpendicularly to a plane dielectric boundary, as shown in Figure 3.7, with the flat top

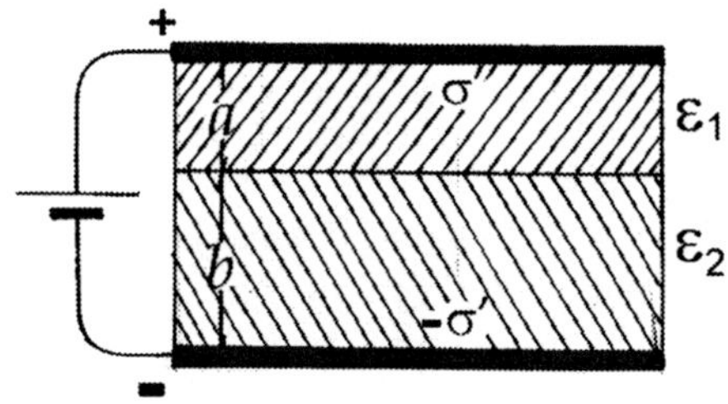

FIGURE 3.9. A charged capacitor filled by parallel layers of two dielectric materials.

and bottom in dielectric media 1 and 2, respectively. It can then be written as

$$\oint_{\text{box}} \boldsymbol{D}.\mathrm{d}\boldsymbol{S} = D_{n1}S_1 - D_{n2}S_2 = 0, \quad \text{where} \quad S_1 = S_2.$$

Hence,

$$\varepsilon_1 E_{n1} = \varepsilon_2 E_{n2}, \tag{3.13}$$

implying that normal components of $\boldsymbol{D}$ are continuous, although the corresponding $\boldsymbol{E}$ is discontinuous across the boundary, because of $\varepsilon_1 \neq \varepsilon_2$. As remarked, this result is also due to the fact that no free charges can, normally, exist on the dielectric boundary.

Denoting angles between $\boldsymbol{E}$ and $\boldsymbol{n}$ in two media as θ_1 and θ_2, as indicated in Figure 3.8, we have $\tan\theta_1 = E_{t1}/E_{n1}$ and $\tan\theta_2 = E_{t2}/E_{n2}$. Using (3.17) and (3.18), we obtain

$$\frac{\tan\theta_1}{\tan\theta_2} = \frac{\varepsilon_1}{\varepsilon_2}, \tag{3.14}$$

indicating that field-lines bend directions at the boundary, except for $\theta_1 = \theta_2 = 0$.

If $\boldsymbol{E} \| \boldsymbol{n}$ particularly, E_n is discontinuous, and $E_t = 0$, thereby characterizing the boundary as an equipotential surface.

The conditions (3.17) and (3.18) can generally be applied to static boundaries between different nonconducting dielectric media, signifying the continuity of $\boldsymbol{E}_t$ and $\boldsymbol{D}_n$. In contrast, on a conducting surface characterized by $\boldsymbol{E} = 0$, and so $\boldsymbol{D} = 0$, the relation $\boldsymbol{D}_n = \sigma$ signifies the discontinuity on a conducting surface. Also, as the tangential component is zero outside the conductor as well as inside, E_t appears to be continuous.

Example. A Planar Dielectric Boundary in a Capacitor.

Consider two dielectric layers of thickness a and b with different dielectric constants ε_1 and ε_2 filling the gap d of a parallel-plate capacitor that is charged by a battery, as shown in Figure 3.9.

In this case, these two layers behave like a single medium of an effective dielectric constant ε' and thickness $a + b = d$, exhibiting induced charges $\mp Q_P$ on surfaces facing the charged capacitor plates at $\pm Q$. On the boundary surface between different dielectric media there appears no macroscopic charge, so that the flux from charges $\pm Q$ penetrates continuously into the media, i.e., $D_{n1} = D_{n2}$, in which another flux of lines from the polarized charges $\mp Q_P$ exists in relation with $P_{n1} = P_{n2}$. Here, we have $D_{n1} - P_{n1} = 0$ and $D_{n2} - P_{n2} = 0$, and so $\varepsilon_1 E_{n1} = \varepsilon_2 E_{n2} = 0$, accordingly $E_{n1} = E_{n2} = 0$ on the boundary. On the other hand, there is no tangential field in the media, so that the boundary surface should be equipotential, ignoring the gap between the media 1 and 2. Thus, signifying surfaces by potentials, V_+, V_1, and V_-, the capacitor can be regarded as equivalent to two capacitors filled with dielectric media ε_1 and ε_2 that are connected in series. Namely, we can write

$$C_1 = \frac{\varepsilon_1 A}{a}, \quad C_2 = \frac{\varepsilon_2 A}{b}, \quad C = \frac{\varepsilon' A}{d} \quad \text{and} \quad \frac{1}{C} = \frac{1}{C_1} + \frac{1}{C_2},$$

and

$$\varepsilon' = \frac{1}{\dfrac{a}{\varepsilon_1} + \dfrac{b}{\varepsilon_2}}.$$

Further, writing that

$$V_+ - V_- = \frac{Q}{C} = E(a+b) \quad \text{and} \quad Q - Q_P = \varepsilon_o E,$$

we obtain

$$Q_P = Q\left(1 + \varepsilon_o \frac{\dfrac{a}{\varepsilon_1} + \dfrac{b}{\varepsilon_2}}{a+b}\right).$$

4
The Laplace–Poisson Equations

4.1. The Electrostatic Potential

One of the basic properties of the electric field is the force on a small electric charge brought in from outside the field or on a hypothetical charge of 1 C that can be assumed as if present inside the field. On the other hand, the field as a whole is depicted by distributed lines originating from electric charges. Represented by two vectors $\boldsymbol{E}$ and $\boldsymbol{D}$, the basic laws of electrostatics are expressed in *integral form* as

$$\oint_{\Gamma} \boldsymbol{E}.\mathrm{d}\boldsymbol{s} = 0 \tag{3.15}$$

and

$$\oint_{S} \boldsymbol{D}.\mathrm{d}\boldsymbol{S} = 0. \tag{3.13}$$

The integral in (3.15) is taken along a closed path Γ, whereas in (3.13) $\boldsymbol{D}$ is integrated over a closed surface S, thereby expressing the static laws within the path Γ and surface S, which are selected for mathematical convenience. Accordingly, it is also significant to formulate the laws in *differential form* to deal with local properties of the field at a given position, as specified by the boundary conditions.

First, we consider the *conservative* nature of the field vector $\boldsymbol{E} = \boldsymbol{E}(\boldsymbol{r})$, where $\boldsymbol{r}$ is an arbitrary position vector in the field. Mathematically, the vanishing integral in (3.15) signifies that $\boldsymbol{E}.\mathrm{d}\boldsymbol{s}$ is a *perfect differential*, so that we can write

$$-\boldsymbol{E}.\mathrm{d}\boldsymbol{s} = \mathrm{d}V,$$

with a scalar V, where the negative sign is attached for convenience. Here, the function $V(\boldsymbol{r})$ is called the *potential*, that is, a function of a position $\boldsymbol{r}$. The line element ds is taken along an arbitrary curve Γ for integration, but can be considered as an arbitrary displacement $d\boldsymbol{r}$ in the field. Hence, using rectangular coordinates $\boldsymbol{r} = (x, y, z)$, we can write that

$$-E_x dx - E_y dy - E_z dz = \frac{\partial V}{\partial x} dx + \frac{\partial V}{\partial y} dy + \frac{\partial V}{\partial z} dz,$$

therefore,

$$E_x = -\frac{\partial V}{\partial x}, \quad E_y = -\frac{\partial V}{\partial y} E_x, \quad \text{and} \quad E_z = -\frac{\partial V}{\partial z}. \tag{4.1a}$$

The set of three differential operations $(\partial V/\partial x, \partial V/\partial y, \partial V/\partial z)$ is a vector operator called the *gradient*, by which (4.1a) can be written as grad V or ∇V. Namely,

$$\boldsymbol{E} = -\text{grad } V = -\nabla V. \tag{4.1b}$$

The differential operator ∇, known as "nabla" (after a Greek word ναβλα), is also read "del," and often used in place of "grad."

It is noted that the equation $V(\boldsymbol{r}) =$ const represents a surface in the field in three dimensions. Characterized by a constant potential, such a surface is called an *equipotential* surface. Considering two close points $\boldsymbol{r}$ and $\boldsymbol{r}'$ on such a $V =$ constant surface, the difference vector $\boldsymbol{r} - \boldsymbol{r}' = d\boldsymbol{r}$ indicates a tangential direction τ to the surface. Therefore, from the relation $V(\boldsymbol{r}) - V(\boldsymbol{r}') \approx \text{grad}_\tau V . d\boldsymbol{r} = 0$, we obtain the relation $\text{grad}_\tau V \perp \tau$. Accordingly, field-lines of vector $\boldsymbol{E}$ are always perpendicular to the tangent of an equipotential surface, i.e., $\boldsymbol{E} \perp \tau$. The static field can therefore be mapped by a number of equipotential surfaces of constant V, similar to *altitude contours* in a geographical map.

4.2. The Gauss Theorem in Differential Form

The charge Q_{inside} in (3.13) is the total amount of the charges enclosed in a Gaussian surface S. Taking a differential volume dv at a point $\boldsymbol{r}$ inside S, we can express the local charge as $\rho(\boldsymbol{r})\, dv$, where $\rho(\boldsymbol{r})$ is the volume charge density at $\boldsymbol{r}$; and hence $Q_{\text{inside}} = \int_v \rho(\boldsymbol{r}) dv$. The volume element dv at a point $\boldsymbol{r}$ is often written as $d^3\boldsymbol{r}$ if it is necessary to specify the point's position in three-dimensional space. The flux is expressed by a *surface integral* as in (3.13), which must be converted to a *volume integral* in order for it to be compared with the charge density $\rho(\boldsymbol{r})$ distributed in S. Such a conversion can be performed mathematically by the *Gauss theorem.*

For simplicity, we use here rectangular coordinates to derive the Gauss theorem, as shown in Fig. 4.1(a).

We consider a volume element $dv = dx dy dz$ in cube shape, where three pairs of rectangular planes of area $dy dz$, $dz dx$, and $dx dy$ are facing along the x, y, and z directions, respectively, constituting a closed cubical surface dS covering the volume dv. For such a small volume, the flux on the left side of (3.13) can be

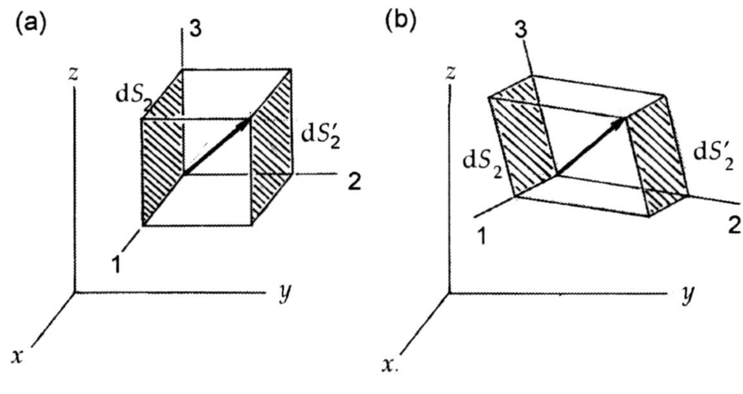

FIGURE 4.1. (a) A differential volume element in rectangular coordinates. (b) An element in curvilinear coordinates (u_1, u_2, u_3).

specially expressed as

$$\begin{aligned}\sum_{dS} \boldsymbol{D}.\mathrm{d}\boldsymbol{S} &= -D_x \mathrm{d}S_x + D_{x+\mathrm{d}x}\mathrm{d}S_x - D_y\mathrm{d}S_y + D_{y+\mathrm{d}y}\mathrm{d}S_y - D_z\mathrm{d}S_z + D_{z+\mathrm{d}z}\mathrm{d}S_z \\ &= (D_{x+\mathrm{d}x} - D_x)\,\mathrm{d}y\mathrm{d}z + \left(D_{y+\mathrm{d}y} - D_y\right)\mathrm{d}z\mathrm{d}x + (D_{z+\mathrm{d}z} - D_z)\,\mathrm{d}x\mathrm{d}y \\ &= \left(\frac{\partial D_x}{\partial x} + \frac{\partial D_y}{\partial y} + \frac{\partial D_z}{\partial z}\right)\mathrm{d}x\mathrm{d}y\mathrm{d}z,\end{aligned}$$

whereas the charge on the right side of (3.13) is given by

$$Q_{\text{inside}} = \rho(x, y, z)\,\mathrm{d}x\mathrm{d}y\mathrm{d}z.$$

Therefore, the flux law (3.13) can be written as

$$\frac{\partial D_x}{\partial x} + \frac{\partial D_y}{\partial y} + \frac{\partial D_z}{\partial z} = \rho\,(x, y, z)\,. \tag{4.2a}$$

Using the vector differential operator $\boldsymbol{\nabla}$, the left hand side expression in (4.2a) can be expressed by the scalar product of $\boldsymbol{\nabla}.\boldsymbol{D}$, called the *divergence* of $\boldsymbol{D}$, and therefore we obtain

$$\boldsymbol{\nabla}.\boldsymbol{D}(\boldsymbol{r}) = \rho(\boldsymbol{r}) \tag{4.2b}$$

or

$$\operatorname{div} \boldsymbol{D}(\boldsymbol{r}) = \rho(\boldsymbol{r}). \tag{4.2c}$$

If the space is empty or filled with a uniform dielectric medium, we have a relation $\boldsymbol{D} = \varepsilon_o \boldsymbol{E}$ or $\boldsymbol{D} = \varepsilon \boldsymbol{E}$, respectively, and (4.2b) and (4.2c) can be written for $\boldsymbol{E}$ as

$$\boldsymbol{\nabla}.\boldsymbol{E} = \operatorname{div}\boldsymbol{E} = \rho(\boldsymbol{r})/\varepsilon_o \quad \text{or} \quad \rho(\boldsymbol{r})/\varepsilon. \tag{4.2d}$$

In (4.2d), the expressions in empty and filled spaces are identical, except for the dielectric constants, and hence it is sufficient to discuss the equation for ε_o for an empty field. The field in normal media can be described sufficiently by one of these vectors $\boldsymbol{D}$ or $\boldsymbol{E}$. In contrast, the potential function $V(\boldsymbol{r})$ derived from (4.1a and b) is a scalar quantity, making mathematical analysis significantly simpler. Combining (4.1a) and (4.1b) with (4.2d), we can write the equation for the potential as

$$-\boldsymbol{\nabla}.\boldsymbol{\nabla} V(\boldsymbol{r}) = -\text{div.grad } V(\boldsymbol{r}) = \frac{\rho(\boldsymbol{r})}{\varepsilon},$$

where

$$\boldsymbol{\nabla}.\boldsymbol{\nabla} = \text{div.grad} = \frac{\partial^2}{\partial x^2} + \frac{\partial^2}{\partial y^2} + \frac{\partial^2}{\partial z^2}$$

is a scalar operator called a *Laplacian*, and often expressed by $\boldsymbol{\nabla}.\boldsymbol{\nabla} = \boldsymbol{\nabla}^2 = \boldsymbol{\Delta}$.

The equation for $V(\boldsymbol{r})$ is an *inhomogeneous* differential equation when $\rho(\boldsymbol{r}) \neq 0$, but becomes a *homogeneous* equation if $\rho(\boldsymbol{r}) = 0$. In the former case

$$\boldsymbol{\nabla}^2 V(\boldsymbol{r}) = -\frac{\rho(\boldsymbol{r})}{\varepsilon} \tag{4.3}$$

is called the *Poisson equation*, whereas the homogeneous equation in the latter, i.e.,

$$\boldsymbol{\nabla}^2 V(\boldsymbol{r}) = 0, \tag{4.4}$$

is the *Laplace equation*. Representing the basic laws of electrostatics, these equations are to be solved for the potential $V(\boldsymbol{r})$, subject to given boundary conditions.

4.3. Curvilinear Coordinates (1)

For electrostatic problems in most practical cases, symmetry of boundaries plays a significant role in obtaining solutions of the Laplace equation. In Section 4.2 rectangular coordinates were used for simplicity; however, for practical problems basic equations may have to be expressed in polar, cylindrical, or other coordinates, depending on the symmetry of a problem.

Polar and cylindrical coordinates are specific *curvilinear coordinates* that can generally be written as u_1, u_2, and u_3. Nevertheless, these coordinates are also functions of rectangular coordinates x, y, and z; a point $P(x, y, z)$ in space can also be specified by (u_1, u_2, u_3). An important feature of these curvilinear systems is that basic coordinate surfaces $(0, u_2, u_3)$, $(u_1, 0, u_3)$ and $(u_1, u_2, 0)$ are mutually orthogonal, and a differential displacement $d\boldsymbol{r}$ can be expressed by the *metric*

$$d\boldsymbol{r}^2 = h_1^2 du_1^2 + h_2^2 du_2^2 + h_3^2 du_3^2, \tag{4.5}$$

representing the orthogonal space. In general, the coordinate transformation is specified by the metric $d\boldsymbol{r}^2 = \Sigma g_{ij} u_i u_j$ expressed with coefficients $g_{ij} = h_i h_j$, which constitute a *metric tensor* (g_{ij}).

Judging from (4.5), in the orthogonal space components of $d\boldsymbol{r}$ are given by components $h_1 du_1$, $h_2 du_2$, and $h_3 du_3$, where h_1, h_2, and h_3 can be considered factors for adjusting coordinates u_1, u_2, and u_3 to linear dimensions. The gradient of a scalar function V can therefore be expressed by components

$$E_1 = -\frac{1}{h_1}\frac{\partial V}{\partial u_1}, \quad E_2 = -\frac{1}{h_2}\frac{\partial V}{\partial u_2} \quad \text{and} \quad E_3 = -\frac{1}{h_3}\frac{\partial V}{\partial u_3}. \tag{4.6}$$

For example, with rectangular coordinates (x, y, z) equation (4.5) for the metric can be expressed as

$$\mathrm{d}r^2 = \mathrm{d}x^2 + \mathrm{d}y^2 + \mathrm{d}z^2,$$

where

$$u_1 = x, \; u_2 = y, \; u_3 = z$$

and

$$h_1 = h_2 = h_3 = 1.$$

Hence, the gradient vector is given by (4.1a).

For polar coordinates (r, θ, φ), on the other hand, we have

$$\mathrm{d}\boldsymbol{r}^2 = \mathrm{d}r^2 + r^2\mathrm{d}\theta^2 + r^2\sin^2\theta\,\mathrm{d}\varphi^2,$$

for which

$$u_1 = r, \; u_2 = \theta, \; u_3 = \varphi$$

and

$$h_1 = 1, \; h_2 = r, \; h_3 = r\sin\theta.$$

In this case, from (4.6) the vector $\boldsymbol{E}$ has components

$$E_r = -\frac{\partial V}{\partial r}, \quad E_\theta = -\frac{1}{r}\frac{\partial V}{\partial \theta}, \quad \text{and} \quad E_\varphi = -\frac{1}{r\sin\theta}\frac{\partial V}{\partial \varphi}. \tag{4.7}$$

Similarly, with cylindrical coordinates (ρ, θ, z) we have

$$\mathrm{d}\boldsymbol{r}^2 = \mathrm{d}\rho^2 + \rho^2\mathrm{d}\theta^2 + \mathrm{d}z^2,$$

for which

$$u_1 = \rho, \; u_2 = \theta, \; u_3 = z$$

and

$$h_1 = 1, \; h_2 = \rho, \; h_3 = 1.$$

The vector $\boldsymbol{E}$ is then expressed by

$$E_\rho = -\frac{\partial V}{\partial \rho}, \quad E_\theta = -\frac{1}{\rho}\frac{\partial V}{\partial \theta} \quad \text{and} \quad E_z = -\frac{\partial V}{\partial z}. \tag{4.8}$$

When converting a surface integral $\int_S \boldsymbol{D}.\mathrm{d}\boldsymbol{S}$ into a volume integral, the volume element needs to be written by curvilinear coordinates as $\mathrm{d}v = (h_1\mathrm{d}u_1)(h_2\mathrm{d}u_2)(h_3\mathrm{d}u_3)$. Similar to a volume element $\mathrm{d}v = \mathrm{d}x\mathrm{d}y\mathrm{d}z$ in rectangular coordinates, the cube-like elemental volume $\mathrm{d}v$ has three pairs of surfaces—($\mathrm{d}S_1$, $\mathrm{d}S_1'$), ($\mathrm{d}S_2$, $\mathrm{d}S_2'$), and ($\mathrm{d}S_3$, $\mathrm{d}S_3'$)—that are perpendicular to the u_1, u_2, and u_3 lines, respectively, as shown in Fig. 4.1(b).

In this case, the flux $\mathrm{d}\Phi$ determined by $\mathrm{d}\boldsymbol{S}$ can be calculated as

$$\begin{aligned}\mathrm{d}\Phi &= \Sigma \boldsymbol{D}.\mathrm{d}\boldsymbol{S}\\ &= (-D_1\mathrm{d}S_1 + D_1'\mathrm{d}S_1') + (-D_2\mathrm{d}S_2 + D_2'\mathrm{d}S_2') + (-D_3\mathrm{d}S_3 + D_3'\mathrm{d}S_3'),\end{aligned} \tag{4.9}$$

where

$$\begin{aligned}\mathrm{d}S_1' &= h_2'h_3'\mathrm{d}u_2\mathrm{d}u_3\\ &= \left(h_2 + \frac{\partial h_2}{\partial u_1}\mathrm{d}u_1\right)\left(h_3 + \frac{\partial h_3}{\partial u_1}\right)\mathrm{d}u_2\mathrm{d}u_3, \quad D_1' = D_1 + \frac{\partial D_1}{\partial u_1}\mathrm{d}u_1, \text{ etc.}\end{aligned}$$

Therefore, we can write

$$\begin{aligned}D_1'\mathrm{d}S_1' &= D_1h_2h_3\mathrm{d}u_2\mathrm{d}u_3 + \left(\frac{\partial D_1}{\partial u_1}h_2h_3 + D_1\frac{\partial h_2}{\partial u_1}h_3 + D_1h_2\frac{\partial h_3}{\partial u_1}\right)\mathrm{d}u_1\mathrm{d}u_2\mathrm{d}u_3\\ &= D_1\mathrm{d}S_1 + \frac{\partial(D_1h_2h_3)}{\partial u_1}\mathrm{d}u_1\mathrm{d}u_2\mathrm{d}u_3,\end{aligned}$$

from which we derive

$$D_1'\mathrm{d}S_1' - D_1\mathrm{d}S_1 = \frac{1}{h_1h_2h_3}\frac{\partial(D_1h_2h_3)}{\partial u_1}\mathrm{d}v.$$

Similarly,

$$D_2'\mathrm{d}S_2' - D_2\mathrm{d}S_2 = \frac{1}{h_1h_2h_3}\frac{\partial(D_2h_3h_1)}{\partial u_2}\mathrm{d}v$$

and

$$D_3'\mathrm{d}S_3' - D_3\mathrm{d}S_3 = \frac{1}{h_1h_2h_3}\frac{\partial(D_3h_1h_2)}{\partial u_3}\mathrm{d}v.$$

Accordingly, we arrive at the expression

$$\mathrm{d}\Phi = \frac{1}{h_1h_2h_3}\left\{\frac{\partial(D_1h_2h_3)}{\partial u_1} + \frac{\partial(D_2h_3h_1)}{\partial u_2} + \frac{\partial(D_3h_1h_2)}{\partial u_3}\right\}\mathrm{d}v = (\operatorname{div}\boldsymbol{D})\,\mathrm{d}v.$$

Therefore, the total flux integrated over a closed surface S can be expressed by

$$\oint_S \boldsymbol{D}.\mathrm{d}\boldsymbol{S} = \int_v \operatorname{div}\boldsymbol{D}\,\mathrm{d}v, \tag{4.10}$$

where

$$\operatorname{div}\boldsymbol{D} = \frac{1}{h_1h_2h_3}\left\{\frac{\partial(D_1h_2h_3)}{\partial u_1} + \frac{\partial(D_2h_3h_1)}{\partial u_2} + \frac{\partial(D_3h_1h_2)}{\partial u_3}\right\}. \tag{4.11}$$

Equation (4.10) is the conversion formula between surface and volume integrals, which is known as the *Gauss theorem*. Equation (4.10) gives the general expression of div $\boldsymbol{D}$ in curvilinear coordinates u_1, u_2, and u_3. Using (4.10), div

$\boldsymbol{D}$ can be expressed in representative curvilinear coordinates. First, in rectangular coordinates as

$$\operatorname{div} \boldsymbol{D} = \frac{\partial D_x}{\partial x} + \frac{\partial D_y}{\partial y} + \frac{\partial D_z}{\partial z}. \tag{4.11a}$$

In polar coordinates as

$$\operatorname{div} \boldsymbol{D} = \frac{1}{r^2 \sin\theta} \left\{ \frac{\partial \left(r^2 \sin\theta D_r\right)}{\partial r} + \frac{\partial \left(r \sin\theta D_\theta\right)}{\partial \theta} + \frac{\partial \left(r D_\varphi\right)}{\partial \varphi} \right\}. \tag{4.11b}$$

In cylindrical coordinates as

$$\operatorname{div} \boldsymbol{D} = \frac{1}{\rho} \frac{\partial \left(\rho D_\rho\right)}{\partial \rho} + \frac{1}{\rho} \frac{\partial D_\theta}{\partial \theta} + \frac{\partial D_z}{\partial z}. \tag{4.11c}$$

The Laplacian operator can also be expressed as

$$\boldsymbol{\nabla}^2 = \frac{\partial^2}{\partial x^2} + \frac{\partial^2}{\partial y^2} + \frac{\partial^2}{\partial z^2}, \tag{4.12a}$$

$$\boldsymbol{\nabla}^2 = \frac{1}{r^2} \frac{\partial}{\partial r} \left(r^2 \frac{\partial}{\partial r} \right) + \frac{1}{r^2 \sin\theta} \frac{\partial}{\partial \theta} \left(\sin\theta \frac{\partial}{\partial \theta} \right) + \frac{1}{r^2 \sin^2} \frac{\partial^2}{\partial \varphi^2}, \tag{4.12b}$$

$$\boldsymbol{\nabla}^2 = \frac{1}{\rho} \frac{\partial}{\partial \rho} \left(\rho \frac{\partial}{\partial \rho} \right) + \frac{1}{\rho^2} \frac{\partial^2}{\partial \theta^2} + \frac{\partial^2}{\partial z^2}, \tag{4.12c}$$

for rectangular, polar, and cylindrical systems, respectively.

4.4. The Laplace–Poisson Equations

A basic mathematical problem in electrostatics is to solve the Laplace equation for the potential function when a charge distribution is specified on the boundary. If there are charges on conducting boundaries and in the field we have to deal with the Poisson equation at these charges' locations. For a complete solution of a static problem, both these equations are often required under given boundary conditions. Mathematically, such a problem can be discussed with the *Green's function method*, which is generally applicable to other boundary problems than those encountered in electrostatics. In this section, the Green's function method is outlined so we may discuss its mathematical consequences for static problems, although some results are inferable with respect to physical ideas.

4.4.1. Boundary Conditions

In Section 3.4 static boundary conditions were discussed using the field vectors $\boldsymbol{E}$ and $\boldsymbol{D}$. For a conducting boundary the conditions are generally specified by (3.16) as

$$E_t = 0 \quad \text{and} \quad D_n = \sigma,$$

which may alternatively be expressed with the potential function in the field as

$$(V)_S = 0 \quad \text{and} \quad -\varepsilon_o \left(\frac{\partial V}{\partial n}\right)_S = \sigma, \tag{4.13}$$

where the suffix S indicates the specific value on the boundary surface.

For dielectric boundaries, the conditions (3.17) can be expressed as

$$\left(\frac{\partial V}{\partial t}\right)_{1S} = \left(\frac{\partial V}{\partial t}\right)_{2S} \quad \text{and} \quad \varepsilon_1 \left(\frac{\partial V}{\partial n}\right)_{1S} = \varepsilon_2 \left(\frac{\partial V}{\partial n}\right)_{2S}, \tag{4.14}$$

which can also be specified by continuities of $(V)_S$ and $\varepsilon(\partial V/\partial n)_S$ across the boundary. On a dielectric boundary, due to different ε_1 and ε_2, $(\partial V/\partial n)_S$ cannot be continuous in general; however the condition can be specified by V and $\partial V/\partial n$ at a point on the surface S. However, such a dielectric boundary may not necessarily be a field boundary a specific if we consider that all the boundary points are physically characterized by a common property. As discussed in Section 3.4, a planar dielectric boundary in a uniform electric field is regarded as an equipotential only if it is parallel with the plates, which is, however, example. Nevertheless, a practical field is confined to an empty or material-filled space with conducting bounds S specified by continuous $(V)_S$ and discontinuous $\varepsilon(\partial V/\partial n)_S$. Known as *Dirichlet's conditions*, such boundaries are one type considered in the mathematical theory.

Although primarily "closed," practical boundary surface S may have an "open" area S' as well. If a small hole exists on a plate of a capacitor, for example, the field can no longer be uniform in the vicinity of the hole, from where some field-lines may "leak out." Such a hole area S' in an empty capacitor should be considered as distinct from closed area S, the former being physically characterized by the absence of surface charge. On S', inside and outside fields are connected by the same medium of ε_o, where $(\partial V/\partial n)_{S'}$ is continuous and expressed as continuous D_n across the area S'. In a general theory of boundary problems, such a boundary as specified by continuous $(\partial V/\partial n)_{S'}$ is called *Neumann's boundary*. Although somewhat specific from a physical viewpoint, a general boundary is regarded as consisting of such surfaces as S and S', on which Dirichlet's and Neumann's conditions can be imposed, respectively. In Subsection 4.4.3, the boundary problem for the Laplace-Poisson equations is discussed with Green's function, where boundary surfaces S and S' of both types are considered. In the next subsection, we pay attention to the uniqueness theorem of the potential prior to discussing the method of Green's function in Subsection 4.4.3.

4.4.2. Uniqueness Theorem

Static potentials given by solutions of the Laplace–Poisson equations are uniquely determined with the boundary conditions on S and S' characterized by continuous

$(V)_S$ and $(\partial V/\partial n)_{S'}$, respectively, constituting a statement known as the *uniqueness theorem*. Here the theorem is proved, thereby assuring that the Laplace-Poisson equations have unique solutions in a given field.

Assuming that two different potentials V and V' obey the same Laplace–Poisson equations, we can prove that the difference function $\varphi = V - V'$ can take no values other than zero at all given points. For that purpose we consider a trial vector function $\varphi\nabla\varphi$, which is then subjected to the Gauss theorem. For such a function we can write

$$\int_v \operatorname{div}(\varphi\nabla\varphi)\,\mathrm{d}v = \int_{S+S'} \varphi\nabla\varphi.\mathrm{d}\left(\boldsymbol{S}+\boldsymbol{S}'\right) = \int_{S+S'} \varphi\frac{\partial\varphi}{\partial n}\mathrm{d}\left(S+S'\right) = 0,$$

because $\varphi = 0$ on S and $\partial\varphi/\partial n = 0$ on S'. On the other hand,

$$\operatorname{div}(\varphi\nabla\varphi) = \varphi\operatorname{div}(\nabla\varphi) + (\nabla\varphi)^2 = (\nabla\varphi)^2,$$

because of the identity $\operatorname{div}(\nabla\varphi) \equiv 0$, and hence

$$\int_v (\nabla\varphi)^2\,\mathrm{d}v = 0.$$

Noting that the integrand $(\nabla\varphi)^2$ can be either positive or zero, this integral becomes zero only when $\nabla\varphi = 0$, or $\varphi =$ onstant. Therefore, $V - V'$ is physically insignificant constant, assuring that the potential $V(\boldsymbol{r})$ can always be unique at a given position $\boldsymbol{r}$.

4.4.3. Green's Function Method

In this subsection we discuss a general problem for the potential $V(\boldsymbol{r})$ in a field of distributed charges on the boundary surface S. We consider that the distribution can primarily be specified by a volume density $\rho(\boldsymbol{r}')$ at a point $\boldsymbol{r}'$ in the surface region, although a surface density can be redefined as surface charges on a mathematical boundary. By considering $\rho(\boldsymbol{r}')$, the potential should obey the Poisson equation

$$\nabla^2 V(\boldsymbol{r}') = -\frac{\rho\left(\boldsymbol{r}'\right)}{\varepsilon_o}. \tag{4.15}$$

Assuming highly localized charges, the density may be expressed as $\rho(\boldsymbol{r}-\boldsymbol{r}') = \rho\,\delta(\boldsymbol{r}-\boldsymbol{r}')$ using a delta function $\delta(\boldsymbol{r}-\boldsymbol{r}')$ to specify a finite ρ located only at $\boldsymbol{r} = \boldsymbol{r}'$, whereas 0 at any other point $\boldsymbol{r} \neq \boldsymbol{r}'$.

We define a function $G(\boldsymbol{r}-\boldsymbol{r}')$ that satisfies the equation

$$\nabla^2 G(\boldsymbol{r}-\boldsymbol{r}') = -\frac{\delta\left(\boldsymbol{r}-\boldsymbol{r}'\right)}{\varepsilon_o} \tag{4.16}$$

and is called the *Green function*. The equation (4.16) is the Laplace equation at all points $\boldsymbol{r}$ other than $\boldsymbol{r}'$, but it is the Poison equation at $\boldsymbol{r} = \boldsymbol{r}'$. Considering ρ as an isolated point charge, (4.16) can be solved for $G(\boldsymbol{r}-\boldsymbol{r}')$ to be expressed in the form

$$R(|\boldsymbol{r}-\boldsymbol{r}'|)\Omega(\omega),$$

where R and Ω are functions of the distance $|\boldsymbol{r}-\boldsymbol{r}'|$ and the solid angle ω subtending the area of charge $\delta(\boldsymbol{r}-\boldsymbol{r}')$, respectively. In this case, Ω, is independent of ω by symmetry, and hence (4.16) can be simplified as

$$\nabla^2 R(r) = \frac{1}{r^2}\frac{\mathrm{d}}{\mathrm{d}r}\left(r^2\frac{\mathrm{d}R}{\mathrm{d}r}\right) = 0,$$

where $|\boldsymbol{r}-\boldsymbol{r}'|$ is replaced by r for brevity. Integrating this, we obtain the solution $R(r) = -c_1/r + c_2$, where the constants $c_1$and c_2 can be determined by the boundary conditions as $r \to 0$ and ∞, respectively. Namely,

$$\lim_{r\to 0}\left(-r^2\frac{\mathrm{d}R}{\mathrm{d}r}\right) = \frac{1}{4\pi\varepsilon_o} \quad \text{and} \quad \lim_{r\to\infty} R(r) = 0,$$

resulting in $c_1 = 1/4\pi\varepsilon_o$ and $c_2 = 0$. Therefore, the Green function can be expressed as

$$G(\boldsymbol{r}-\boldsymbol{r}') = \frac{1}{4\pi\varepsilon_o}\frac{\delta\left(\boldsymbol{r}-\boldsymbol{r}'\right)}{|\boldsymbol{r}-\boldsymbol{r}'|}. \tag{4.17}$$

It is noted that the function (4.17) is symmetric by exchanging $\boldsymbol{r}$ and $\boldsymbol{r}'$, whereas the vector $\nabla G(\boldsymbol{r}-\boldsymbol{r}')$ is anti-symmetric in relation to the variable for differentiation: $\boldsymbol{r}$ or $\boldsymbol{r}'$, i.e.,

$$\nabla_{\boldsymbol{r}} G(\boldsymbol{r}-\boldsymbol{r}') = -\nabla_{\boldsymbol{r}'} G(\boldsymbol{r}-\boldsymbol{r}'), \tag{4.18}$$

where the suffixes indicates variables for differentiation.

Acting on the local potential $V(\boldsymbol{r}')$, the Green function can be considered as a *propagator* to shift it to a separate point $\boldsymbol{r}$ in the field, for which we define a scalar function $G(\boldsymbol{r}-\boldsymbol{r}')V(\boldsymbol{r}')$ to satisfy the Poisson equation (4.15), thereby deriving an expression for the potential $V(\boldsymbol{r})$ at $\boldsymbol{r}$. In this method, a field vector written as $\boldsymbol{F}(\boldsymbol{r}) = -\nabla_{\boldsymbol{r}}\{G(\boldsymbol{r}-\boldsymbol{r}')V(\boldsymbol{r}')\}$ is considered for further analysis. For such a field vector $\boldsymbol{F}(\boldsymbol{r})$, the flux density at $(\boldsymbol{r}')$ on the boundary surface can be constructed as follows and evaluated in the limit of $\boldsymbol{r} \to (\boldsymbol{r}')$. That is,

$$\begin{aligned}\oint_S \boldsymbol{F}.\mathrm{d}\boldsymbol{S} &= \oint_S \lim_{\boldsymbol{r}\to\boldsymbol{r}'} \nabla_{\boldsymbol{r}}\left\{G\left(\boldsymbol{r}-\boldsymbol{r}'\right)V\left(\boldsymbol{r}'\right)\right\}.\mathrm{d}\boldsymbol{S}\\ &= \oint_S \lim_{\boldsymbol{r}\to\boldsymbol{r}'}\left\{G\left(\boldsymbol{r}-\boldsymbol{r}'\right)\nabla_{\boldsymbol{r}}V\left(\boldsymbol{r}'\right) + V\left(\boldsymbol{r}'\right)\nabla_{\boldsymbol{r}}G\left(\boldsymbol{r}-\boldsymbol{r}'\right)\right\}.\mathrm{d}\boldsymbol{S}.\end{aligned}$$

Using (4.17) and $\lim_{\boldsymbol{r}\to\boldsymbol{r}'}\nabla_{\boldsymbol{r}}V(\boldsymbol{r}')\cdot\mathrm{d}\boldsymbol{S} = (\partial V/\partial n)|\boldsymbol{r}-\boldsymbol{r}'|\mathrm{d}S$ in this expression, the first term becomes $\oint_S \frac{\partial V}{\partial n}\mathrm{d}S$ in the limit of $\boldsymbol{r}\to\boldsymbol{r}'$, and the second term vanishes because of $\lim_{\boldsymbol{r}\to\boldsymbol{r}'}\nabla_{\boldsymbol{r}}G\left(\boldsymbol{r}-\boldsymbol{r}'\right) = 0$. On the other hand, by the Gauss theorem we have

$$\oint_S \boldsymbol{F}.\mathrm{d}\mathbf{S} = \int_v (\nabla_{\boldsymbol{r}}.\boldsymbol{F})\,\mathrm{d}v = \int_v \nabla_r.\nabla_r \left\{G\left(\boldsymbol{r}-\boldsymbol{r}'\right)V\left(r'\right)\right\}\mathrm{d}v$$
$$= \int_v \nabla_{\boldsymbol{r}'}.\nabla_{\boldsymbol{r}'} \left\{G\left(\boldsymbol{r}-\boldsymbol{r}'\right)V\left(\boldsymbol{r}'\right)\right\}\mathrm{d}v$$
$$= \int_v V\left(\boldsymbol{r}'\right)\nabla^2_{\boldsymbol{r}'}G\left(\boldsymbol{r}-\boldsymbol{r}'\right)\mathrm{d}v + \int_v G\left(\boldsymbol{r}-\boldsymbol{r}'\right)\nabla^2_{\boldsymbol{r}'}V\left(\boldsymbol{r}'\right)\mathrm{d}v$$
$$= -\int_v G\left(\boldsymbol{r}-\boldsymbol{r}'\right)\frac{\rho\left(\boldsymbol{r}'\right)}{\varepsilon_o}\mathrm{d}v - \int_v V\left(\boldsymbol{r}'\right)\frac{\delta\left(\boldsymbol{r}-\boldsymbol{r}'\right)}{\varepsilon_o}\mathrm{d}v,$$

in which the second term of the last expression can be simplified as $V(\boldsymbol{r})/\varepsilon_o$. Here, $\oint_S \boldsymbol{F}\cdot\mathrm{d}S$ should be equal to $\varepsilon_o \oint_S \frac{\partial V}{\partial n}\mathrm{d}S$, and we finally obtain

$$V(\boldsymbol{r}) = \int_v \rho\left(\boldsymbol{r}'\right)G\left(\boldsymbol{r}-\boldsymbol{r}'\right)\mathrm{d}v + \varepsilon_o\left(\frac{\partial V}{\partial n}\right)_S,$$
$$= \frac{1}{4\pi\varepsilon_o}\int_{v'} \frac{\rho\left(\boldsymbol{r}'\right)\delta\left(\boldsymbol{r}-\boldsymbol{r}'\right)}{\left|\boldsymbol{r}-\boldsymbol{r}'\right|}\mathrm{d}v' + \varepsilon_o\left(\frac{\partial V}{\partial n}\right)_S. \tag{4.19}$$

It is noted that the integral in (4.19) is to be carried out for the volume of distributed charges v', and hence the first term expresses a *Coulomb potential* due to the total charge that looks like a point charge at a distant point $r \gg r'$. Also, we can apply formula (4.19) directly to individual charges in a group of charges, which can be added to obtain the total potential.

4.5. Simple Examples

In this section we discuss simple applications of the Laplace–Poisson equations and the Coulomb potential in fields of high symmetry, applications that require only elementary mathematical analysis. Problems for advanced mathematical analysis will be discussed in Chapter 5.

Example 1. A Long Cylindrical Capacitor.

The electric field and potential in a cylindrical capacitor were already discussed in Section 3.4. In this example the same problem is discussed with the Laplace equation as the mathematical exercise. The charge density on the conductor is expressed by a given charge per length, i.e., $\lambda = Q/l$, where the field is radial and perpendicular to the conductor and expressed as a function of radius ρ only if the length is sufficiently long, where $\rho = a$ and b are radii of inner and outer conducting surfaces. By virtue of cylindrical symmetry, the coordinates θ and z

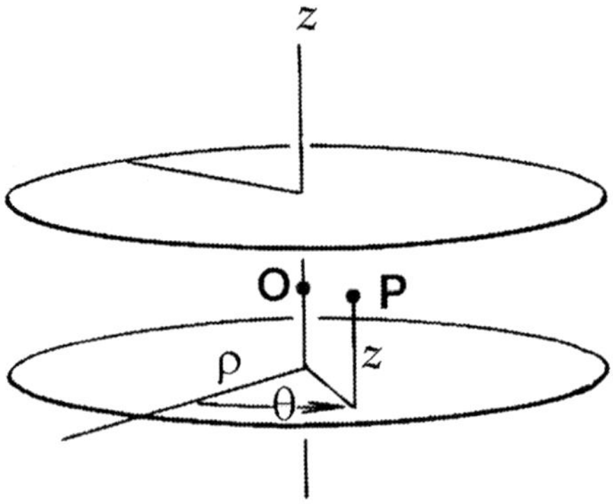

FIGURE 4.2. A circular parallel-plate condenser. The point P is specified by cylindrical coordinates (ρ, θ, z). The origin O is set at the center.

are not necessary, and hence from (4.12c) the Laplace equation can be written as

$$\frac{1}{\rho}\frac{\mathrm{d}}{\mathrm{d}\rho}\left(\rho\frac{\mathrm{d}V}{\mathrm{d}\rho}\right) = 0.$$

This can be integrated as

$$\frac{\mathrm{d}V}{\mathrm{d}\rho} = \frac{c}{\rho} \quad \text{and} \quad V = c\ln\rho + c' \quad \text{for} \quad a \le \rho \le b.$$

The integration constant c can be determined by the boundary condition on the conductor surface $\rho = a$, that is, $(D_\rho)_{\rho=a} = \sigma_a$ (the surface charge density). Here, $\sigma_a = \lambda/2\pi a$, and from the relations $-\varepsilon_o(\mathrm{d}V/\mathrm{d}\rho)_{\rho=a} = \sigma_a$, and $-\varepsilon_o(\mathrm{d}V/\mathrm{d}\rho)_{\rho=b} = \sigma_b$ we obtain

$$c = \lambda/2\pi\varepsilon_o.$$

The other constant, c', can be obtained from the relation $(V)_{\rho=a} = c\ln a + c'$, but is left undetermined as insignificant because only the potential difference is physically meaningful. Hence,

$$V(\rho) = \frac{\lambda}{2\pi\varepsilon_o}\ln\rho + c \quad \text{and} \quad E_\rho g - \frac{\partial V}{\partial\rho} = \frac{\lambda}{2\pi\varepsilon_o\rho}.$$

Exercise. Using the Laplace equation, obtain the expressions for the potential and electric field of radii a and b, respectively, in a spherical capacitor.

Example 2. A Parallel-Plate Capacitor of Circular Plates.

In a parallel-plate capacitor of gap d and circular area A of radius R, i.e., $A = \pi R^2$, we consider R not to be significantly larger than d. However, note that the capacitor is symmetric around the z-axis, as shown in Fig. 4.2, and hence the Laplace equation can be expressed with cylindrical coordinates ρ, θ, and z, where the angle θ is unnecessary for circular symmetry. That is,

$$\rho\frac{\partial}{\partial\rho}\left(\rho\frac{\partial V}{\partial\rho}\right) + \rho^2\frac{\partial^2 V}{\partial\rho^2} = 0.$$

Assuming that $V(\rho, z) = P(\rho)Z(z)$, the equation can be written as a sum of two separated terms of ρ only and of z only, that is,

$$\frac{1}{\rho P}\frac{\mathrm{d}}{\mathrm{d}\rho}\left(\rho\frac{\mathrm{d}P}{\mathrm{d}\rho}\right) + \frac{1}{Z}\frac{\mathrm{d}^2 Z}{\mathrm{d}z^2} = 0.$$

Therefore, by introducing a constant α^2, we can separate the equation into two equations

$$\frac{1}{Z}\left(\frac{\mathrm{d}^2 Z}{\mathrm{d}z^2}\right) = \alpha^2 \qquad \text{(i)}$$

and

$$\frac{1}{\rho P}\frac{\mathrm{d}}{\mathrm{d}\rho}\left(\rho\frac{\mathrm{d}P}{\mathrm{d}\rho}\right) = -\alpha^2, \qquad \text{(ii)}$$

Rewriting (ii), we obtain

$$\frac{\mathrm{d}^2 P}{\mathrm{d}\rho^2} + \frac{1}{\rho}\frac{\mathrm{d}P}{\mathrm{d}\rho} + \alpha^2 P = 0,$$

that is, the Bessel equation, whose solution is expressed by

$$P(\rho) = P(0)\, J_o(\alpha\rho).$$

Equation (i) has a solution generally given by $\exp(\pm\alpha z)$, but we take their linear combination to write

$$Z(z) = Z(0)\sinh(\alpha z)$$

to satisfy the boundary conditions on the plates at $z = \pm\frac{1}{2}d$. Denoting the potential difference between the plates by ΔV, the solution of the Laplace equation can then be expressed as

$$V(\rho, z) = \frac{1}{2}\Delta V\, J_o(\alpha\rho)\sinh(\alpha z), \qquad \text{where} \quad \Delta V = 4P(0)Z(0).$$

The boundary conditions at $z = \pm\frac{1}{2}d$ are

$$-\varepsilon_o\left(\frac{\partial(\rho, z)}{\partial z}\right)_{z=\pm\frac{1}{2}d} = \pm\frac{\varepsilon_o}{2\alpha}\Delta V\cosh\left(\frac{1}{2}\alpha d\right) J_o(\alpha\rho) = \pm\sigma(\rho),$$

which are not constant as in a capacitor of $A = \infty$, but approaching constant values as $\alpha \to 0$. It is noted that ρ and z become independent in this limit.

4.6. The Coulomb Potential

Examples for the Coulomb potential (4.19) are discussed in this section using the point-charge model to calculate for static fields. Although only mathematically justifiable, a point charge can be considered as a solution (4.19) of the Laplace equation, which is used as a classical model to represent charged particles.

Distributed charges are normally described in terms of densities $\rho(\boldsymbol{r}')$ in volume, $\sigma(\boldsymbol{r}')$ in surface area and $\lambda(\boldsymbol{r}')$ in line, and the potentials at a position $\boldsymbol{r}$ are expressed respectively as

$$V(\boldsymbol{r}-\boldsymbol{r}') = \frac{1}{4\pi\varepsilon_o}\int_{v'} \frac{\rho\left(\boldsymbol{r}'\right) d^3\boldsymbol{r}'}{|\boldsymbol{r}-\boldsymbol{r}'|},$$

$$= \frac{1}{4\pi\varepsilon_o}\int_{S'} \frac{\sigma\left(\boldsymbol{r}'\right) d^2\boldsymbol{r}'}{|\boldsymbol{r}-\boldsymbol{r}'|},$$

and

$$= \frac{1}{4\pi\varepsilon_o}\int_{L'} \frac{\lambda\left(\boldsymbol{r}'\right) d^1\boldsymbol{r}'}{|\boldsymbol{r}-\boldsymbol{r}'|}. \tag{4.20}$$

Example 1. A Uniformly Charged Straight Conductor.

Figure 4.3 shows a long conductor along the z-axis. If it is uniformly charged the length should be infinitely long. Also, it is noted that such a line conductor is considered as one conductor of a cylindrical capacitor, where the second one is placed at infinity. Disregarding effects from the second conductor, the Coulomb potential arising from the long conductor can be calculated at a distance r perpendicular to the conductor.

As illustrated in the figure, the electric field vector due to charge elements λdz_1 and λdz_2 located at $\pm z$ becomes perpendicular to the line by symmetry as the z

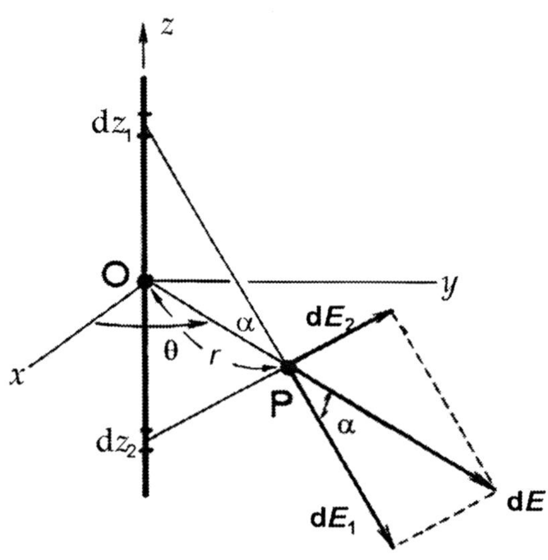

FIGURE 4.3. Calculating the $\boldsymbol{E}$ vector of a long charged rod.

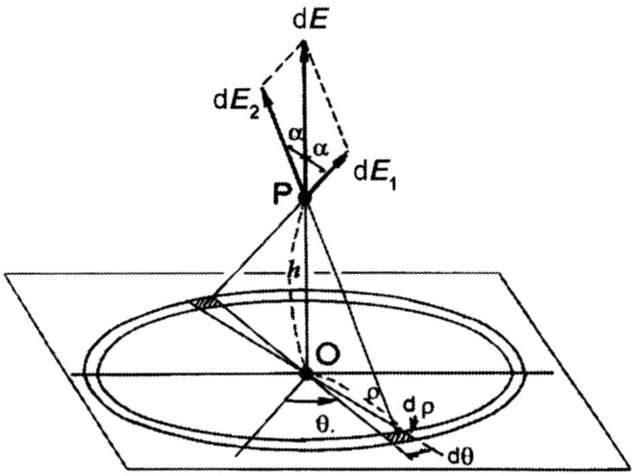

FIGURE 4.4. Calculating the $\boldsymbol{E}$ vector of a charged plate of infinite area.

components of $\mathrm{d}\boldsymbol{E}_1$ and $\mathrm{d}\boldsymbol{E}_2$ are cancelled. Therefore,

$$|\mathrm{d}\boldsymbol{E}| = \frac{1}{4\pi\varepsilon_o}\int_0^\infty \frac{2\lambda\,\mathrm{d}z}{z^2+r^2}\cos\alpha, \quad \text{where} \quad \cos\alpha = \frac{r}{\sqrt{z^2+r^2}}.$$

Hence, we obtain

$$|\boldsymbol{E}| = E_r = \frac{\lambda}{2\pi\varepsilon_o}\int_0^\infty \frac{r\mathrm{d}z}{\left(z^2+r^2\right)^{\frac{3}{2}}} = \frac{\lambda}{2\pi\varepsilon_o r},$$

and the logarithmic potential

$$V(r) = \frac{\lambda}{2\pi\varepsilon_o}\ln r + \text{const.}$$

Example 2. A Charged Surface of a Conductor.

Figure 4.4 shows a flat conducting surface, where the charge distribution is considered to be uniform. However, this assumption is valid only if the surface extends to infinity. Besides, $\boldsymbol{E} = 0$ inside the conductor, as characterized by a constant V that can be equal to 0 if the area is infinitely large. In this case, the field vector is strictly perpendicular to the surface because of the symmetry and of a constant magnitude as derived by the Gauss theorem.

As illustrated in the figure, at a point $P(0, 0, h)$ we consider all surface charges in a ring area between circles of radii ρ and $\rho + \mathrm{d}\rho$, and notice that field elements $\mathrm{d}\boldsymbol{E}_1$ and $\mathrm{d}\boldsymbol{E}_2$ due to differential charges on areas $\rho\mathrm{d}\rho\mathrm{d}\theta$ at θ and $\theta + \pi$ have horizontal components that are cancelled by symmetry and signified only by vertical components. Thus, the net contribution is determined by the integral

$$E_h = \int_S \mathrm{d}E_h = \frac{\sigma}{\varepsilon_o}\int_0^\infty \frac{r\mathrm{d}r}{\left(r^2+h^2\right)^{\frac{3}{2}}} = \frac{\sigma}{\varepsilon_o}\left|-\frac{h}{\sqrt{r^2+h^2}}\right|_0^\infty = \frac{\sigma}{\varepsilon_o}, \quad \text{and} \quad D_h = \sigma,$$

which is the same result as obtained previously with the Gauss theorem.

4.7. Point Charges and the Superposition Principle

Equations (4.20) permit us to interpret distributed charges at $\boldsymbol{r}'$ as a group of point charges that are individually responsible for the potential at $\boldsymbol{r}$, and their combined result is expressed in integral form. We realize that such an additive property of static potentials is simply the mathematical consequence of a linear differential equation. On the other hand, physically it is due to algebraically additive charges as postulated early on, from which the law of additive field vectors can arise. However, the nature of electricity is still not quite substantiated in contemporary physics, so that it is fair to mention that the *superposition principle* governs the nature of electromagnetic fields.

If charges are confined to a small isolated sphere, the formula (4.20) allows us to consider it at a distant point as a point charge from which radial field-lines emerge or converge. On the other hand, if another charge is present nearby, the field can no longer be radial, as expressed by a superposition of two fields. At this point we realize that those charges cannot form a stable static configuration, unless supported by a hidden mechanism. Nevertheless, disregarding the stability, such clustered charges are discussed as an exercise for electrical states of matter.

Guided by the superposition principle, we can use formulas

$$V(\boldsymbol{r}) = \frac{1}{4\pi\varepsilon_o}\sum_i \frac{q_i}{|\boldsymbol{r}-\boldsymbol{r}_i|} \tag{4.21}$$

for a system of point charges q_i located at $\boldsymbol{r}_i$, where $i = 1, 2, \ldots,$ and calculate the electric field vector with and $\boldsymbol{E}(\boldsymbol{r}) = -\boldsymbol{\nabla}V(\boldsymbol{r})$.

4.7.1. An Electric Image

Figure 4.5(a) shows a field pattern of two point charges of an equal q, which is compared with a pattern of charges $+q$ and $-q$ sketched in Fig. 4.5(b), where these charges are separated by a distance $2h$. The field-lines deviate considerably from radial symmetry in their vicinity; between charges the density of field-line charges is low in the former, compared with a high density in the latter, arising from repulsive and attractive forces, respectively. Although unstable, both represent cases for some electrical states of matter each providing a useful model for distributed charge on conductors, polarization of insulators, and dipolar molecules.

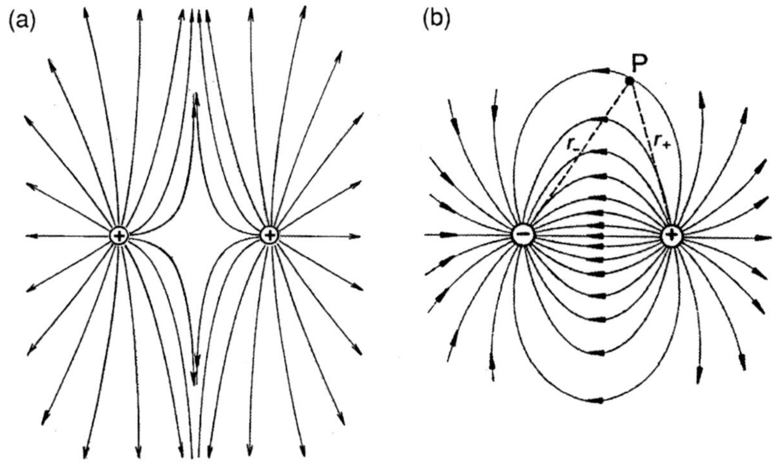

FIGURE 4.5. (a) Field-lines due to two identical point charges separated by a short distance. (b) Field-lines due to point charges of an equal magnitude and opposite signs.

At a point P, the potential of two charges $+q$ and $-q$ is written as

$$V(r_+, r_-) = \frac{1}{4\pi\varepsilon_o}\left(\frac{q}{r_+} - \frac{q}{r_-}\right), \tag{4.22}$$

where r_+ and r_- are the distances from $+q$ and $-q$, respectively.

Among the many field-lines sketched in Fig. 4.5(b) it is noted that a plane bisecting the distance between q and $-q$ is equipotential, because $r_+ = r_-$. Therefore, we can replace it by a conducting sheet without modifying the original field-lines on the right side. Further, noticeable is that the field pattern is unchanged by removing $-q$ from the field. Thus, two opposite charges q and $-q$ are equivalent to a single charge q and a conducting plane. We can consider that the conducting plane is represented by $-q$, which is called an *image charge* on the plane. Figure 4.6(a) shows the field-lines when a positive q is placed at a distance h from a conducting surface. Here, lines of the image charge $-q$ are drawn with dotted curves.

Field-lines of E arising from the charge q should be terminated on the conducting surface, where the induced charge density can be calculated with the diagram shown in Fig. 4.5(b). At a point distant r on the plane from the center between q and its image $-q$, their field vectors $\boldsymbol{E}_+$ and $\boldsymbol{E}_-$ can be added to obtain $\boldsymbol{E}$ that is perpendicular to the surface. That is,

$$|\boldsymbol{E}_+| = |\boldsymbol{E}_-| = \frac{q}{4\pi\varepsilon_o\left(h^2 + r^2\right)},$$

and

$$E = 2\,|\boldsymbol{E}_+|\cos\alpha \quad \text{where} \quad \cos\alpha = \frac{h}{\sqrt{h^2 + r^2}}.$$

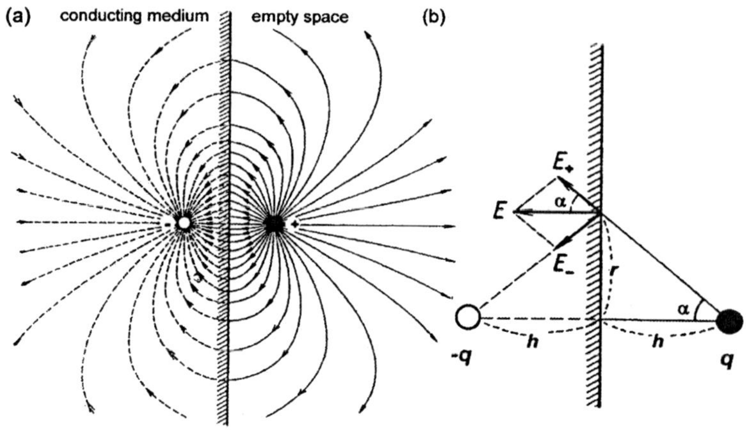

FIGURE 4.6. (a) A point charge q and its electric image $-q$ on infinite conducting plane. (b) Calculating the image force.

This perpendicular field should be related to an induced charge density, which can be written as

$$-\sigma(r) = -\varepsilon_o E(r) = -\frac{qh}{2\pi\left(h^2 + r^2\right)^{\frac{3}{2}}},$$

indicating maximum at $r = 0$. The total amount of induced charges on the surface can be calculated by integrating $-\sigma(r)$ over the entire area, that is,

$$-\int_0^\infty \sigma(r)\, r\mathrm{d}r \int_0^{2\pi} \mathrm{d}\theta = \frac{qh}{2\pi}\int_0^\infty \frac{r\mathrm{d}r}{\left(h^2 + r^2\right)^{\frac{3}{2}}} \times 2\pi = -\frac{qh}{2\pi}\frac{1}{h}2\pi = -q,$$

which is exactly equal to the image charge. Owing to the negative sign of the image charge or total induced charge, the charge q is attracted by the conducting plate. The force is expressed as the Coulomb force between point charges q and $-q$ and called an *image force*. That is,

$$F = -\frac{1}{4\pi\varepsilon_o}\frac{q^2}{(2h)^2}. \tag{4.23}$$

4.7.2. Electric Dipole Moment

Asymmetric molecules are often represented electrically by a pair of charges q and $-q$ that are separated by a small finite distance l. Such a separation l is so small in size that the pair of charges appears as if neutral as observed at a very distant point. However, at a closer distance, the pair of charges exhibits explicit field-lines, called a *dipolar* field. Such a distance r for a predominantly dipolar

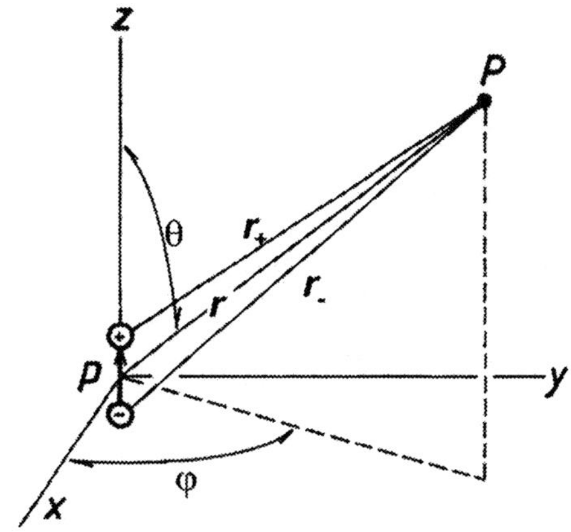

FIGURE 4.7. Calculation of the dipolar potential at a distant point $P(r, \theta, \varphi)$.

field is significantly larger than l, where the pair of charges is characterized by the product ql called an electric *dipole.*

Figure 4.7 shows such a model of a dipole located at the origin of a coordinate system, and along the z axis. It is usually free to rotate, and hence represented conveniently by a vector quantity defined by

$$\boldsymbol{p} = q\boldsymbol{l}, \tag{4.24}$$

where the vector $\boldsymbol{l}$ signifies the direction from $-q$ to q, and the vector $\boldsymbol{p}$ is referred to as the *dipole moment.* The MKSA unit of a dipole moment is C-m, which however is too large for molecular dipoles of the order of 10^{-29} C-m. In molecular physics, we traditionally use the unit called "Debye" defined as equal to 10^{-18} esu-cm. Typical examples are $\boldsymbol{p} = 0.12$ Debye for CO, and 1.86 Debye for H_2O.

The dipole potential can be obtained with a model in Fig. 4.5(b) by considering $2h = l$. The potential can be calculated from (4.22) approximately for $r \gg l$. Noted that

$$r_\pm^2 = r^2 + \left(\frac{l}{2}\right)^2 \mp 2r\left(\frac{l}{2}\right)\cos\theta,$$

we expand $\dfrac{1}{r_\pm}$ with respect to a small $\dfrac{l}{r}$, and derive

$$\frac{1}{r_\pm} = \frac{1}{r\sqrt{1 + \left(\dfrac{l}{r}\right)^2 \mp \dfrac{l}{r}\cos\theta}} \approx \frac{1}{r}\left(1 \pm \frac{l}{2r}\cos\theta + \cdots\right) \quad \text{for} \quad r \gg l.$$

Therefore, the potential can be approximately expressed as

$$V(r_+, r_-) \approx \frac{1}{4\pi\varepsilon_o}\frac{ql}{r^2}\cos\theta = \frac{1}{4\pi\varepsilon_o}\frac{p\cos\theta}{r^2}.$$

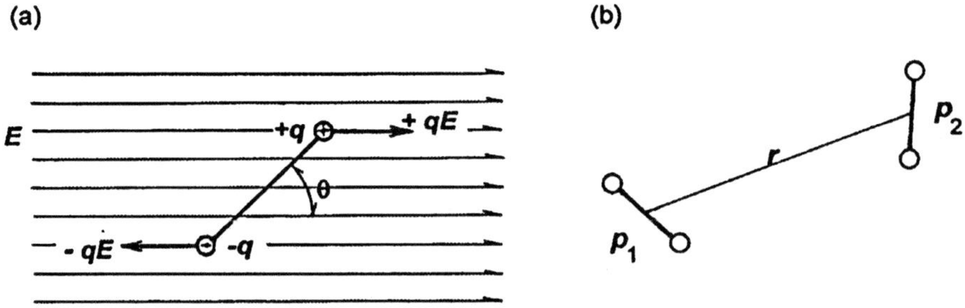

FIGURE 4.8. (a) An electric dipole moment in a uniform electric field $\boldsymbol{E}$. (b) Dipole-dipole interaction.

Since we defined the dipole moment as a vector, the dipole potential can be written as

$$V_1(r) = \frac{1}{4\pi\varepsilon_o} \frac{\boldsymbol{p}.\boldsymbol{r}}{r^3}, \tag{4.25}$$

where the index 1 is attached to $V_1(r)$ as related to the order in the series expansion of potentials, as the potential will be expressed in higher approximation in later discussions.

If an electric dipole $\boldsymbol{p}$ can be in rotational motion like a molecular dipole, external field $\boldsymbol{E}$ exerts a torque. Figure 4.8(a) shows such a dipole in a uniform field, where a couple of forces $+q\boldsymbol{E}$ and $-q\boldsymbol{E}$ is responsible for rotating the dipole moment. Accordingly, the dipole has a potential energy for rotation

$$U = -\boldsymbol{p}.\boldsymbol{E} = -pE\cos\theta \tag{4.26a}$$

as related to the torque

$$\tau = \boldsymbol{p} \times \boldsymbol{E} \quad \text{where} \quad |\tau| = pE\sin\theta. \tag{4.26b}$$

Here θ is the angle between the vectors $\boldsymbol{p}$ and $\boldsymbol{E}$.

If $\boldsymbol{E}$ is not uniform over the dipole moment along a direction x, we need to consider a force related to the field gradient in addition to the torque given by (4.26b). Such a force is responsible for translational motion of the dipole. If the field $\boldsymbol{E}$ varies as a function of position x, a field gradient arises from the difference between fields at q and $-q$, i.e.,

$$E(x + \mathrm{d}x) - E(x) = (\partial E/\partial x)\,\mathrm{d}x,$$

where $\mathrm{d}x = l\cos\theta$, and therefore the net force is given by

$$F = q\frac{\partial E}{\partial x} l\cos\theta = p\cos\theta\frac{\partial E}{\partial x}. \tag{4.26c}$$

4.7.3. The Dipole-Dipole Interaction

In dielectric materials, polar molecules can be represented by their dipole moments. In this case, the mutual interaction energy arises from dipole-dipole interactions. Considering two dipole moments $\boldsymbol{p}_1$ and $\boldsymbol{p}_2$ that can be rotated at fixed positions as shown in Fig. 4.8(b), the interaction potential can be calculated for a given distance. The distance between $\boldsymbol{p}_1$ and $\boldsymbol{p}_2$ can be expressed as a vector $\boldsymbol{r}$, although the direction is not significant for the mutual interaction energy. Nevertheless, directions of interacting dipole moments are significantly related to their distance from one another. On the other hand, the interaction energy should be unchanged by inverting $\boldsymbol{r}$ to $-\boldsymbol{r}$, as consistent with Newton's action-reaction principle.

The interaction energy can be calculated for $\boldsymbol{p}_1$ to be in the dipole field of $\boldsymbol{p}_2$ or for $\boldsymbol{p}_2$ in the field of $\boldsymbol{p}_1$. In either case, we combine (4.25) and (4.26a) as derived in the following. The electric field originating from $\boldsymbol{p}_2$ at the position of $\boldsymbol{p}_1$ can be expressed as

$$\boldsymbol{E}_{12} = -\frac{1}{4\pi\varepsilon_o}\boldsymbol{\nabla}_1\frac{\boldsymbol{p}_2.\boldsymbol{r}}{r^3} = \frac{3\left(\boldsymbol{p}_2.\boldsymbol{r}\right)\boldsymbol{r} - \boldsymbol{p}_2 r^2}{4\pi\varepsilon_o r^5},$$

and hence $\boldsymbol{p}_1$ has an energy

$$U_{12} = -\boldsymbol{p}_1.\boldsymbol{E}_{12} = \frac{1}{4\pi\varepsilon_o}\left[\frac{(\boldsymbol{p}_1.\boldsymbol{p}_2)r^2 - 3(\boldsymbol{p}_1.\boldsymbol{r})(\boldsymbol{p}_2.\boldsymbol{r})}{r^5}\right]. \tag{4.27}$$

Similarly, we obtain the expression for $U_{21} = -\boldsymbol{p}_2.\boldsymbol{E}_{21}$ that is identical to (4.27). Equation (4.27) is the formula for a dipole-dipole interaction.

5
The Legendre Expansion of Potentials

5.1. The Laplace Equation in Spherical Coordinates

In electrostatics a basic problem is to find the potential function for a given charge distribution that can be solved with the Laplace-Poisson equations under specified boundary conditions. Although the cause for distributed charges is unspecified in the given system, here we are only concerned about the electrostatic problem. For many applications we are interested in the potential at a distant point from distributed charges in a small region. The corresponding field is often dominated by deviations from spherical symmetry, a deviation that can be conveniently viewed with polar coordinates with respect to the center of distribution. Such a charge distribution is often described with respect to a *unique* direction for deformation that arises from an internal origin.

The Laplace equation $\nabla^2 V = 0$ is a *partial differential equation* to which the method of separating variables is applied, reducing it to a set of *ordinary differential equations* for the individual curvilinear coordinates. In polar coordinates r, θ, and φ, the Laplace equation can be expressed as

$$\frac{1}{r^2}\frac{\partial}{\partial r}\left(r^2\frac{\partial V}{\partial r}\right)+\frac{1}{r^2\sin\theta}\frac{\partial}{\partial\theta}\left(\sin\theta\frac{\partial V}{\partial\theta}\right)+\frac{1}{r^2\sin^2\theta}\frac{\partial^2 V}{\partial\varphi^2}=0. \tag{5.1}$$

Assuming $V(r,\theta,\varphi) = R(r)\Theta(\theta)\Phi(\varphi)$, where the factors R, Θ and Φ are functions of r, θ, and φ, respectively, equation (5.1) can be written as

$$\frac{1}{r^2R}\frac{\mathrm{d}}{\mathrm{d}r}\left(r^2\frac{\mathrm{d}R}{\mathrm{d}r}\right)+\frac{1}{r^2\Theta\sin\theta}\frac{\mathrm{d}}{\mathrm{d}\theta}\left(\sin\theta\frac{\mathrm{d}\Theta}{\mathrm{d}\theta}\right)+\frac{1}{r^2\Phi\sin^2\varphi}\frac{\mathrm{d}^2\Phi}{\mathrm{d}\varphi^2}=0. \tag{i}$$

First, we let

$$\frac{\mathrm{d}^2\Phi}{\mathrm{d}\varphi^2}=-\mathrm{m}^2\Phi, \tag{ii}$$

from which we have

$$\Phi(\varphi)=\Phi_o\exp(\pm i\mathrm{m}\varphi).$$

In order for the potential to assume a unique value, we must consider that $\Phi(\varphi) = \Phi(\varphi + 2\pi)$, so that m is determined from $\exp(\pm 2m\pi i) = 1$. Therefore, m must be either 0 or $\pm$ an integer, i.e., $m = 0, \pm 1, \pm 2, \ldots$.

Next, in (i) we set

$$\frac{1}{R}\frac{d}{dr}\left(r^2\frac{dR}{dr}\right) = k, \tag{iii}$$

in order to obtain a radial function that can be expressed as $R(r) = Ar^n$, where n is an integer. Then, from (iii) k should be related to n as given by

$$k = n(n+1). \tag{iv}$$

It is noted that (iv) can also be fulfilled by another integer $n' = -(n+1)$, as we can write $k = n'(n'+1)$. Therefore, generally, the function $R(r)$ can be expressed as

$$R(r) = A_n r^n + \frac{B_n}{r^{n+1}}, \tag{v}$$

where A_n and B_n are constants for $R(r)$ to be determined as a linear combination of these terms of r^n and $1/r^{n+1}$, as required by boundary conditions. For the positive r, it is sufficient to consider that $n = 0, 1, 2, 3, \ldots$.

Using (ii) and (iv) in the relation (i), we obtain the equation for $\Theta(\theta)$ expressed as

$$n(n+1) + \frac{1}{\Theta\sin\theta}\frac{d}{d\theta}\left(\sin\theta\frac{d\Theta}{d\theta}\right) - \frac{m^2}{\sin^2\theta} = 0,$$

and hence the function $\Theta(\theta)$ depends generally on indexes n and m. However, noting that the specific case of $m = 0$ signifies the potential independent of the angle φ, such an axially symmetric function is specified only by the index n. Therefore, the function Θ for $m = 0$ is specially written as $P_n(\theta)$. For the axial function $P_n(\theta)$, called the *Legendre function*, we have a differential equation

$$\frac{1}{\sin\theta}\frac{d}{d\theta}\left(\sin\theta\frac{dP_n}{d\theta}\right) + n(n+1)P_n = 0, \tag{5.2a}$$

which can also be expressed in terms of a variable $\mu = \cos\theta$; that is,

$$\frac{d}{d\mu}\left\{\left(1-\mu^2\right)\frac{dP_n}{d\mu}\right\} + n(n+1)P_n = 0 \tag{5.2b}$$

Noting that the equation (5.2b) is unchanged when n is replaced by $-(n+1)$, we have the relation

$$P_n(\cos\theta) = P_{-n-1}(\cos\theta) \tag{5.3}$$

Combining $P_n(\cos\theta)$ with the radial function $R(r)$, the axial potential can be expressed as

$$V(r,\theta)=\sum_n\left(A_n r^n+\frac{B_n}{r^{n+1}}\right)P_n(\cos\theta), \tag{5.4}$$

which is a linear combination of all solutions indexed by integers n. The Legendre function is specified by an index n. The first term in the brackets of (5.4) is zero at $r=0$, while diverging as $r\to\infty$, whereas the second term is singular at $r=0$ but finite at all r. Therefore, these terms are considered to represent different regions in the field, and on their boundaries coefficients A_n and B_n should be determined for each potential signified by n of the Legendre function.

5.2. Series Expansion of the Coulomb Potential

In the previous section the Laplace equation is separated into ordinary differential equations, for which the origin of polar coordinates is unspecified. In Chapter 4, on the other hand, the solution (4.19) is obtained with respect to conducting boundaries on which charges are distributed, and where charges and the point of observation are specified by vectors $\boldsymbol{r}'$ and $\boldsymbol{r}$, respectively. The Coulomb potential at $\boldsymbol{r}$ is thus expressed as

$$V(\boldsymbol{r})=\frac{1}{4\pi\varepsilon_o}\int_{v(\boldsymbol{r}_o)}\frac{\rho(\boldsymbol{r}_o)\,d^3\boldsymbol{r}_o}{|\boldsymbol{r}-\boldsymbol{r}_o|},$$

where $\boldsymbol{r}'$ and dv' in (4.19) are replaced by $\boldsymbol{r}_o$ and $d^3\boldsymbol{r}_o$ to specify the source in the following discussion. A general situation for calculating the potential is illustrated in Figure 5.1, where we have the geometrical relation $(\boldsymbol{r}-\boldsymbol{r}_o)^2=r^2+r_o^2-2rr_o\cos\theta$. Dividing the space into two regions characterized by $r>r_o$ and

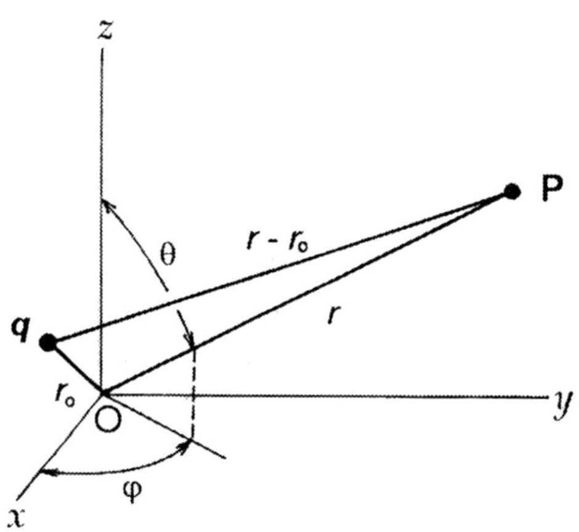

FIGURE 5.1. A point charge q located near the origin O and a distant point P showing the relation between coordinates.

$r < r_o$, we have

$$\frac{1}{|\boldsymbol{r}-\boldsymbol{r}_o|} = \frac{1}{r\sqrt{1-2\mu s+\mu^2}}, \qquad \text{where} \quad s = \frac{r_o}{r},$$

and

$$\frac{1}{|\boldsymbol{r}-\boldsymbol{r}_o|} = \frac{1}{r_o\sqrt{1-2\mu s+s^2}}, \qquad \text{where} \quad s = \frac{r}{r_o},$$

respectively, where the factor $(1-2\mu s+s^2)^{-1}$, known as the *generating function*, can be expanded into power series for a small variable s. That is,

$$\frac{1}{\sqrt{1-2\mu s+s^2}} = 1+\mu s+\frac{1}{2}(3\mu^2-1)s^2+\frac{3}{2}(5\mu^3-3\mu)s^3 + \frac{1}{8}(35\mu^4-30\mu^2+3)s^4+\ldots,$$

where the coefficients can be identified as Legendre's functions $P_n(\mu)$ for $n = 0, 1, 2, 3, \ldots$. Leaving the proof to Section 5.3, we write the series expansion as

$$\frac{1}{\sqrt{1-2\mu s+s^2}} = \sum_n P_n(\mu)\, s^n, \tag{5.5}$$

where

$$P_0(\mu) = 1,\ P_1(\mu) = \mu,\ P_2(\mu) = 1(3\mu^2-1),\ P_3(\mu) = \partial(5\mu^3-3\mu), \ldots .$$

Using $P_n(\mu)$, the Coulomb potential for the region $r > r_o$ can be expressed by

$$V(\boldsymbol{r}) = \frac{1}{4\pi\varepsilon_o}\sum_n \frac{P_n(\mu)}{r^{n+1}} \int_{v(r_o)} \rho(\boldsymbol{r}_o)\, r_o^n \mathrm{d}^3\boldsymbol{r}_o, \tag{5.6a}$$

whereas, for $r < r_o$,

$$V(\boldsymbol{r}) = \frac{1}{4\pi\varepsilon_o}\sum_n r^n P_n(\mu) \int_{v(r_o)} \frac{\rho(\boldsymbol{r}_o)}{r_o^{n+1}} \mathrm{d}^3\boldsymbol{r}_o. \tag{5.6b}$$

The potential $V(\boldsymbol{r})$ of (5.6a) and (b) are expressed as a superposition of component potentials $V_n(\boldsymbol{r})$ for $n = 0, 1, 2, \ldots$, that is, $V(\boldsymbol{r}) = \Sigma_n V_n(\boldsymbol{r})$.

In the region for $r > r_o$ in particular, for $n = 0$,

$$V_o(\boldsymbol{r}) = \frac{1}{4\pi\varepsilon_o}\frac{Q}{r}, \text{ where } \quad Q = \int_{v(\boldsymbol{r}_o)} \rho(\boldsymbol{r}_o)\boldsymbol{r}_o d^3\boldsymbol{r}_o$$

represents a point charge model effectively for the total charge Q. For $n = 1$,

$$V_1(\boldsymbol{r}) = \frac{1}{4\pi\varepsilon_o}\frac{p\cos\theta}{r^2}, \qquad \text{where} \quad \boldsymbol{p} = \int_{v(\boldsymbol{r}_o)} \rho(\boldsymbol{r}_o)\boldsymbol{r}_o \mathrm{d}^3\boldsymbol{r}_o$$

expresses the dipole moment associated with distributed charges.

5.3. Legendre's Polynomials

We discuss the Legendre function briefly in this mathematical section. Here, for the function $P_n(\mu)$ defined as expansion coefficients in (5.5), we show the proof that $P_n(\mu)$ satisfies the equation (5.2b), and the recurrence formula.

Differentiating (5.5) with respect to the variable s, we obtain

$$\frac{\mu - s}{\left(1 - 2\mu s + s^2\right)^{\frac{3}{2}}} = \sum_n n s^{n-1} P_n(\mu).$$

Using (5.5) in this expression, we can write

$$(\mu - s) \sum_n P_n(\mu) s^n = \sum_n \left(n s^{n-1} + n s^{n+1} - 2n s^n\right) P_n(\mu)$$

and, hence,

$$\sum_n (2n+1) \mu s^n P_n(\mu) = \sum_n (n+1) s^{n+1} P_n(\mu) + \sum_n n s^{n-1} P_n(\mu)$$

We can compare terms of s^n on both sides of this relation and obtain a *recurrence formula*,

$$(2n+1) \mu P_n(\mu) = (n+1) P_{n+1}(\mu) + n P_{n-1}(\mu). \tag{5.7}$$

For positive values, $k = n(n+1) \geq 0$, the order n can be equal to $-1, 0, 1, 2, \ldots$, therefore from (5.7) we can write, successively, that

$$\begin{aligned}
P_{-1}(\mu) &= 0,\ P_0(\mu) = 1,\ P_1(\mu) = \mu P_0(\mu) = \mu, \\
2P_2(\mu) &= 3\mu P_1(\mu) - P_0(\mu) = 3\mu^2 - 1, \\
P_3(\mu) &= 5\mu P_2(\mu) - 2P_1(\mu) = \frac{1}{2}(15\mu^3 - 9\mu), \ldots .
\end{aligned}$$

Thus, the expansion coefficients $P_n(\mu)$ in the series (5.5), known as the *Legendre polynomials*, are related by the recurrence formula (5.7).

Next, we differentiate (5.5) with respect to the variable μ, and obtain

$$\frac{s}{\left(1 - 2\mu s + s^2\right)^{\frac{3}{2}}} = \sum_n s^n P_n'(\mu), \quad \text{where} \quad P_n'(\mu) = \frac{dP_n(\mu)}{d\mu}.$$

Hence,

$$s\left(\sum_n s^n P_n(\mu)\right) = \left(1 - 2\mu s + s^2\right) \sum_n s^n P_n'(\mu).$$

Equating the coefficient of s^n on the left side to those on the right, we obtain another recurrence formula,

$$P_n(\mu) = P_{n+1}'(\mu) + P_{n-1}'(\mu) - 2\mu P_n'(\mu). \tag{5.8}$$

Next, we derive the Legendre equation (5.2b) using these recurrence formulas to show that $P_n(\mu)$ is the solution. Differentiating (5.7) with respect to μ, we obtain

$$(2n+1)\left\{P_n(\mu) + \mu P_n'(\mu)\right\} = (n+1) P_{n+1}'(\mu) + n P_{n-1}'(\mu),$$

and so

$$(2n+1)\,P_n(\mu) = (n+1)\,P'_{n+1}(\mu) + n P'_{n-1}(\mu) - (2n+1)\,\mu P'_n(\mu)$$

$$= n\left\{P'_{n+1}(\mu) + P'_{n-1}(\mu) - 2\mu P'_n(\mu)\right\} + P'_{n+1}(\mu) - \mu P'_n(\mu)$$

or

$$(2n+1)P_n(\mu) = (n+1)\left\{P'_{n+1}(\mu) + P'_{n-1}(\mu) - 2\mu P'_n(\mu)\right\} \\ - P'_{n-1}(\mu) + \mu P'_n(\mu).$$

Using (5.7), we can replace the quantity in the curly brackets by $P_n(\mu)$, and

$$(n+1)\,P_n(\mu) = P'_{n+1}(\mu) - \mu P'_n(\mu) \quad \text{or} \quad n P_n(\mu) = -P'_{n-1}(\mu) + \mu P'_n(\mu).$$

The first expression can be written as $n P_{n-1}(\mu) = P'_n(\mu) - \mu P'_{n-1}(\mu)$ by changing n to $n-1$. Eliminating $P'_{n-1}(\mu)$ from this and the second expression, we obtain

$$n P_{n-1}(\mu) = \mu n P_n(\mu) + \left(1-\mu^2\right) P'_n(\mu).$$

Therefore,

$$\frac{d}{d\mu}\left\{\left(1-\mu^2\right) P'_n(\mu)\right\} = n\left\{P'_{n-1}(\mu) - \mu P'_n(\mu)\right\} - n P_n(\mu) = -n(n+1)\,P_n(\mu),$$

which is the Legendre equation (5.2b).

5.4. A Conducting Sphere in a Uniform Field

A conducting sphere is polarized in a uniform field, which becomes axially symmetric in the vicinity of the field, although almost unchanged at distant places. Induced charges appear on the spherical surface, producing an axial field that modifies the original uniform field.

For such a problem, polar coordinates r, θ, and φ can be used with respect to the direction of the uniform field passing through the center of the sphere. In this case, at the distant point the field E_o is constant and expressed by the potential $V = -E_o z$, that is,

$$V(r,\theta) = -E_o r\cos\theta \quad \text{for} \quad r \to \infty, \tag{i}$$

which is regarded as the condition for the potential to satisfy in the distant region from the sphere. On the other hand, the potential must also be constant on the conducting spherical surface of radius R, and hence

$$V(R,\theta) = V_o(\text{constant}) \tag{ii}$$

is the other boundary condition to be satisfied, and it can also be expressed by

$$E_\theta = 0 \quad \text{and} \quad \varepsilon_o E_r = \sigma_P \quad \text{at} \quad r = R. \tag{iii}$$

Noting that the net field is axially symmetrical and independent of φ, the potential is generally expressed by (5.4), as it is a solution of the Laplace equation. However, due to the condition (i), the potential should be signified by the factor $\cos\theta$, and only the term of $P_1(\cos\theta) = \cos\theta$ in the series is significant. Therefore, the net potential can be expressed as

$$V(r,\theta) = \left(A_1 r + \frac{B_1}{r^2}\right)\cos\theta. \qquad \text{(iv)}$$

Using the condition (i) at $r \to \infty$ in this potential, we obtain $A_1 = -E_o$. From (iv), we have

$$E_r = -\frac{\partial V}{\partial r} = -\left(A_1 - \frac{2B_1}{r^3}\right)\cos\theta \quad \text{and} \quad E_\theta = -\frac{1}{r}\frac{\partial V}{\partial\theta} = \left(A_1 + \frac{B_1}{r^3}\right)\sin\theta,$$

to which we can apply the condition (iii), and obtain

$$(E_\theta)_{r=R} = \left(A_1 + \frac{B_1}{R^3}\right)\sin\theta = 0 \text{ and } \varepsilon_o (E_r)_{r=R} = \varepsilon_o\left(A_1 - \frac{2B_1}{R^3}\right)\cos\theta = \sigma_P.$$

From the first relation we obtain that $B_1 = -A_1 R^3 = E_o R^3$, and from the second we can express the polarized charge density on the sphere as

$$\sigma_P = 3\varepsilon_o E_o \cos\theta.$$

Induced surface charges are therefore distributed as proportional to $\cos\theta$, giving maximum at $\theta = 0$ and minimum at $\frac{1}{2}\pi$. Also notable is that the field E_r is zero at the point ($r = R$, $\theta = \frac{1}{2}\pi$, corresponding to the minimum charge density. Figure 5.2 illustrates such field-lines in the vicinity of a polarized sphere. As

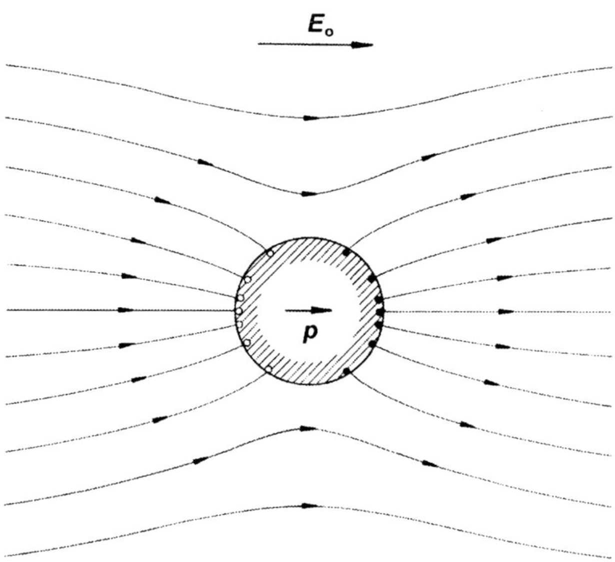

FIGURE 5.2. Field-lines of a polarized sphere in a uniform electric field E_o.

indicated by (iv), the potential is given by a superposition of a uniform potential $-E_o r\cos\theta$ plus a dipole potential $V_1 = p/4\pi\varepsilon_o r^2$, where the dipole moment is defined as $p = (4\pi\varepsilon_o)E_o R^3$.

5.5. A Dielectric Sphere in a Uniform Field

In this section, we discuss a dielectric sphere that is polarized by a uniform electric field. We assume that the material is of a uniform density and characterized by a dielectric constant ε. As is true for a conducting case, such a sphere is in axial symmetry, however unlike in a conductor, there is an electric field inside the sphere. We describe the field by coordinates r and θ with respect to the center of the sphere, referring to the direction of applied field expressed at distant points by the potential

$$V = -E_o z = -E_o r\cos\theta.$$

The potential $V(r,\theta)$ can be expressed outside and inside the sphere as

$$V_{out}(r,\theta) = -E_o r\cos\theta + \sum_n \frac{B_n}{r^{n+1}} P_n(\cos\theta) \quad \text{for} \quad r > R$$

and

$$V_{in}(r,\theta) = \sum_n A_n r^n P_n(\cos\theta) \quad \text{for} \quad r < R,$$

respectively. The boundary conditions on the spherical surface are

$$V_{out}(R,\theta) = V_{in}(R,\theta)$$

and

$$-\varepsilon_o\left(\frac{\partial V_{out}}{\partial r}\right)_{r=R} = -\varepsilon\left(\frac{\partial V_{in}}{\partial r}\right)_{r=R}$$

We can find that the coefficients A_n and B_n for $n > 1$ cannot satisfy these boundary conditions simultaneously, and, consequently, only A_1 and B_1 need be determined, as in the case of a conducting sphere. Thus, for $n = 1$ the conditions are written as

$$-E_o R + \frac{B_1}{R^2} = A_1 R \quad \text{and} \quad -\varepsilon_o\left(E_o\frac{2B_1}{R^3}\right) = \varepsilon A_1.$$

Solving these for A_1 and B_1, we obtain

$$A_1 = \frac{3\varepsilon_o E_o}{\varepsilon + 2\varepsilon_o} \quad \text{and} \quad B_1 = \frac{\varepsilon - \varepsilon_o}{\varepsilon + 2\varepsilon_o} E_o R^3.$$

Hence

$$V_{out}(r,\theta) = -E_o r\cos\theta + \frac{\varepsilon - \varepsilon_o}{\varepsilon + 2\varepsilon_o}\frac{E_o R^3\cos\theta}{r^2}$$

and

$$V_{\text{in}}(r,\theta) = -\frac{3\varepsilon_o}{\varepsilon + 2\varepsilon_o} E_o r \cos\theta.$$

The potential $V_{\text{out}}(r,\theta)$ is composed of uniform and dipolar terms, whereas the potential $V_{\text{in}}(r,\theta)$ gives a uniform field.

Replacing $r\cos\theta$ in $V_{\text{in}}(r,\theta)$ by z,

$$E_z = -\frac{\partial V_{\text{in}}(z)}{\partial z} = \frac{3\varepsilon_o}{\varepsilon + 2\varepsilon_o} E_o,$$

which is constant, but not equal to E_o unless $\varepsilon = \varepsilon_o$. Further, the induced surface charge density on the sphere can be calculated as

$$\sigma_P = -\varepsilon_o \left(\frac{\partial V_{\text{out}}(r,\theta)}{\partial r}\right)_{r=R} + \varepsilon \left(\frac{\partial V_{\text{in}}(r,\theta)}{\partial r}\right)_{r=R} = \frac{2(\varepsilon - \varepsilon_o)}{\varepsilon + 2\varepsilon_o} E_o \cos\theta,$$

which is a maximum at $\theta = 0$ and minimum at $\frac{1}{2}\pi$. From $V_{\text{out}}(r,\theta)$, the dipole moment induced in a dielectric sphere can be written as

$$p = 4\pi\varepsilon_o \frac{\kappa - 1}{\kappa + 2} E_o R^3, \quad \text{where} \quad \kappa = \frac{\varepsilon}{\varepsilon_o}.$$

Also, the electric field outside the sphere can be calculated as

$$E_r - \frac{\partial V_{\text{out}}(r,\theta)}{\partial r} = \left\{1 + \frac{2(\kappa - 1)}{\kappa + 2}\frac{R^3}{r^3}\right\} E_o \cos\theta$$

$$E_\theta = -\frac{1}{r}\frac{\partial V_{\text{out}}(r,\theta)}{\partial \theta} = \left\{1 - \frac{2(\kappa - 1)}{\kappa + 2}\frac{R^3}{r^3}\right\} E_o \sin\theta.$$

We note that at $\theta = \frac{1}{2}\pi$, while $E_r = 0$, the field is dominated by E_θ, which is less than 1; this is because $\kappa > 1$ signifies a low flux density in this vicinity, as sketched in Figure 5.3(a).

Exercise. A Spherical Cavity in a Dielectric Medium.

Consider a spherical cavity of radius R in a medium of dielectric constant $\varepsilon = \kappa\varepsilon_o$, as shown in Figure 5.3(b). In this case, the cavity is polarized by an applied field in the medium. Obtain the expressions for the electric field vectors in the medium as well as in the cavity, and explain that the field pattern is as illustrated in the figure.

5.6. A Point Charge Near a Grounded Conducting Sphere

In Section 4.7 we discussed an electric image $-q$ of a point charge q on a conducting plane of infinite area. The potential of the plane is regarded as zero because it is extended to infinity. In this section we discuss a conducting sphere of radius R in the field of a point charge q_1, to see if an image charge $-q_2$ can represent the

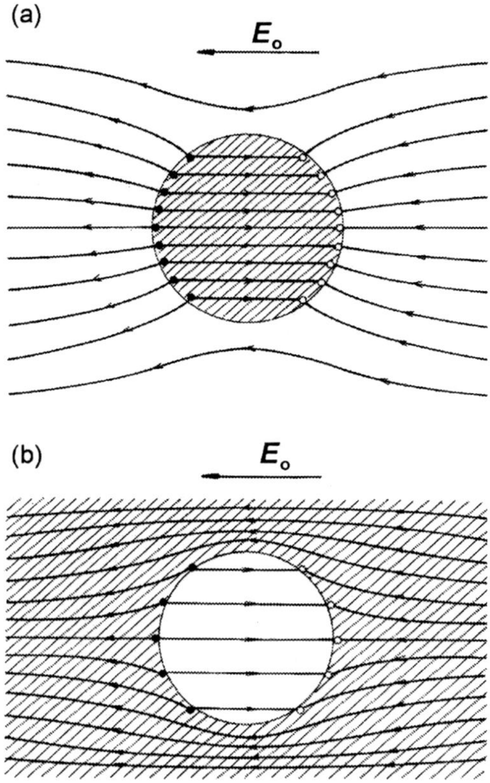

FIGURE 5.3. A dielectric sphere in a uniform electric field E_o. (a) $\varepsilon_{inside} > \varepsilon_{outside}$, (b) $\varepsilon_{inside} < \varepsilon_{outside}$.

sphere. However, for such an image to exist it is realized that the sphere should be grounded to maintain the potential at zero. Otherwise, the sphere is polarized, and not represented by a dipole moment. On the grounded sphere, the induced charges can of one kind reside, keeping the other kind away at infinity, so that the sphere can effectively be represented by a charge $-q_2$. Figure 5.4(b) shows a sketch of field-lines near the sphere and q_1, where some of the lines from q_1 are ended at the conducting surface, whereas other lines are extended to infinity. Hence, the magnitude of the image charge cannot be of the same as q_1.

We calculate the potential $V(r_1, r_2)$ at a point P, where r_1 and r_2 are the distances from a given q_1 and a hypothetical q_2, as shown in Figure 5.4(b). Taking P at exactly on a surface point of the sphere, we have

$$V(\mathrm{P}) = V(r_1, r_2) = \frac{1}{4\pi\varepsilon_o}\left(\frac{q_1}{r_1} - \frac{q_2}{r_2}\right) = 0.$$

Indicating the positions of q_1 and q_2 as $z = a$ and b, respectively, referring to the rectangular axes x, y, and z shown in the figure,

$$r_1^2 = x^2 + y^2 + (z - a)^2 = r^2 - 2az + a^2$$

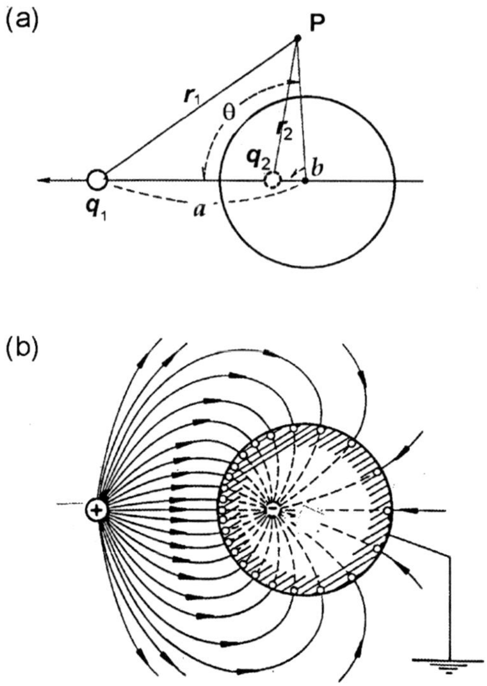

FIGURE 5.4. A point charge q_1 and a grounded conducting sphere. (a) The electric image q_2. (b) Field-lines due to q_1 and its image q_2.

and

$$r_2^2 = x^2 + y^2 + (z - b)^2 = r^2 - 2bz + b^2,$$

where $r = (x^2 + y^2 + z^2)^1$ is the radial distance from the center of the sphere. At all points on the sphere $r = R$, $V(P) = 0$, and hence

$$\frac{q_1}{\sqrt{R^2 - 2az + a^2}} = \frac{q_2}{\sqrt{R^2 - 2bz + b^2}},$$

from which we obtain

$$\left(q_1^2 - q_2^2\right) R^2 - 2z\left(q_1^2 b - q_2^2 a\right) + q_1^2 b^2 - q_2^2 a^2 = 0.$$

In order for this equation to be held at all points on the surface, the second term on the right should vanish regardless of z, and therefore

$$q_1^2 b - q_2^2 a = 0 \quad \text{and} \quad R^2 = \frac{q_2^2 a^2 - q_1^2 b^2}{q_1^2 - q_2^2}.$$

Accordingly, for the image charge $-q_2$ we have

$$b = \frac{R^2}{a} \quad \text{and} \quad q_2 = q_1\sqrt{\frac{b}{a}},$$

indicating $q_2 < q_1$.

The induced charge density can be calculated with the potentials of point charges q_1 and q_2, that is, $\sigma(\theta) = -\varepsilon_o(\partial V/\partial r)_{r=R}$. Writing $\alpha = R/a$ for convenience, the result can be expressed as

$$\sigma(\theta) = -\frac{q_1\alpha(1-\alpha)}{4\pi R^2\left(1-2\alpha\cos\theta-\alpha^2\right)^{\frac{3}{2}}},$$

giving distributed densities as a function of θ, as sketched in Figure 5.4(b). The total induced charge can be obtained by

$$\int_{-1}^{+1}\sigma(\theta)\,2\pi R^2 \mathrm{d}(\cos\theta) = -\alpha q_1 = q_2,$$

as expected. Further, the force between q_1 and $-q_2$ is attractive, and expressed by

$$F = -\frac{1}{4\pi\varepsilon_o}\frac{q_1 q_2}{(a-b)^2} = -\frac{q_1^2}{4\pi\varepsilon_o R^2}\frac{\alpha^3}{\left(1-\alpha^2\right)^2}.$$

5.7. A Simple Quadrupole

As a simple example of a *quadrupole*, we consider a spherical charge Q that is distorted by four external point charges, two positive and two negative, located close to Q on the x- and y-axis, respectively, as illustrated in Figure 5.5.

This is a model for an ionic charge distorted at a lattice site of a rectangular crystal, where the deformed charge is characterized by four small point charges $+q$, $-q$, $+q$ and $-q$ to express a distortion from the spherical Q, as schematically shown in the figure. Assuming that such charge displacements occur in the xy-plane, the potential of such a quadrupole can be expressed with polar coordinates r, θ, and φ with respect to the z-axis. For convenience, we denote these charge deviations

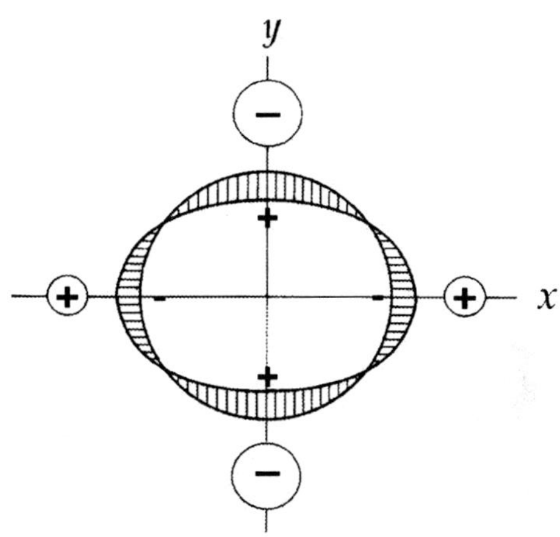

FIGURE 5.5. A simple quadrupole moment.

as $q_i(i = 1, 2, 3, 4)$ that are located at $\boldsymbol{a}_i$: $(a, \frac{1}{2}\pi, \varphi_i)$, where $\varphi_i = 0, \frac{1}{2}\pi, \pi, \frac{3}{2}\pi$, respectively. For a point $P(r, \theta, \varphi)$ of observation, we use angles α_i defined by $\cos\alpha_i = (\boldsymbol{r}.\boldsymbol{a}_i)/ra_i$. In this case, we have

$$\cos\alpha_1 = -\cos\alpha_3 = \sin\theta\cos\varphi \quad \text{and} \quad \cos\alpha_2 = -\cos\alpha_4 = \sin\theta\sin\varphi$$

and the potential at P can be written for $r \gg a_i$ as

$$V(r, \theta, \varphi) = \frac{1}{4\pi\varepsilon_o}\sum_n \frac{a^n}{r^{n+1}}\{P_n(\cos\alpha_1) - P_n(\cos\alpha_2) + P_n(\cos\alpha_3) - P_n(\cos\alpha_4)\}.$$

By the relation (5.3),

$$P_n(\cos\alpha_3) = P_n(-\cos\alpha_1) = (-1)^n P_n(\cos\alpha_1) \text{ and } P_n(\cos\alpha_4) = (-1)^n P_n(\cos\alpha_2)$$

Therefore, the $n = 0$ and 1 terms can be disregarded, and the $n = 2$ terms are significant for the quadrupole potential, that is

$$\begin{aligned} V_2(r, \theta, \varphi) &= \frac{qa^2}{2\pi\varepsilon_o r^3}\{P_2(\sin\theta\cos\varphi) - P_2(\sin\theta\sin\varphi)\} \\ &= \frac{qa^2}{2\pi\varepsilon_o r^3}\left(\frac{3}{2}\sin^2\theta\cos 2\varphi\right) = \frac{1}{4\pi\varepsilon_o}\frac{3qa^2}{r^3}\sin^2\theta\cos 2\varphi. \end{aligned}$$

Exercises.

1. In the above model the quadrupole may be considered as consisting of two dipoles of $(q, -q)$ and $l = \sqrt{2}a$. Show such a model of two dipoles from which the same expression for V_2 can be obtained.
2. Assuming that four charges q in the model are all of the same sign, obtain the expression of the potential V for $r \gg a$.

5.8. Associated Legendre Polynomials

The Coulomb potential (4.19) has been expanded into the Legendre series. We discussed the cases of $n = 0$ and $n = 1$ for a point charge and dipole; further, a simple quadrupole potential was described as given by $n = 2$ in the previous section. Cases for $n > 2$ can be discussed, also, however, no significant applications to practical problems can be found. In contrast, functions P_1 and P_2 are of specific importance in applications to molecular dipoles, nuclear quadrupoles, antenna design, and many other problems.

In symmetry lower than uniaxial, a charge system is signified not only by θ but also by the *azimuthal* angle φ, and the potential has a factor $\Phi(\varphi) = \Phi_o\exp(im\varphi)$, where m can take zero and $\pm$ integer values. For such a potential, the angular factor is fully expressed as $P_n(\cos\theta)\exp(im\varphi) = P_n^m(\theta, \varphi)$, which is called the *associated Legendre function*. The function $P_n^m(\theta, \varphi)$ is determined as solutions of the differential equation

$$\sin\theta\frac{d}{d\theta}\left(\sin\theta\frac{dP_n^m}{d\theta}\right)+\left\{n(n+1)\sin^2\theta-m^2\right\}P_n^m=0, \tag{5.9a}$$

or using $\mu = \cos\theta$ instead of θ,

$$\frac{d}{d\mu}\left\{\left(1-\mu^2\right)\frac{dP_n^m}{d\mu}\right\}+\left\{n(n+1)-\frac{m^2}{1-\mu^2}\right\}P_n^m=0, \tag{5.9b}$$

which is called the *associated Legendre equation*. The function $P_n^m(\mu, \varphi)$ has singularities at $\mu = \pm 1$ if n and m are integers and $n \geq |m|$, constituting physically meaningful solutions in the ranges of $0 \leq \theta \leq \pi$ and $0 \leq \varphi \leq 2\pi$. Mathematically, there is another solution known as the second kind solution, but we disregard it, since only the $P_n^m(\mu, \varphi)$ are meaningful. Leaving the mathematical detail to reference books, here we discuss only $P_n^m(\theta, \varphi)$, called the first kind, for potential problems.

The associated function $P_n^m(\mu, \varphi)$ can be derived from the Legendre polynomial $P_n(\mu)$ by differentiations, i.e.,

$$P_n^{|m|}(\mu)=(-1)^{|m|}\left(1-\mu^2\right)^{|m|/2}\frac{d^{|m|}P_n(\mu)}{d\mu^{|m|}}. \tag{5.10}$$

In the Appendix, a proof is given for (5.10) to satisfy the associated Legendre equation (5.9b), showing the orthogonality relation

$$\int_{-1}^{+1} P_n^{|m|}(\mu)\,P_{n'}^{|m|}(\mu)\,d\mu=\frac{2}{2n+1}\frac{(n+|m|)!}{(n-|m|)!}\delta_{n,n'}, \tag{5.11}$$

where $\delta_{n,n'} = 1$ if $n = n'$, otherwise $\delta_{n,n'} = 0$ for $n \neq n'$, and clearly $n \geq |m|$ for the associated function to be real. The normalized Legendre functions, called *spherical harmonics*, are then expressed as

$$Y_{n,m}(\theta,\varphi)=\sqrt{\frac{(2n+1)(n-m)!}{4\pi(n+m)!}}\,P_n^m(\cos\theta)\exp(im\varphi), \tag{5.12}$$

for which the orthogonality relation can be written as

$$\int_0^{2\pi}\int_0^{\pi} Y_{n,m}(\theta,\varphi)\,Y_{n',m'}(\theta,\varphi)^*\sin\theta\,d\theta\,d\varphi=0. \tag{5.13}$$

It is realized that these spherical harmonics for $m = n, n-1, n-2, \ldots, -n$ constitute an *orthonormal* set of functions, so that Legendre's function $P_n(\cos\theta)$ can be expressed as a linear combination of these spherical harmonics for m. In this connection, we can derive a formula, known as the *addition theorem*, which is important for a *multipole* expansion of the potential, to be discussed in Section 5.9. Considering an angle γ between two given directions (θ, φ) and (θ', φ'), we look for the formula for $P_n(\cos\gamma)$ to be expressed by these angles. Corresponding to the cosine law, i.e.,

$$\cos\gamma = \cos\theta\cos\theta' + \sin\theta\sin\theta'\cos(\varphi - \varphi'),$$

we can prove the relation

$$P_n(\cos\gamma) = \frac{4\pi}{2n+1}\sum_{m=+n}^{-n} Y_{n,m}(\theta,\varphi)\,Y_{n,m}{}^*(\theta',\varphi'). \tag{5.14}$$

Proof: First, we write that

$$P_n(\cos\gamma) = \sum_{m=n}^{-n} A_{nm}\,P_n^m(\cos\theta)\exp(im\varphi), \tag{i}$$

where

$$A_{nm} = \frac{2n+1}{4\pi}\frac{(n-|m|)!}{(n+|m|)!}\int_0^{2\pi} d\varphi \int_{+1}^{-1} d(\cos\theta) P_n(\cos\gamma)\,P_n^{|m|}(\cos\theta)\exp(i\,|m|\,\varphi).$$

Considering a direction (γ, ϕ), we can also write that

$$P_n^{|m|}(\cos\theta)\exp(i\,|m|\,\varphi) = \sum_{m'=+n}^{-n} B_{nm'}\,P_n^{|m'|}(\cos\gamma)\exp\left(i\,|m'|\,\phi\right), \tag{ii}$$

where

$$B_{nm'} = \frac{2n+1}{4\pi}\frac{\left(n-|m'|\right)!}{(n+|m'|)!}\int_0^{2\pi} d\phi \int_{+1}^{-1} d(\cos\gamma)\,P_n^{|m|}(\cos\theta)\exp(i\,|m|\,\varphi)\,P_n^{|m'|}$$
$$\times(\cos\phi)\exp\left(i\,(m')\,\phi\right). \tag{iii}$$

If we set $\gamma = 0$ in (ii), we have $\theta = \theta'$ and $\varphi = \varphi'$, and then from (ii)

$$P_n^m(\cos\theta')\exp(im\varphi') = \sum_{m'=+n}^{-n} B_{nm'}\,P_n^{m'}(1)\exp\left(im'\phi\right) = B_{n0}\,P_n^0(1) = B_{n0}.$$

Writing (iii) for $m' = 0$, we also obtain

$$B_{n0} = \frac{2n+1}{4\pi}\int_0^{2\pi} d\phi \int_{+1}^{-1} d(\cos\gamma)\,P_n(\cos\gamma)\left\{P_n^m(\cos\theta)\exp(im\varphi)\right\},$$

and the relation between them indicates that $P_n(\cos\gamma)$ is the coefficient of the spherical harmonic $Y_{n,m}(\theta', \varphi')$ when $Y_{n,m}(\theta, \varphi)$ is expressed as a linear expansion with respect to θ' and φ'. Thus, taking the normalization factor into account, we arrive at the addition theorem for spherical harmonics (5.14).

5.9. Multipole Potentials

The potential function satisfying the Laplace equation can be expressed generally in two regions $r > r_o$ and $r < r_o$: Using (5.14), for the first case, we have

$$V(r,\theta,\varphi) = \frac{1}{4\pi\varepsilon_o}\sum_n \frac{1}{r^{n+1}}\sum_m M_{n,m}Y_{n,m}(\theta,\varphi), \tag{5.15a}$$

where the *multipole potentials* are expressed as

$$M_{n,m} = \frac{4\pi}{2n+1}\int_{v(r_o)} \rho(\boldsymbol{r}_o)\, r_o^n Y_{n,m}{}^*(\theta,\varphi)\mathrm{d}^3\boldsymbol{r}_o. \tag{5.15b}$$

For the second case,

$$V(r,\theta,\varphi) = \frac{1}{4\pi\varepsilon_o}\sum_n r^n N_{n,m}Y_{n,m}(\theta,\varphi), \tag{5.16a}$$

where

$$N_{n,m} = \frac{4\pi}{2n+1}\int_{v(r_o)} \frac{\rho(\boldsymbol{r}_o)}{r_o^n} Y_{n,m}{}^*(\theta,\varphi)\mathrm{d}^3\boldsymbol{r}_o. \tag{5.16b}$$

Nevertheless, we are only concerned about the first region $r > r_o$, where a charge cloud can be expressed in terms of *multipoles*. Equation (5.15a) indicates that the potential is given by a superposition of potentials characterized by indexes $\boldsymbol{n}$, and hence written as

$$V(r,\theta,\varphi) = \sum_{n=0}^{\infty} V_n(r,\theta,\varphi), \tag{5.17}$$

where

$$\begin{aligned}
V_o &= \frac{1}{4\pi\varepsilon_o r}M_{0,0},\\
V_1 &= \frac{1}{4\pi\varepsilon_o r^2}\left(M_{1,1}Y_{1,1} + M_{1,0}Y_{1,0} + M_{1,-1}Y_{1,-1}\right),\\
V_2 &= \frac{1}{4\pi\varepsilon_o r^3}\left(M_{2,2}Y_{2,2} + M_{2,1}Y_{2,1} + M_{2,0}Y_{2,0} + M_{2,-1}Y_{2,-1} + M_{2,-2}Y_{2,-2}\right),
\end{aligned}$$

and so on. The term V_o is a *point-charge potential*, where $M_{0,0} = \int_v \rho(\boldsymbol{r}_o)\mathrm{d}^3\boldsymbol{r}_o$ represents a charge q. For the potential V_1,

$$Y_{1,\pm1} = \pm\sqrt{\frac{3}{8\pi}}\sin\theta\exp(\pm i\varphi) = \pm\sqrt{\frac{3}{8\pi}}\frac{x\pm iy}{r} \quad \text{and}$$

$$Y_{1,0} = \sqrt{\frac{3}{4\pi}}\cos\theta = \sqrt{\frac{3}{4\pi}}\frac{z}{r},$$

which are direction cosines of the vector $\boldsymbol{r}$, and hence

$$M_{1,\pm 1} = \frac{4\pi}{3}\sqrt{\frac{3}{8\pi}} \int\limits_{v(\boldsymbol{r}_o)} \rho\,(\boldsymbol{r}_o)\,(x_o \mp i y_o)\,\mathrm{d}^3\boldsymbol{r}_o$$

and

$$M_{1,0} = \frac{4\pi}{3}\sqrt{\frac{3}{8\pi}} \int\limits_{v(\boldsymbol{r}_o)} \rho\,(\boldsymbol{r}_o)\, z_o \mathrm{d}^3\boldsymbol{r}_o,$$

where the integrals are components of a dipole moment $\boldsymbol{p} = \int_v \rho\,(\boldsymbol{r}_o)\boldsymbol{r}_o \mathrm{d}^3\boldsymbol{r}_o$.

For the potential V_2 we have five spherical harmonics for $n = 2$ and $m = 2, 1, 0, -1, -2$, which are

$$Y_{2,\pm 2} = \frac{1}{2}\sqrt{\frac{15}{8\pi}}\frac{(x \pm i y)^2}{r^2},$$

$$Y_{2,\pm 1} = \pm\sqrt{\frac{15}{8\pi}}\frac{(x \pm i y)\, z}{r^2}, \qquad Y_{2,0} = \frac{1}{2}\sqrt{\frac{5}{4\pi}}\frac{2z^2 - x^2 - y^2}{r^2},$$

and

$$M_{2,\pm 2} = \frac{2\pi}{5}\sqrt{\frac{15}{8\pi}} \int\limits_{v(\boldsymbol{r}_o)} \rho\,(\boldsymbol{r}_o)\,(x_o \mp i y_o)^2\,\mathrm{d}^3\boldsymbol{r}_o,$$

$$M_{2,\pm 1} = \pm\frac{4\pi}{5}\sqrt{\frac{15}{8\pi}} \int\limits_{v(\boldsymbol{r}_o)} \rho\,(\boldsymbol{r}_o)\,(x_o \mp i y_o)\, z_o \mathrm{d}^3\boldsymbol{r}_o,$$

$$M_{2,0} = \frac{2\pi}{5}\sqrt{\frac{5}{4\pi}} \int\limits_{v(\boldsymbol{r}_o)} \rho\,(\boldsymbol{r}_o)\left(2z_o^2 - x_o^2 - y_o^2\right)\mathrm{d}^3\boldsymbol{r}_o.$$

Using these results, the potential V_2 can be expressed in a quadratic form,

$$V_2 = \frac{1}{8\pi\varepsilon_o r^5}\left(Q_{xx}x^2 + Q_{yy}y^2 + Q_{zz}z^2 + 2Q_{yz}yz + 2Q_{zx}zx + 2Q_{xy}xy\right)$$

where

$$Q_{xx} = \int\limits_{v(\boldsymbol{r}_o)} \rho\,(\boldsymbol{r}_o)\left(2x_o^2 - y_o^2 - z_o^2\right)\mathrm{d}^3\boldsymbol{r}_o,\ \ Q_{xy} = \int\limits_{v(\boldsymbol{r}_o)} \rho\,(\boldsymbol{r}_o)\, x_o y_o \mathrm{d}^3\boldsymbol{r}_o,\ \text{etc.}$$

If these quantities Q_{ij}, where $\mathrm{i,j} = x, y, z$, are considered elements of a tensor $\mathbf{Q}$, the potential V_2 can be written as

$$V_2 = \frac{1}{4\pi\varepsilon_o}\frac{\langle \boldsymbol{r}\,|\mathbf{Q}|\,\boldsymbol{r}\rangle}{2r^5}, \tag{5.18}$$

where $\langle r|$ and $|r\rangle$ are a row matrix (x, y, z) and a column matrix $\begin{pmatrix} x \\ y \\ z \end{pmatrix}$, respectively, and $\mathbf{Q}$ can be called a *quadrupole tensor*. It is significant that $\mathbf{Q}$ defined above is a symmetric tensor characterized by $Q_{\mathrm{ij}} = Q_{\mathrm{ji}}$, and that it is traceless, as V_2 should satisfy the Laplace equation $\nabla^2 V_2 = 0$, that is,

$$\text{trace}\,\mathbf{Q} = Q_{xx} + Q_{yy} + Q_{zz} = 0. \tag{5.19}$$

A distorted charge cloud can therefore be specified by a *quadrupole moment* for the potential V_2. The symmetric character of $\mathbf{Q}$ originates from the basic law of electrostatics, which corresponds microscopically to the fact that a charge cloud should be in a stable state. In Section 5.7, we discussed a simple quadruple of four point charges as a mathematical example, but it may, however, comprise a physically feasible state of a charged cluster. However, if it is asymmetrical by some reason, the cluster cannot be characterized by a quadrupole.

Further, a quadrupole may not only be a clustered point charge, but it may arise also from continuously distributed charges. In the following, we discuss the potential of a charged ring.

Example. Potentials of a Charged Ring.

Figure 5.6 shows a charged ring of radius a of thin wire.

The charge q is assumed to be uniformly distributed at a density $q\mathrm{d}\phi/2\pi$. Taking the origin of polar coordinates as the center of the ring, the potential at $\mathrm{P} = (r, \theta, \varphi)$

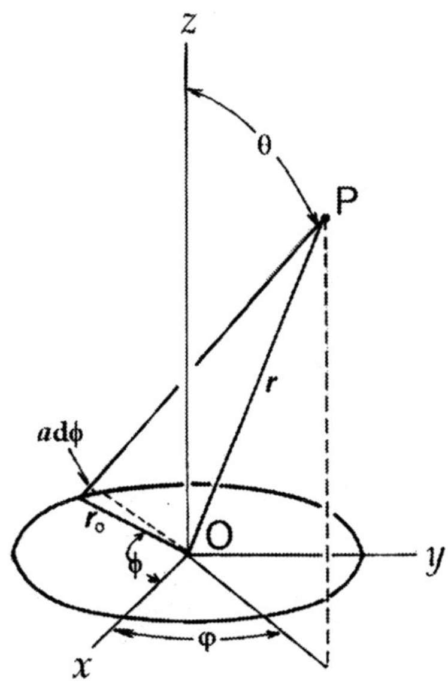

FIGURE 5.6. Calculating the potential due to a charged ring at a distant point $\mathrm{P}(r, \theta, \varphi)$.

is expressed as

$$V(r,\theta) = \frac{1}{4\pi\varepsilon_o}\frac{q}{2\pi}\int_0^{2\pi}\frac{d\varphi}{|\boldsymbol{r} - \boldsymbol{r}_o|},$$

where $\boldsymbol{r}$ and $\boldsymbol{r}_o$ are position vectors of P and a charged element $d\varphi$ at $|\boldsymbol{r}_o| = a$, respectively. Here,

$$\begin{aligned}(\boldsymbol{r} - \boldsymbol{r}_o)^2 &= r^2 - 2ra\sin\theta\cos(\varphi - \phi) + a^2\\ &= r^2\left\{1 - \frac{2a}{r}\sin\theta\cos(\varphi - \phi) + \left(\frac{a}{r}\right)^2\right\},\end{aligned}$$

and, therefore,

$$\begin{aligned}\frac{1}{|\boldsymbol{r} - \boldsymbol{r}_o|} = \frac{1}{r}\bigg\{&1 + \frac{a}{r}\sin\theta\cos(\varphi - \phi) - \frac{1}{2}\left(\frac{a}{r}\right)^2\\ &+ \frac{3}{2}\left(\frac{a}{r}\right)^2\sin^2\theta\cos^2(\varphi - \phi) + \ldots\bigg\}.\end{aligned}$$

Noting that $\int_0^{2\pi}\cos(\varphi - \phi)\,d\phi = 0$ and $\int_0^{2\pi}\cos^2(\varphi - \phi)\,d\phi = \frac{1}{2}$, the potential can be expressed as

$$\begin{aligned}V(\theta,\varphi) &= \frac{q}{4\pi\varepsilon_o r}\left\{1 - \frac{1}{2}\left(\frac{a}{r}\right)^2 + \frac{3}{4}\left(\frac{a}{r}\right)^2\sin^2\theta + \ldots\right\}\\ &= \frac{q}{4\pi\varepsilon_o r} - \frac{qa^2}{2\pi\varepsilon_o r^3}P_2(\cos\theta) + \ldots.\end{aligned}$$

The second term can be interpreted as a quadrupole potential that is generally expressed as (5.18). In the present case, the $\mathbf{Q}$ tensor must be axially symmetrical along the z direction, and $Q_{xx} = Q_{yy} = -1Q_{zz}$, and hence by comparison we obtain

$$\langle\boldsymbol{r}\,|\mathbf{Q}|\,\boldsymbol{r}\rangle = Q_{zz}\frac{3z^2 - r^2}{2} \quad \text{and} \quad Q_{zz} = -qa^2.$$

Part 2
Electromagnetism

6
The Ampère Law

6.1. Introduction

Humans have utilized magnetized materials such as iron, magnetite, hematite, and other minerals for centuries. In particular, a thin iron bar was utilized as a compass for navigation. To explain the terrestrial magnetism, Gilbert (ca. 1600) first considered the planet Earth as a giant magnet. Since that time, magnetism became a subject for serious scientific studies, bringing about discoveries of many other magnetic materials. Due to their attractive and repulsive nature, each magnet was characterized by two distinct "poles," named north and south poles (N- and S-poles), according to the way they interacted. These poles were recognized to exist always in pairs, and they were understood to be inseparable, unlike electric charges. Further, because of their delocalization on magnetized surfaces, polar charges were not easily measurable. Nevertheless, Coulomb (1785) defined magnetic poles according to the force between them in an idealized case, analogous to the static force between electric charges.

Using a small magnet, Oersted (1820) identified the magnetic effect of an electric current, an effect that was then formulated by Ampère (1822). Studies on *electromagnetism* began after Faraday's discovery of *electromagnetic induction* (1831), a phenomenon that was later theorized by Maxwell (1873) with the concept of "fields." The static electric field and the steady magnetic field were investigated exclusive of one another (though we now know the relation between the charges and currents for electromagnetism). It was only after Einstein's discovery of the relativity principle (1905) that such a relation became properly interpreted, thereby recognizing the Maxwell theory as a relativistic law.

Laws of electromagnetism are formulated in Maxwell's equations, although the origin of magnetism was not properly understood until Heisenberg's (1928) interpretation of the magnetic interaction as an "exchange interaction" between electron spins. Today, such a quantum character of magnetic interactions is well established, while no single magnetic pole has yet been isolated. However, the magnetic origin of an electron spin must be considered a fact in nature, and presently the absence of a single magnetic pole constitutes one of the laws of electromagnetism.

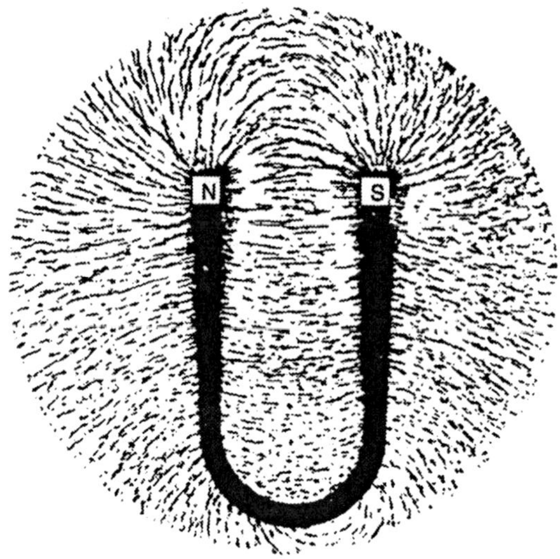

FIGURE 6.1. Magnetic field-lines of a horseshoe magnet.

Figure 6.1 shows a typical magnetic field of a horseshoe magnet, where the field-lines are visualized by a pattern formed by iron powder sprinkled over the magnet. Here, powder particles behave like small needles of a magnetic moment $\boldsymbol{p}_m$, being aligned along the magnetic field $\boldsymbol{H}$ by a torque $\boldsymbol{p}_\mathrm{m} \times \boldsymbol{H}$. Such a torque on a small compass needle is used to measure the field strength $|\boldsymbol{H}|$, whereas such a needle indicates the direction of $\boldsymbol{H}$. We can recognize a vector character of the magnetic field from such a primitive experiment.

6.2. The Ampère Law

Oersted discovered that a small compass needle showed a deflection in the vicinity of a wire carrying an electric current, which was the same magnetic effect as observed near a magnetized material. Figure 6.2(a) shows a photograph of field-lines exhibited by iron powder in the vicinity of a straight steady current. It is significant that magnetic field-lines due to a current are in "closed" loops, in contrast to those near a magnet as shown in Figure 6.1. In Figure 6.2(b) a field pattern around a ring current is shown. The patterns are mostly closed but not exactly circular. Lines around a current exhibit a *vortex*-like pattern, which are different from those observed near a magnet. In such a magnetic field the field pattern varies depending on the curved path of a current, however appearing close to circular near the current.

The Ampère law states that the magnetic field intensity, as measured by a small compass at a distance r from a straight current I, is described as

$$|\boldsymbol{H}| \propto \frac{I}{r}, \tag{6.1}$$

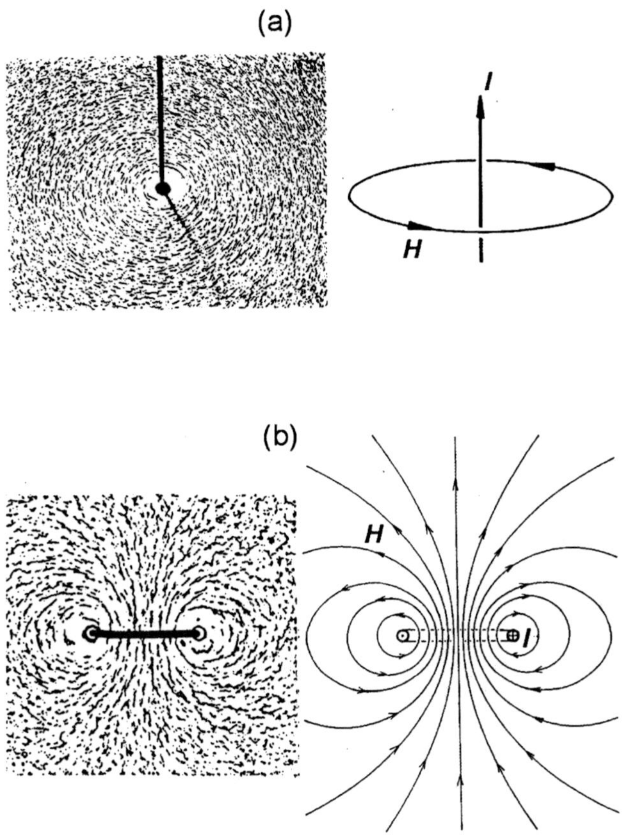

FIGURE 6.2. (a) Magnetic field-lines of a straight current, showing the relation between the current I and the $\boldsymbol{H}$ vector in the right-hand screw rule. (b) Magnetic field-lines of a ring current.

for which the directions of I and the vector $\boldsymbol{H}$ are related by the *right-hand screw* rule, as shown in Figure 6.2(a). Referring to a conventional machine screw, the rule indicates the relation between clockwise rotation and advancing directions for $\boldsymbol{H}$ and I. For a given I the magnitude $|\boldsymbol{H}|$ is constant on a circle of radius r, but decreasing as proportional to $1/r$ toward distant points.

Assuming the factor 2π in (6.1) the relation $\boldsymbol{H}(2\pi r) \propto I$ can be interpreted as an integral $\oint_\Gamma \boldsymbol{H}.\mathrm{d}\boldsymbol{s}$ performed on a circular path Γ of radius r is equal to I. The nonzero integral $\oint_\Gamma \boldsymbol{H}.\mathrm{d}\boldsymbol{s}$ indicates a significantly different feature of magnetic field $\boldsymbol{H}$ from a conservative electric field $\boldsymbol{E}$ characterized by $\oint_\Gamma \boldsymbol{E}.\mathrm{d}\boldsymbol{s} = 0$. In MKSA system, the unit for $\boldsymbol{H}$ is defined by (6.1) where the proportionality constant is set equal to $1/2\pi$, that is, A-m^{-1}.

The Ampère law can thus be expressed in MKS units as

$$\oint_{\Gamma} \boldsymbol{H}.\mathrm{d}\boldsymbol{s} = I, \tag{6.2}$$

which is the integral form, referring to a closed path Γ that encloses the current I. It is noted that mathematically the path Γ does not necessarily need to be a field-line for (6.2) to express the Ampère law. Choosing Γ to exclude I, equation (6.2) can be written as $\oint_{\Gamma} \boldsymbol{H}.\mathrm{d}\boldsymbol{s} = 0$, which however does not imply that $\boldsymbol{H}$ is conservative in this case.

For a magnet the magnetism originates from an entirely different microscopic mechanism unrelated to currents. In this context, the unit of $\boldsymbol{H}$ defined above appears to be illogical for magnets. However, a magnetic induction effect occurs with magnets, allowing such a definition of current-related unit for fields of both kinds. As will be discussed in the next chapter, the vector $\boldsymbol{H}$ by itself is not sufficient for characterizing a magnetic field; for the full description of this field the density of field-lines is also required. Accordingly, we retain the unit A-m^{-1} for $\boldsymbol{H}$.

The line integral of (6.2) in the field of a straight current is shown schematically in Figure 6.3(a) and (b).

Here, the integration can be performed along radial and angular zigzag steps, where the vector $\boldsymbol{H}$ is perpendicular and parallel to small successive displacements $\mathrm{d}s_r$ and $\mathrm{d}s_\theta$, respectively. Figure 6.3(a) shows a case where the closed path of integration Γ encloses the current I, whereas in Figure 6.3(b) I flows completely outside of Γ. Noted that $H_r = 0$, and $\boldsymbol{H}$ is signified only by the component H_θ, the integral is contributed by $H_\theta(r\mathrm{d}\theta) = (I/2\pi r)(r\mathrm{d}\theta) = I\mathrm{d}\theta/2\pi$. Therefore, in

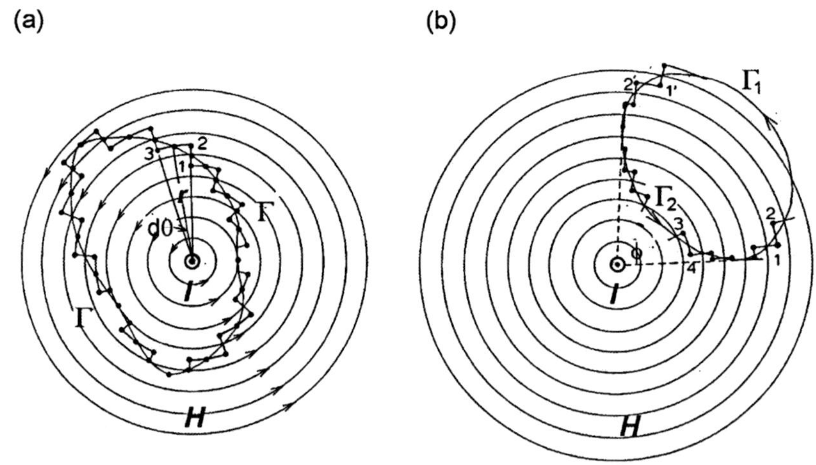

FIGURE 6.3. Calculating a line integral $\oint_{\Gamma} \boldsymbol{H}.\mathrm{d}\boldsymbol{S} = 0$ in a magnetic field of a straight current. (a) The current I threaded in the path Γ; (b) I located outside Γu

the former case,

$$\oint_{\Gamma} \boldsymbol{H}.\mathrm{d}\boldsymbol{s} = \frac{I}{2\pi}\int_{0}^{2\pi} \mathrm{d}\theta = I,$$

when I is inside Γ. If no current is inside, divide the path Γ into two parts, Γ_1 and Γ_2, as in Figure 6.3(b), for which the subtending angles are opposite, i.e., $\mathrm{d}\theta_1 = -\mathrm{d}\theta_2$. Hence, we obtain

$$\oint_{\Gamma_1+\Gamma_2} \boldsymbol{H}.\mathrm{d}\boldsymbol{s} = 0.$$

6.3. A Long Solenoid

As an example of the Ampère law, we consider a coiled wire wound in many turns, which is a basic magnetic device called a *solenoid* serving as a practical component to control currents in electrical circuits. Figures 6.4(a) and (b) illustrate the typical pattern of the magnetic field, where a circular current in each turn produces a magnetic field as in Figure 6.4(a), consequently providing a nearly uniform magnetic field in a solenoid. Here, the direction of $\boldsymbol{H}$ in the coil is related to the current as specified by the right-hand rule. Inside a tightly wound coil in a long length, the magnetic field can be idealized as uniform in parallel with the

(a)

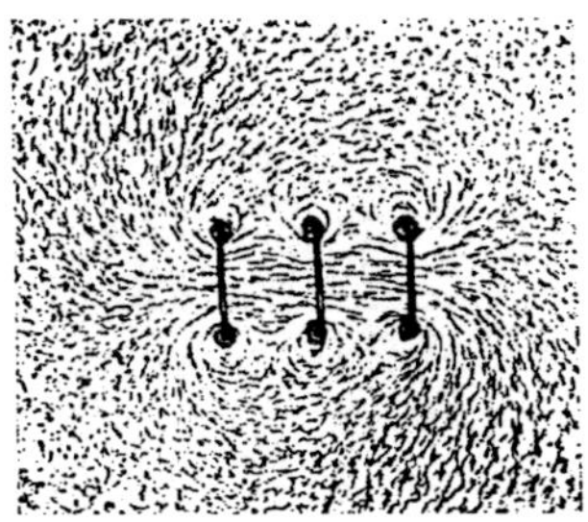

(b)

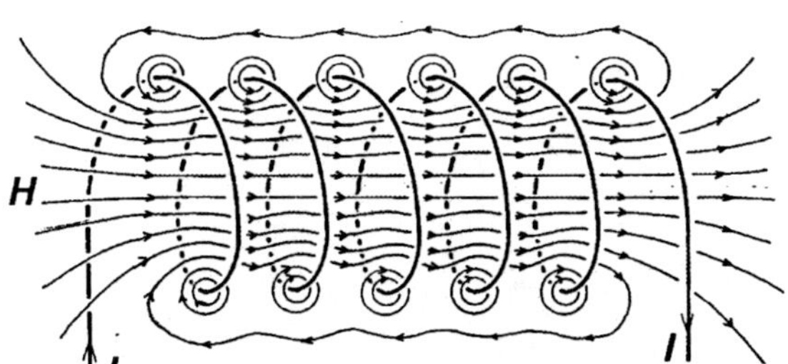

FIGURE 6.4. (a) Magnetic field-lines of a coil. (b) The current-field relation in a coil, an exaggerated view.

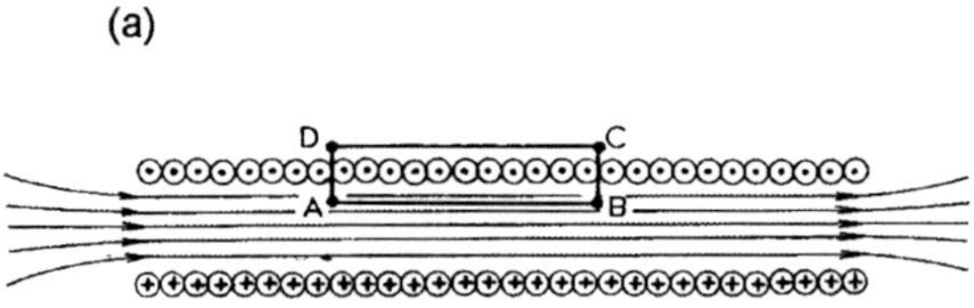

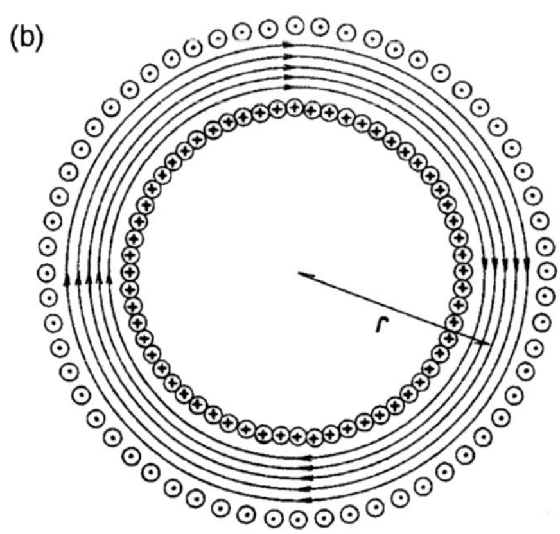

FIGURE 6.5. (a) A uniform solenoid, where the Ampère law is applied to the rectangular path A-B-C-D-A. (b) A uniform toroid.

coil axis by ignoring diverted field-lines near both ends. A long solenoid bent in circular form with two ends joined together, as shown in Figure 6.5(b), is called a *toroid* or *toroidal* coil, where all parallel field-lines are confined to inside space completely.

In such idealized coils, the uniform field $\boldsymbol{H}$ can be related by a simple rule to the responsible current I. Consider a rectangular path ABCD as shown in Figure 6.5(a), where AB and CD are parallel to the coil, whereas BC and DA are perpendicular, crossing the layer of currents.

Along the latter paths, $\boldsymbol{H} \perp \mathrm{d}\boldsymbol{s}$, i.e., $\boldsymbol{H}.\mathrm{d}\boldsymbol{s} = 0$, and along the outside CD, $\boldsymbol{H} = 0$. Therefore, in (6.2) applied to ABCD, the path integral is contributed only from AB, and (6.2) can be expressed as

$$Hl_{\mathrm{AB}} = IN_{\mathrm{AB}},$$

where l_{AB} and N_{AB} are the length and the number of turns between A and B. Writing $n = N_{\mathrm{AB}}/l_{\mathrm{AB}}$ as the density of turn number, the uniform field H is given by

$$H = nI. \tag{6.3}$$

It is noted that in such an idealized coil, a significantly large magnitude of H can be obtained with a large n by increasing current I.

For a toroid of the turn number N, shown in Figure 6.5(b), we can write a relation similar to (6.3) if the radius r is sufficiently large. In this case, H can be considered as constant circularly, for which the Ampère law can be written as $H(2\pi r) = NI$, and hence $H = NI/2\pi r$. In such a solenoid or toroid, there is a large H as given by a large nI or NI, respectively, similar to the relation between the electric flux density D and the condensed charges $\pm\sigma$ in a capacitor.

6.4. Stokes' Theorem

Macroscopic electric currents can be attributed to moving charged particles in a conductor. Although invisible, such a current is evidenced by heat generated in a conducting material as well as by the magnetic field produced by the motion. In fact, it was Rowland (1848) who demonstrated magnetically that a charged body in motion is equivalent to a current (See Section 6.7). Effects of a current such as heat, magnetic field, and chemical reactions in electrolysis were regarded as visual evidence for an invisible current. Conduction current can thus be identified geometrically along with a curve of the wire, which is, however, not the same for moving charge carriers .

The current I is defined by the total charge passing through the area S across the passage per second, where the density at a local point is often significant. Namely, $I = \int_S \boldsymbol{j}.\mathrm{d}\boldsymbol{S}$, where $\boldsymbol{j}$ represents the *current density* on differential area $\mathrm{d}\boldsymbol{S}$ and determines the integral $\oint_\Gamma \boldsymbol{H}.\mathrm{d}\boldsymbol{s}$ for the magnetic field $\boldsymbol{H}$ along the rim Γ of the area S. However, to deal with local properties, we need to convert the Ampère law into a differential form. Stokes' theorem is required for such a conversion, as we now explain.

In this section, we derive Stokes' theorem in rectangular coordinates, for simplicity, then in curvilinear coordinates, as discussed in the next section. First, we write

$$\oint_\Gamma \boldsymbol{H}.\mathrm{d}\boldsymbol{s} = \oint_\Gamma \left(H_x\mathrm{d}x + H_y\mathrm{d}y + H_z\mathrm{d}z\right)$$

$$= \frac{1}{2}\left[\oint_{C_z}\left(H_x\mathrm{d}x + H_y\mathrm{d}y\right) + \oint_{C_x}\left(H_y\mathrm{d}y + H_z\mathrm{d}z\right) + \oint_{C_y}\left(H_z\mathrm{d}z + H_x\mathrm{d}x\right)\right] \quad \text{(i)}$$

where terms on the right can be integrated along closed paths C_z, C_x, and C_y that are projections of the curve C onto planes *xy*, *yz*, and *zx*, respectively. Figure 6.6 shows C_z on the *xy*-plane, for example, where the integral can be calculated approximately as the sum of integrals over such closed square paths marked as *abcda*. In fact, both clockwise and counter clockwise integrals give the same results, making 1 the front factor of the square brackets in (i), instead of $\frac{1}{2}$, if we

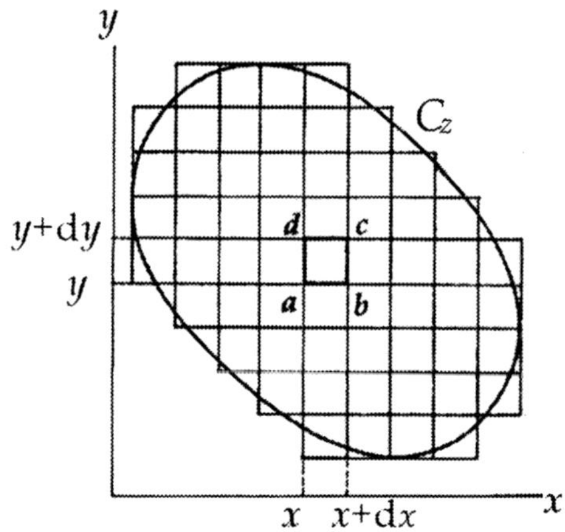

FIGURE 6.6. A component path C_z projected on the xy-plane for Stokes' theorem. The square path a-b-c-d-a represents a differential area for integration.

take both integrations into account. Physically, however, we note that only one of these can be associated with a given current direction, as determined by the right-hand screw rule.

We integrate the first term counter clockwise along a square a-b-c-d-a, that is,

$$\oint_{abcda} (H_x \mathrm{d}x + H_y \mathrm{d}y) = \int_a^b H_x \mathrm{d}x + \int_b^c H_y \mathrm{d}y + \int_c^d H_x \mathrm{d}x + \int_d^a H_y \mathrm{d}y,$$

where integrals on the right can be evaluated approximately by the values of integrands at these lower limits multiplied by the ranges. Namely,

$$\int_a^b H_x \mathrm{d}x = H_x \mathrm{d}x, \int_c^d H_x \mathrm{d}x = \left(H_x + \frac{\partial H_x}{\partial y}\mathrm{d}y\right)(-\mathrm{d}x)$$

$$\int_b^c H_y \mathrm{d}y = -H_y \mathrm{d}y \int_d^a H_y \mathrm{d}y = \left(H_y + \frac{\partial H_y}{\partial x}\mathrm{d}x\right) dy$$

Therefore,

$$\oint_{C_z} (H_x \mathrm{d}x + H_y \mathrm{d}y) \approx \sum_{C_z} \oint_{abcda} (\ldots\ldots) = \iint_{S_z} \left(\frac{\partial H_y}{\partial x} - \frac{\partial H_x}{\partial y}\right) \mathrm{d}x\mathrm{d}y,$$

where S_z is the area bordered by C_z. By similar calculations, we obtain

$$\oint_{C_x} (H_y \mathrm{d}y + H_z \mathrm{d}z) = \iint_{S_x} \left(\frac{\partial H_z}{\partial y} - \frac{\partial H_y}{\partial z}\right) \mathrm{d}y\mathrm{d}z$$

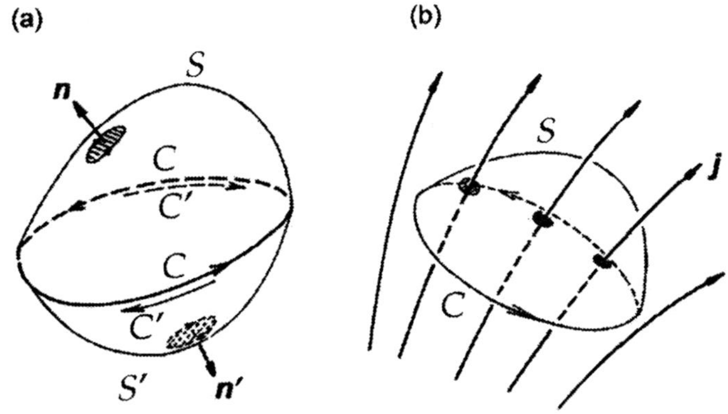

FIGURE 6.7. A convex surface $\boldsymbol{S}$ and its periphery C for calculating the current $I = \int_S \boldsymbol{j}.\mathrm{d}\boldsymbol{S}$, where the normal $\boldsymbol{n}$ and the direction along C are related by the right-screwhand rule. (a) $(\boldsymbol{n}, C)$ is opposite to $(\boldsymbol{n}', C')$. (b) Total current determined by S.

and

$$\oint_{C_y} (H_z \mathrm{d}z + H_x \mathrm{d}x) = \iint_{S_y} \left(\frac{\partial H_x}{\partial z} - \frac{\partial H_z}{\partial x} \right) \mathrm{d}z\mathrm{d}x.$$

In these results, we can consider that differential areas $\mathrm{d}y\mathrm{d}z$, $\mathrm{d}z\mathrm{d}x$, and $\mathrm{d}x\mathrm{d}y$ to be components of a vector area element $\mathrm{d}\boldsymbol{S} = \mathrm{d}S\boldsymbol{n}$, where the unit vector $\boldsymbol{n}$ represents a normal to a *convex* area subtending the closed path C. In this case, using the right-hand screwrule, we can define the normal **n** as associated with the current direction along the integration path C. In this context, the vector area $\mathrm{d}\boldsymbol{S}$ can represent only one-side of the area $\mathrm{d}S$ specified by the vector $\boldsymbol{n}$, which, consequently, allows us to disregard the factor 1 from (i). Figure 6.7(a) shows the right-hand screwrelation between $\boldsymbol{n}$ to the convex area S with respect to C, whereas $\boldsymbol{n}'$, S' and C' are in opposite relation. In either case we notice that both clockwise and counter clockwise integrations have the same contribution to the integral $\oint_C \boldsymbol{H}.\mathrm{d}\boldsymbol{s}$.

We define a vector quantity of rectangular components

$$\frac{\partial H_z}{\partial y} - \frac{\partial H_y}{\partial z}, \quad \frac{\partial H_x}{\partial z} - \frac{\partial H_z}{\partial x}, \quad \text{and} \quad \frac{\partial H_y}{\partial x} - \frac{\partial H_x}{\partial y}$$

as "curl $\boldsymbol{H}$," with which Stokes' theorem can be expressed as

$$\oint_C \boldsymbol{H}.\mathrm{d}\boldsymbol{s} = \int_S (\mathrm{curl}\,\boldsymbol{H}).\mathrm{d}\boldsymbol{S}, \tag{6.4}$$

where $\mathrm{d}\boldsymbol{S}$ is the differential area defined as a vector with the direction $\boldsymbol{n}$ related to C by the right-hand rule.

It is noted that curl $\boldsymbol{H}$ is identical to the cross-product between ∇ and $\boldsymbol{H}$, and written as

$$\text{curl}\,\boldsymbol{H} = \nabla\times\boldsymbol{H} = \begin{vmatrix} \boldsymbol{i} & \boldsymbol{j} & \boldsymbol{k} \\ \dfrac{\partial}{\partial x} & \dfrac{\partial}{\partial y} & \dfrac{\partial}{\partial z} \\ H_x & H_y & H_z \end{vmatrix}, \tag{6.5}$$

where $\boldsymbol{i}, \boldsymbol{j}$, and $\boldsymbol{k}$ are unit vectors along the x-, y-, and z-axes, respectively.

6.5. Curvilinear Coordinates (2)

In Section 6.4, we expressed the differential operator curl $\boldsymbol{H}$ in rectangular coordinates. For practical applications, however, it is often necessary to write it in curvilinear coordinates, as we discussed for other operators, e.g., div and ∇^2, in Section 4.3. However, the operator curl $\equiv \nabla\times$ is a little more complex, and hence we elaborate on it in this section for future applications.

The rectangular coordinates x, y, and z can be expressed as functions of the corresponding curvilinear coordinates u_1, u_2, and u_3. We can therefore define the following three vectors defined as the basic vectors for the curvilinear space. That is,

$$\begin{aligned} \mathbf{a}_1 &= \frac{\partial \boldsymbol{r}}{\partial u_1} = \frac{\partial x}{\partial u_1}\boldsymbol{i} + \frac{\partial y}{\partial u_1}\boldsymbol{j} + \frac{\partial z}{\partial u_1}\boldsymbol{k} = h_1\boldsymbol{e}_1, \\ \mathbf{a}_2 &= \frac{\partial \boldsymbol{r}}{\partial u_2} = \frac{\partial x}{\partial u_2}\boldsymbol{i} + \frac{\partial y}{\partial u_2}\boldsymbol{j} + \frac{\partial z}{\partial u_2}\boldsymbol{k} = h_2\boldsymbol{e}_2, \\ \mathbf{a}_3 &= \frac{\partial \boldsymbol{r}}{\partial u_3} = \frac{\partial x}{\partial u_3}\boldsymbol{i} + \frac{\partial y}{\partial u_3}\boldsymbol{j} + \frac{\partial z}{\partial u_3}\boldsymbol{k} = h_3\boldsymbol{e}_3, \end{aligned} \tag{6.6}$$

where

$$h_1^2 = \left(\frac{\partial x}{\partial u_1}\right)^2 + \left(\frac{\partial y}{\partial u_1}\right)^2 + \left(\frac{\partial z}{\partial u_1}\right)^2, \quad h_2^2 = \left(\frac{\partial x}{\partial u_2}\right)^2 + \left(\frac{\partial y}{\partial u_2}\right)^2 + \left(\frac{\partial z}{\partial u_2}\right)^2,$$

$$h_3^2 = \left(\frac{\partial x}{\partial u_3}\right)^2 + \left(\frac{\partial y}{\partial u_3}\right)^2 + \left(\frac{\partial z}{\partial u_3}\right)^2.$$

With these expressions a differential displacement d$\boldsymbol{r}$ can be written as

$$\mathrm{d}\boldsymbol{r} = \frac{\partial \boldsymbol{r}}{\partial u_1}\mathrm{d}u_1 + \frac{\partial \boldsymbol{r}}{\partial u_2}\mathrm{d}u_2 + \frac{\partial \boldsymbol{r}}{\partial u_3}\mathrm{d}u_3 = (h_1\mathrm{d}u_1)\boldsymbol{e}_1 + (h_2\mathrm{d}u_2)\boldsymbol{e}_2 + (h_3\mathrm{d}u_3)\boldsymbol{e}_3, \tag{6.7}$$

indicating that $(h_1\mathrm{d}u_1, h_2\mathrm{d}u_2, h_3\mathrm{d}u_3)$ are curvilinear components of d$\boldsymbol{r}$.

For a gradient of a scalar function V, we can write that

$$\frac{\partial V}{\partial u_1} = \frac{\partial V}{\partial x}\frac{\partial x}{\partial u_1} + \frac{\partial V}{\partial y}\frac{\partial y}{\partial u_1} + \frac{\partial V}{\partial z}\frac{\partial z}{\partial u_1} = h_1 \boldsymbol{e}_1 \cdot \boldsymbol{\nabla} V,$$

$$\frac{\partial V}{\partial u_2} = h_2 \boldsymbol{e}_2 \cdot \boldsymbol{\nabla V} \quad \text{and} \quad \frac{\partial V}{\partial u_3} = h_3 \boldsymbol{e}_3 \cdot \boldsymbol{\nabla V},$$

where $\boldsymbol{\nabla} V = \left(\frac{\partial V}{\partial x}, \frac{\partial V}{\partial y}, \frac{\partial V}{\partial z}\right)$ consistent with (4.6).

To obtain the curvilinear expression for the operator $\boldsymbol{\nabla}\times$, we define the *reciprocal vectors* $\boldsymbol{a}_1$, $\boldsymbol{a}_2$ and $\boldsymbol{a}_3$, corresponding to $\mathbf{a}_1$, $\mathbf{a}_2$, and $\mathbf{a}_3$, respectively. These are related as

$$\boldsymbol{a}_1 \cdot \mathbf{a}_1 = \boldsymbol{a}_2 \cdot \mathbf{a}_2 = \boldsymbol{a}_3 \cdot \mathbf{a}_3 = 1, \quad \text{and} \quad \boldsymbol{a}_1 = \frac{1}{\Omega}(\mathbf{a}_2 \times \mathbf{a}_3),$$

$$\boldsymbol{a}_2 = \frac{1}{\Omega}(\mathbf{a}_3 \times \mathbf{a}_1), \quad \boldsymbol{a}_3 = \frac{1}{\Omega}(\mathbf{a}_2 \times \mathbf{a}_3),$$

where $\Omega = (\mathbf{a}_1\mathbf{a}_2\mathbf{a}_3) = h_1 h_2 h_3 (\boldsymbol{e}_1\boldsymbol{e}_2\boldsymbol{e}_3) = h_1 h_2 h_3$ is the volume of the reciprocal unit cell. It is noted that the gradient operator $\boldsymbol{\nabla}$, therefore, has components in the reciprocal space that are given by

$$\nabla_1 = \boldsymbol{a}_1 \frac{\partial}{\partial u_1}, \; \nabla_2 = \boldsymbol{a}_2 \frac{\partial}{\partial u_2}, \quad \text{and} \quad \nabla_3 = \boldsymbol{a}_3 \frac{\partial}{\partial u_3}. \tag{6.8}$$

Now, for an arbitrary vector $\boldsymbol{F}$, we derive an expression for curl $\boldsymbol{F} = \boldsymbol{\nabla} \times \boldsymbol{F}$. Writing

$$\boldsymbol{F} = F_1 \boldsymbol{a}_1 + F_2 \boldsymbol{a}_2 + F_3 \boldsymbol{a}_3,$$

$$(\boldsymbol{\nabla} \times \boldsymbol{F})_{\mathrm{i}} = \frac{\partial F_{\mathrm{k}}}{\partial u_{\mathrm{j}}}(\boldsymbol{a}_{\mathrm{j}} \times \boldsymbol{a}_{\mathrm{k}}) + \frac{\partial F_{\mathrm{j}}}{\partial u_{\mathrm{k}}}(\boldsymbol{a}_{\mathrm{k}} \times \boldsymbol{a}_{\mathrm{j}}) = \left(\frac{\partial F_{\mathrm{k}}}{\partial u_{\mathrm{j}}} - \frac{\partial F_{\mathrm{j}}}{\partial u_{\mathrm{k}}}\right)(\boldsymbol{a}_{\mathrm{j}} \times \boldsymbol{a}_{\mathrm{k}})$$

$$= \frac{1}{\Omega}\left(\frac{\partial F_{\mathrm{k}}}{\partial u_{\mathrm{j}}} - \frac{\partial F_{\mathrm{j}}}{\partial u_{\mathrm{k}}}\right)\mathbf{a}_{\mathrm{i}},$$

and hence

$$\text{curl } \boldsymbol{F} = \frac{1}{\Omega}\begin{vmatrix} \mathbf{a}_1 & \mathbf{a}_2 & \mathbf{a}_3 \\ \frac{\partial}{\partial u_1} & \frac{\partial}{\partial u_2} & \frac{\partial}{\partial u_3} \\ F_1 & F_2 & F_3 \end{vmatrix}, \tag{6.9a}$$

or

$$\text{curl } \boldsymbol{F} = \frac{1}{h_1 h_2 h_3}\begin{vmatrix} h_1 \boldsymbol{e}_1 & h_2 \boldsymbol{e}_2 & h_3 \boldsymbol{e}_3 \\ \frac{\partial}{\partial u_1} & \frac{\partial}{\partial u_2} & \frac{\partial}{\partial u_3} \\ h_1 F_1 & h_2 F_2 & h_3 F_3 \end{vmatrix}. \tag{6.9b}$$

With polar coordinates, (6.9b) can be written specifically as

$$\operatorname{curl} \boldsymbol{F}(r, \theta, \varphi) = \frac{1}{r^2 \sin\theta} \begin{vmatrix} \boldsymbol{e}_r & r\boldsymbol{e}_\theta & r\sin\theta\boldsymbol{e}_\varphi \\ \dfrac{\partial}{\partial u_1} & \dfrac{\partial}{\partial u_2} & \dfrac{\partial}{\partial u_3} \\ F_r & rF_\theta & r\sin\theta F_\varphi \end{vmatrix} \tag{6.10}$$

6.6. The Ampère Law in Differential Form

A vortex-like magnetic field is produced by an electric current as described by the Ampère law, where current directions are related by the right-hand screw rule. The clockwise direction of the $\boldsymbol{H}$ vector and the related current I correspond to rotating and advancing directions of a right-hand screw. In this context, the current $\boldsymbol{j}$ passing through a closed field-line of $\boldsymbol{H}$ can be expressed as

$$I = \int_S \boldsymbol{j}.\mathrm{d}\boldsymbol{S},$$

where $\boldsymbol{S}$ is a convex surface bordered by a closed curve C, as illustrated in Figure 6.7(b).

Therefore, the Ampère law (6.2), if we consider Γgo be such a closed field-line C, can be written as

$$\oint_C \boldsymbol{H}.\mathrm{d}s = \int_S \boldsymbol{j}.\mathrm{d}\boldsymbol{S}. \tag{6.11}$$

Using Stokes' theorem (6.4), the left side can be replaced by a surface integral, and therefore

$$\int_S (\operatorname{curl} \boldsymbol{H}).\mathrm{d}\boldsymbol{S} = \int_S \boldsymbol{j}.\mathrm{d}\boldsymbol{S},$$

and we obtain

$$\operatorname{curl} \boldsymbol{H}(\boldsymbol{r}) = \boldsymbol{j}(\boldsymbol{r}) \tag{6.12}$$

at a point $\boldsymbol{r}$, which is the Ampère law in differential form.

Equation (6.12) is the law that can be applied locally to an arbitrary position r in the field. Although $\boldsymbol{H} = 0$ at a position $\boldsymbol{r}$ where $\boldsymbol{j} = 0$, there is a point $\boldsymbol{r}'$ in the field where $\boldsymbol{H} \neq 0$, even if $\boldsymbol{j}(\boldsymbol{r}') = 0$. In this case, curl $\boldsymbol{H}(\boldsymbol{r}') = 0$ corresponds to the integral law $\oint_\Gamma \boldsymbol{H}.\mathrm{d}s = 0$, where Γ is chosen arbitrarily by enclosing no currents. Such a magnetic field $\boldsymbol{H}$ appears to be conservative in this region, where a magnetic scalar potential can be defined from $\boldsymbol{H}(r') = -\nabla_\mathrm{m}(\boldsymbol{r}')$. However, in the Ampère law (6.2) the curve Γ must be a field-line C encircling the current j. Along an arbitrary path Γ, physically neither a unit *magnetic charge* could be found to move, as will be discussed in Chapter 8, and hence the field of $\boldsymbol{H}$ is not conservative. Nevertheless, the magnetic potential $V_m(\boldsymbol{r}')$ is a useful mathematical concept for calculation, called a *pseudo-potential.*

As an example, we consider a long cylindrical conductor of circular cross-sectional area of radius a, which extends along the z-axis. Assuming that a current I is flowing steadily along the conductor, we calculate the magnetic field, using first the Ampère law in integral form, $\oint_C \boldsymbol{H} \cdot \mathrm{d}\boldsymbol{s} = I$. We can use cylindrical coordinates r, φ, z, referring z to the cylinder axis. At a point for $r > a$, a circle C of radius r encloses the entire current I inside of Γ, whereas for $r < a$, only $I(r/a)^2$ needs to be considered for the integral on C. Further, considering that only the component H_φ is significant, we can write that $H_\varphi(2\pi r) = I$ and $H_\varphi(2\pi r) = I(r/a)^2$ for $r > a$ and for $r < a$, respectively, and obtain

$$H_\varphi = \frac{I}{2\pi r} \quad (\text{for } r > a) \quad \text{and} \quad H_\varphi = \frac{Ir}{2\pi a^2} \quad (\text{for } r < a).$$

The differential equation (6.12) offers a more accurate analysis for obtaining the correct solution. In terms of the components, (6.12) can be expressed as

$$\frac{1}{r}\frac{\partial H_z}{\partial \varphi} - \frac{\partial H_\varphi}{\partial z} = 0, \quad \frac{\partial H_r}{\partial z} - \frac{\partial H_z}{\partial r} = 0, \quad \text{and} \quad \frac{1}{r}\frac{\partial (rH_\varphi)}{\partial r} - \frac{1}{r}\frac{\partial H_r}{\partial \varphi} = j.$$

As the field components should be independent of φ and z by symmetry, only the derivatives with respect to r are significant in these equations. From the second relation, we obtain $\dfrac{\partial H_z}{\partial r} = 0$, and hence $H_z =$ constant, which eventually leads us to a constant H_r, independent of the current. However, this value cannot be determined by (6.11) only, unless the problem is supplemented by the relation div $\boldsymbol{H} = 0$.

From the third equation, we have

$$\frac{1}{r}\frac{\mathrm{d}(rH_\varphi)}{\mathrm{d}r} = j, \quad \text{where} \quad j = 0 \ (\text{for } r > a) \text{ and } j = \frac{I}{\pi a^2} (\text{for } r < a),$$

which can be integrated as

$$H_\varphi = \frac{c_1}{r} \quad (\text{for } r > a) \quad \text{and} \quad H_\varphi = \frac{Ir}{2\pi a^2} + \frac{c_2}{r} \ (\text{for } r < a),$$

where $c_1 = I/4\pi$ and $c_2 = 0$ were determined by the conditions at $r = a$ and $r = \infty$.

The above example shows clearly that the Ampère law for the $\boldsymbol{H}$ vector is not sufficient to determine the magnetic field of a current, hence we require another law such as div $\boldsymbol{H} = 0$ to complete the description. It is notable from the relation $H_\varphi = I/2\pi r$ for $r > a$ we can define the scalar potential with $\boldsymbol{H} = -\boldsymbol{\nabla} V_m$, namely, $V_m = -\dfrac{I}{2\pi}\varphi$ satisfying

$$H_r = -\frac{\partial V_m}{\partial r} = 0 \quad \text{and} \quad H_\varphi = -\frac{1}{r}\frac{\partial V_m}{\partial \varphi} = \frac{I}{2\pi r}.$$

Here, it is noted that $H_r = 0$ is originated from div $\boldsymbol{H} = 0$, regardless of I.

As indicated by this example, the magnetic field is not signified only by the $\boldsymbol{H}$ vector, but requires an additional relation $\mathrm{div}\boldsymbol{H} = 0$. In fact, this relation for $\boldsymbol{H}$ constitutes another basic law for the absence of magnetic charges in nature, as will be discussed in Chapter 8.

6.7. The Rowland Experiment

It is a well-established fact that an electric current is due to moving charges through matter, however, it was not quite obvious in early days. Hence, the current was defined by means of visible phenomena such as magnetic, thermal, and chemical effects. Although today we have no doubt about moving charged particles, in early days it was questionable whether a charged body in motion was equivalent to an electric current. H. A. Rowland (1878) demonstrated magnetically that a current also arises from moving charges in the observer's frame of reference.

Figure 6.8 illustrates Rowland's experimental set up, as quoted from Pohl's textbook. A disk of an insulator with a metallic rim (A) of about $d \sim 20$ cm in diameter was rotated at a frequency $\nu = 50$ rev per second around the shaft. It was rotated inside a cylindrical metal case (B) at a narrow gap without contact. The metallic rim was charged to about 10^{-7}C by applying a voltage of about 1 kV between the disk and case. The rim had a split ab as shown in the figure, so that

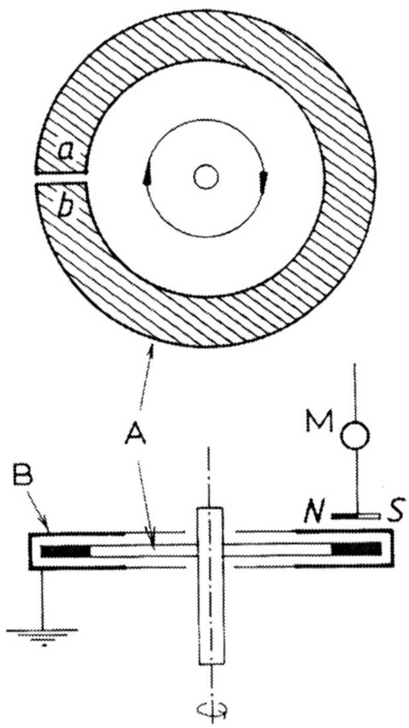

FIGURE 6.8. A set up for the Rowland experiment consisting of an insulating disk with conducting rim (A), a metal case (B) and a small compass with a mirror (M) suspended vertically.

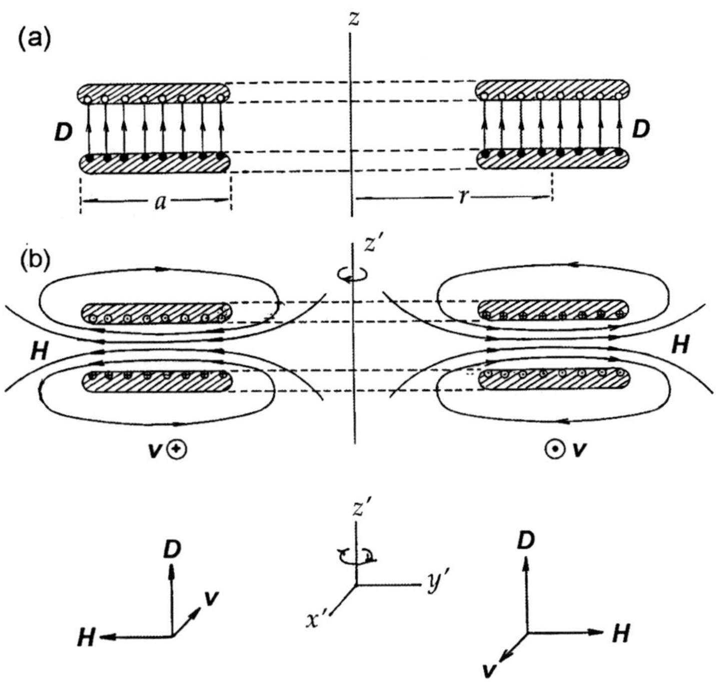

FIGURE 6.9. A rotating ring condenser, where (x', y', z') are the coordinate axes rotating with the condenser at which magnetic field-lines should appear. The fields $\boldsymbol{D}$ and $\boldsymbol{H}$ are related to the linear speed v of rotation as illustrated.

the charge could not be in conducting motion, but could be moved only with the rotating disk.

If such a rotating disk were equivalent to a current in space given by $I = qv$, where $v = (\pi d)\omega/2\pi$ is the speed of motion, a magnetic field H could be expected in the vicinity of the case. In Rowland's experiment such a magnetic field was detected by a small magnetic needle (M) suspended at a close location.

This result can be extended further for investigating the relation between a static electric field and a magnetic field with regard to an observer in relative motion. Figure 6.9 shows a capacitor of parallel disks of a ring-shaped area A with a narrow gap d, which is rotatable around the z-axis. The capacitor can be charged to $\pm q$ before rotation, where the electric field-lines are signified by the density $|\boldsymbol{D}| = q/A$, as illustrated in Figure 6.9(a). On the other hand, when the capacitor rotates at a constant speed, we can consider a pair of ring currents $\pm qv$, where $v = r\omega$ the linear speed of charges, thereby generating a magnetic field $\boldsymbol{H}$ across the gap between the rotating rings.

Referring to the rotation axis, we can set rectangular coordinate axes x, y, and z on the capacitor at rest, and another set of axes x', y' and $z' = z$ in rotational motion. The vector $\boldsymbol{D}$ is signified only by the components D_z and D'_z in these systems, although $D_z \approx D'_z$ unless v is close to the speed of light c, according to the relativity theory. However, a stationary observer should recognize a magnetic field between the disks in the radial direction y', as related to D'_z and v, depending on how v compares with c. As illustrated in the Figure 6.9(b), these factors are related by $H'_y \propto v \times D'_z$, as consistent with Ampère's law, and such a field H'_z should be detectable by a stationary observer, in principle. In contrast, an observer moving with the disks should detect only D_z. Such an attempt may remain a thought experiment, but it gives insight into the relativity theory in relation to dynamic electromagnetism.

7
Magnetic Induction

7.1. Laws of Magnetic Induction

The magnetic field is a property of the space that surrounds an electric current and a magnet. In early physics powdered iron was used to visualize its geometric character, and the distributed field intensities were studied with a compass needle. However, such observations were hardly subject to systematic investigations, until a coiled wire became available for studying a uniform magnetic field at a measurable intensity. Indicated by a ripple pattern of iron particles, the field-lines revealed the distribution in space, which were recognized as significant for describing the nature of a magnetic field. Faraday (1832) made a number of sketches of field-lines emphasizing the significance of a distributed pattern, and he discussed the field with the concept of *flux* of field-lines. In fact, in the previous chapters we used his idea for electric fields for which electric charges on conductors were responsible. For a magnetic field, in contrast, we cannot argue further using electric analogies because magnetic charges are absent in nature. Nevertheless, with the concept of flux Faraday described the induction effect in magnetic fields, which lead him to establish the fundamental laws of electromagnetism. Owing to Faraday's discovery, the flux of magnetic field-lines, as determined by induced voltages, allows a quantitative description of the magnetic field with no responsible magnetic charges.

7.1.1. The Faraday Law

Figure 7.1 shows a basic set up to demonstrate the magnetic induction effect. The coil L in Figure 7.1(a) produces a magnetic field (as sketched in Figure 6.4), where the field-lines are in near parallel inside L, when activated by a steady current. A smaller coil L_0 connected with a galvanometer G can be moved to and from L to detect an induction current. Figure 7.1(b) shows another set up with a permanent magnet, where a similar induction effect can be observed by moving L_0. An induction current in L_0 flows as related to the flux through the cross-sectional area S, and hence the flux can be defined in terms of an observed induction current. A modern version of L_0 with a connected circuit, called a *search coil* and *fluxmeter*, is a basic device for measuring the magnetic field.

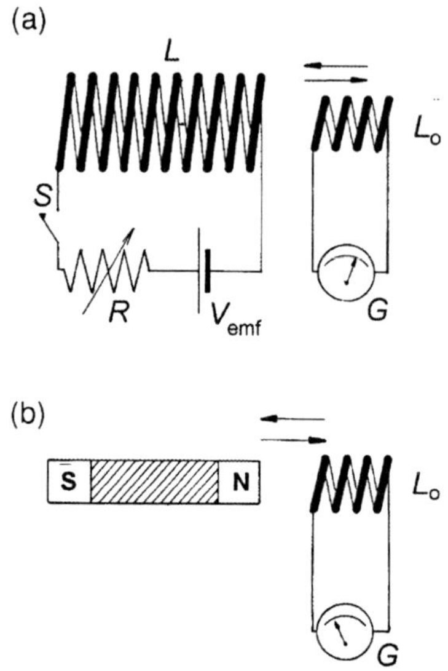

FIGURE 7.1. (a) Magnetic induction between two coils L and L_o. Induction between a magnet and a search coil L_o.

When L_o is moved against L, an induced current I can be observed on G, depending on the electrical resistance of L_o, for which the induced emf (V_{emf}) is considered primarily responsible. The emf voltage varies as a function of the time for the relative motion between L_o and L or the magnet, exhibiting a pulse-shaped peak, as sketched in Figure 7.2, which is higher the faster the movement. The observed V_{emf} depends on other factors of L_o, such as the area S, turn number N, and the closest position. However, most significant is that the voltage pulse defined by $U = \int_0^\infty V_{emf}(t)\,dt$ is independent of the speed, which is therefore considered a variable for expressing the magnetic flux.

The number of field-lines, or the flux, is determined by $\Delta\Phi = \int_S B dS = BS$ per one turn in L_o if the field is uniform, which can then be described by an observed pulse U/N, where N is the number of turns in the search coil L_o. By equating $\Delta\Phi$ to U/N, the flux density B can be defined as $B = U/\mathrm{N}S$. In this definition the unit of B is given by V-sec/m^2 in the MKSA system, a unit called the "tesla." The traditional CGS unit "gauss" is still in use; the conversion to tesla is 1 tesla

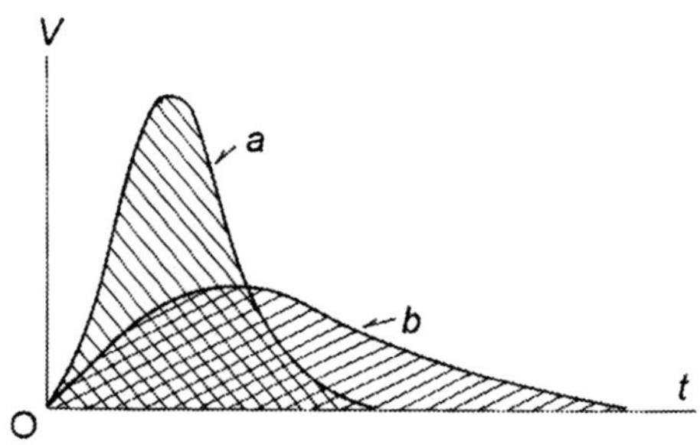

FIGURE 7.2. Examples of induction pulses. a and b correspond to fast and slow movement of a search coil, respectively. Empirically, these areas $\int_0^\infty V dt$ are identical.

$= 10^4$gauss. Also, the flux $\Delta\Phi = BS$ is measured in the practical unit "weber," which is the same as the voltage pulse in V-sec, and accordingly tesla = weber/m^2.

The quantity $\Delta\Phi$ represents the flux of field-lines that is picked up by one turn of a search coil when inserted in the magnetic field. At a given position of L_o, the Faraday law can then be expressed as

$$\Delta\Phi(t) = \int_0^t |V_{\mathrm{emf}}(t)|\,\mathrm{d}t \quad \text{or} \quad \frac{\mathrm{d}\Phi}{\mathrm{d}t} = |V_{\mathrm{emf}}(t)| \tag{7.1}$$

and, hence, the observed $|V_{\mathrm{emf}}(t)|$ represents a changing rate of the flux through the area S of a search coil. It is noted that the formula (7.1) between the magnitude of $V_{\mathrm{emf}}(t)$ and $\mathrm{d}\Phi/\mathrm{d}t$ is not exactly correct for magnetic induction phenomena since the direction of $V_{\mathrm{emf}}(t)$ is not specified. Nevertheless, (7.1) indicates that the magnetic induction is signified by an induced voltage pulse in a moving coil, and is called the *Faraday law*.

7.1.2. The Lenz Law

Lenz also studied magnetic induction and published his interpretation in 1834. Lenz revised the Faraday law by attaching the correct sign to the induced voltage; the revised law is called the *Lenz law*. At his time, electronic motion in metals was not a familiar concept, so that induced emf was not recognized as originating from an electronic mechanism. Nevertheless, Lenz considered a mutual interaction between coils, referring to Newton's *action-reaction principle*.

The induction current flowing in a coil generates heat in L_o, indicating that a change $\Delta\Phi$ in the main coil L can be attributed to a mutual interaction energy between L_o and L. If so, it is logical to consider the *reaction* $-\Delta\Phi$ of L_o to the change $\Delta\Phi$ as originating from such a mutual interaction, and hence the induction current can flow as the reaction $-\Delta\Phi$. In this way (7.1) can be revised as

$$-\Delta\Phi = \int_0^t |V_{\mathrm{emf}}(t)|\,\mathrm{d}t \quad \text{or} \quad -\frac{\mathrm{d}\Phi}{\mathrm{d}t} = |V_{\mathrm{emf}}(t)|, \tag{7.2}$$

which is the *Faraday-Lenz formula*. Originally Lenz stated that the induction current should flow in such a direction as to oppose the change $\Delta\Phi$, which can be precisely expressed in the revised formula (7.2). Figure 7.3 shows the relation between positive directions for $\mathrm{d}\Phi/\mathrm{d}t$ and $V_{\mathrm{emf}} = RI$, where R is the resistance of L_o and the detector, a resistance that is described by the *left-hand screw* rule.

7.1.3. Magnetic Field Vectors

In the above, we considered a search coil L_o of cross-sectional area S, placed perpendicular to the field direction, in which the flux was defined as $\Delta\Phi = BS$. Considering S as a vector $\boldsymbol{S} = S\boldsymbol{n}$, where $\boldsymbol{n}$ is the unit vector to the area, the quantity B can logically be regarded as a vector $\boldsymbol{B}$, signifying the direction of field-lines. In this definition, $|\boldsymbol{B}|\cos\theta$, where $\cos\theta = (\boldsymbol{B}.\boldsymbol{n})/B$, represents the density of field-lines measured as the induced voltage pulse U/N of L_o.

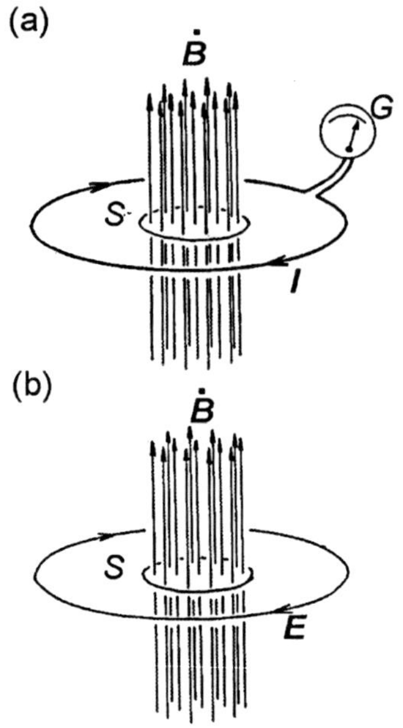

FIGURE 7.3. Magnetic induction. (a) Induction current I as related to $\dot{B}$, (b) dynamic electric field E related to $\dot{B}$.

On the other hand, the magnetic field in the coil L was expressed by another field vector $\boldsymbol{H}$, as stated by the Ampère law. The magnitude is given by $|\boldsymbol{H}| = nI$ if L is sufficiently long, and the direction of $\boldsymbol{H}$ is tangential to the field-line. With these two descriptions combined, the region of "intense" field can be specified by a large value of $|\boldsymbol{B}|$, whereas the direction of field-lines can be specified by the vector $\boldsymbol{H}$.

The uniform field in a long solenoid can be shown by parallel field-lines that are evenly distributed, where the vectors $\boldsymbol{B}$ and $\boldsymbol{H}$ are parallel. For such an empty solenoid, we can write a proportionality relation

$$\boldsymbol{B} = \mu_o \boldsymbol{H}, \tag{7.3}$$

where μ_o is an empirical constant. The value of μ_o can be determined experimentally from the relation $U/NS = \mu_o nI$, and is known as

$$\mu_o = 1.256 \times 10^{-6} = 4\pi \times 10^{-7} \text{ V sec}/A\text{-}m.$$

Considered for representing empty space, like dielectric parameter ε_o in electrostatics, the constant μ_o is required for all calculations with MKSA units.

7.2. Differential Law of Induction and the Dynamic Electric Field

In a magnetic field, field-lines are described by two vectors $\boldsymbol{H}$ and $\boldsymbol{B}$ expressing their intensities and distribution, respectively. In a uniform field inside a long solenoid, the two vectors represent parallel field-lines at a constant density over almost entire volume, where we have the relation $B = \mu_o H$, and the flux is simply expressed by $\Phi = BS$. In a more general field the distributed flux should be related with respect to a closed area S for counting the number of lines within its peripheral curve C, as expressed by $\Phi = \int_S \boldsymbol{B}.\mathrm{d}\boldsymbol{S}$

Generally, such a flux is a function of time t, as indicated by induction experiments. According to the Faraday-Lenz law, an emf (V_{emf}) is induced in a coil by changing magnetic flux Φ through the sectional area S. Such an area S for defining Φ is essential for V_{emf}; on the other hand, due to the emerging potential difference $-V(t) = V_{\mathrm{emf}}$, the corresponding electric field $E(t)$must be considered in the coiled wire, driving charge carriers in motion. In this context, the Faraday law can be re-expressed as

$$-\frac{\mathrm{d}}{\mathrm{d}t}\int_S \boldsymbol{B}.\mathrm{d}\boldsymbol{S} = \oint_C \boldsymbol{E}.\mathrm{d}\boldsymbol{s}, \tag{7.4}$$

which is the Faraday-Lenz formula in integral form. Such an electric field $\boldsymbol{E}(t)$ in (7.4), called the *dynamic electric field*, has been best substantiated in the *betatron accelerator*, where particles of charge e are accelerated by a force $e\boldsymbol{E}(t)$.

Using the Stokes theorem, we convert the integral on the right side of (7.4) to a surface integral, i.e., $\oint_C \boldsymbol{E}.\boldsymbol{ds} = \int_S \mathrm{curl}\,\boldsymbol{E}.\mathrm{d}\boldsymbol{S}$, and hence the equation (7.4) can be reduced to a differential form, that is,

$$-\frac{\partial B}{\partial t} = \mathrm{curl}\,\boldsymbol{E}. \tag{7.5}$$

Such a dynamic electric field is not conservative because curl $\boldsymbol{E} \neq 0$, as indicated by (7.5), or $\oint_C \boldsymbol{E}.\mathrm{d}\boldsymbol{s} \neq 0$, for $\frac{\partial \boldsymbol{B}}{\partial t} \neq 0$. If a small charge $q$is brought into such a field $\boldsymbol{E}$, a characteristic force $\boldsymbol{F} = q\boldsymbol{E}$ can be detected to verify the effect. Such forces should act on charged particles and mobile charge carriers in matter as well, so that a moving conductor in a magnetic field should be subjected to such a force due to the induction effect, which is therefore called a *magnetic force*.

Figure 7.4 illustrates an example of a conducting wire that is moved in a uniform magnetic field against a magnetic force. Consider that $\boldsymbol{B}$ is normal to the page, and a pair of long parallel wires is placed perpendicular to $\boldsymbol{B}$, to which a galvanometer G connected. A metal rod *ab* with an insulating handle P slides along the wires to the right or to the left. The rod and a portion of the wires form a closed loop along which an induction current can flow when the rod is moved. Denoting the length between contact points a and b by l, and the sliding distance of the rod as x from the initial position marked o-o, the magnetic flux in the area swept by the rod is given by $\Phi = Blx$. When *ab* slides at a speed $v = \mathrm{d}x/\mathrm{d}t$, the Faraday-Lenz law describes the induced emf voltage

$$V_{\mathrm{emf}} = -\frac{\mathrm{d}\Phi}{\mathrm{d}t} = -Blv,$$

giving a potential difference $V(t) = -V_{\mathrm{emf}} = Blv$ on the moving rod.

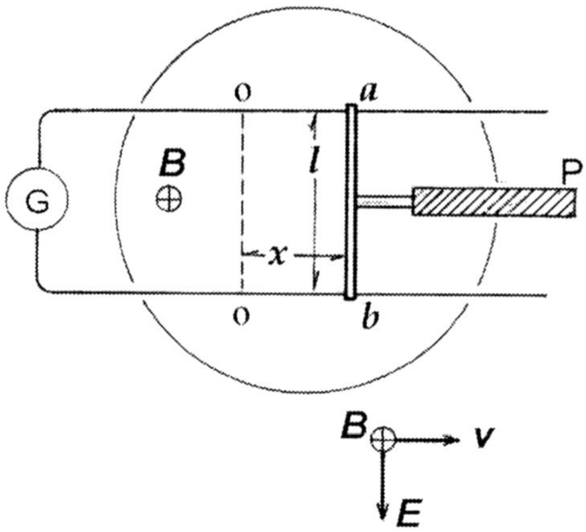

FIGURE 7.4. Magnetic induction with a moving rod ab along parallel wires in a magnetic field $\boldsymbol{B}$. $\boldsymbol{v}$ and $\boldsymbol{E}$ are the speed and induced electric field in ab, respectively.

Accordingly, an electric field, described by $E = -\,\mathrm{d}V/\mathrm{d}l = Bv$, is in the rod between a and b. Taking such directions of v, B, and E into account, we can write a vector relation

$$\boldsymbol{E} = \boldsymbol{v} \times \boldsymbol{B}, \tag{7.6}$$

as shown in the figure, which is detectable with mobile carriers or particles of charges q. On a particle moving at a speed $v \neq 0$ in a magnetic field $\boldsymbol{B}$, the magnetic force can therefore be expressed as

$$\boldsymbol{F}_{\mathrm{L}} = q\boldsymbol{E} = q(\boldsymbol{v} \times \boldsymbol{B}), \tag{7.7}$$

which is known as the *Lorentz force*.

The Lorentz force acts on a charged particle of mass m and charge q in a direction normal to the speed **v**, and hence no acceleration takes place along $\boldsymbol{v}$ in a magnetic field, but deflects from the initial path determined by $\boldsymbol{v}$. Therefore, we can write the equation of motion as

$$m\frac{\mathrm{d}v}{\mathrm{d}t} = 0, \quad m\frac{v^2}{r} = qvB,$$

and therefore

$$r = \frac{mv}{qB}, \tag{7.8}$$

where r is the radius of a circular orbit in a uniform field of $B =$ constant. Such a circular motion of a particle, known as *cyclotron motion*, is shown in Figure 7.5 for a negative electron, $q = -e$. From (7.8), the radius is large for a large momentum mv, but small in a weak magnetic field B.

In a betatron, such a dynamic electric field is utilized for accelerating electrons. Electrons injected normal into a large magnetic field $\boldsymbol{B}$ exhibit a cyclotron motion of radius $r$given by (7.8). The speed v in a circular orbit remains unchanged in a steady magnetic field, but by varying the field at a rate $\mathrm{d}\Phi/\mathrm{d}t$, $\boldsymbol{v}$ can be modulated

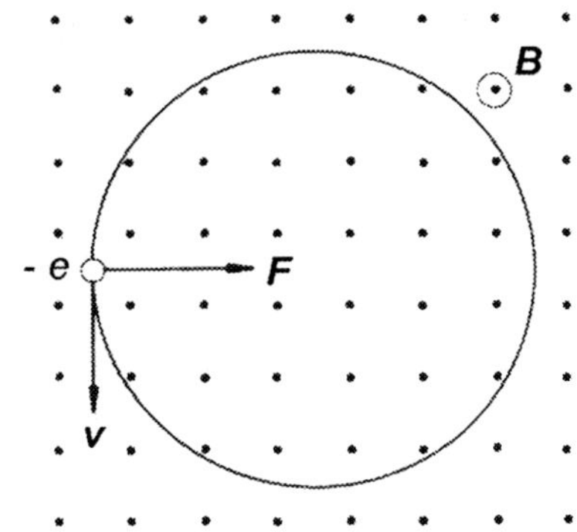

FIGURE 7.5. An electron in a uniform magnetic field $\boldsymbol{B}$. The centripetal force $\boldsymbol{F} = -e(\boldsymbol{v} \times \boldsymbol{B})$ is responsible for the cyclotron motion.

by the dynamic electric emf voltage arising from $d\Phi/dt$. The circular motion and varying field can be synchronized to accelerate electrons. An average rate of 1 weber/10^{-3}-s is technically quite feasible, giving rise to an induced emf of the order of 1 kV, which can accelerate electrons when the cyclotron motion and $d\Phi/dt$ are in phase.

Equation (7.7) can also be expressed for a macroscopic magnetic force on a current flowing in a magnetic field. Consider that a conducting wire of cross-sectional area A and the length l is exposed to a uniform field B. The current on such a wire is given by $I = jA$, where $j = nqv$, and $n = N/(Al)$ are the current density and the number density of carrier particles, respectively. Hence, $qv = I/nA = (I/N)l$. The Lorentz force $\boldsymbol{F}_L$ given by (7.7) acts on all particles in the current, so that the macroscopic force is given by $N\boldsymbol{F}_L = \boldsymbol{F}$. Hence, the force on the conduction wire can be expressed as

$$\boldsymbol{F} = I(\boldsymbol{l} \times \boldsymbol{B}), \tag{7.9}$$

where the vector l indicates the direction of the current I specified by the wire. Figure 7.6 illustrates the relation (7.9).

Because of the induction effect, an amount of work is required for a conducting body to move in a magnetic field, resulting in an induced current and heat in the coil. Such a process can be described by the law of energy conservation; the

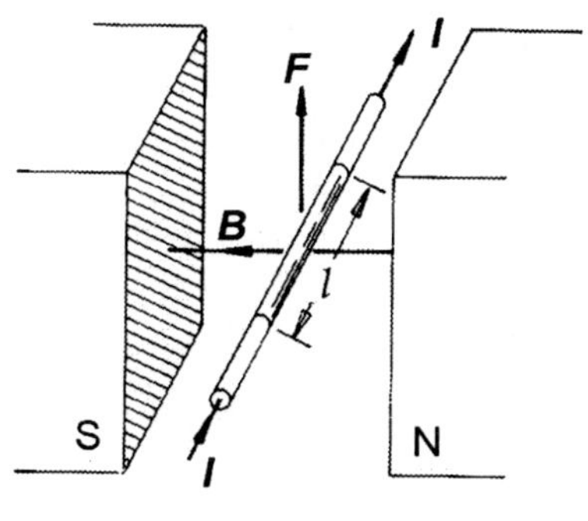

FIGURE 7.6. A magnetic force F on a straight current I in a uniform magnetic field $B(B \perp I)$.

input power to move against the magnetic force is equal to the heat produced in the output. Namely, $P_{\text{in}} = -\boldsymbol{F}.\boldsymbol{v} = -IlBv = (V_{\text{emf}}/R)lBv$, where $V_{\text{emf}} = Blv$ and R is the electrical resistance of the conductor. Therefore, we obtain $P_{\text{in}} = (Blv)^2/R = V_{\text{emf}}^2/R = P_{\text{out}}$.

We discussed a moving rod in the above example, but the induction can generally take place in any conducting body in a magnetic field, where the induction causes a distributed current called an *eddy current*. For example, in a simple pendulum of a metallic bob in a magnetic field, the amplitude is decreased with time, as induced eddy currents generate heat in the bob.

7.3. Magnetic Moments

A magnetic force on a current given by (7.9) has numerous engineering applications: alternating current (AC) generators, AC and DC motors, metering devices, and many others. A coiled wire in compact size is suitable for the rotation in a limited space between magnetic poles, converting electric energies to mechanical ones via a magnetic induction effect. Such a rotating coil in a magnetic field may be described conveniently in terms of a *magnetic moment*.

Although rectangular and circular coils are used in many practical applications, a coil in arbitrary shape can be discussed for a general description. As we discussed for the Stokes theorem, the area of a given coil can be divided into many small rectangular areas as shown in Figure 6.6, where each one is regarded as an elementary rectangular coil. Assuming that all of these currents flow around small coils in the same direction with an equal magnitude, their integrated result is equivalent to a current around the peripheral rim. Therefore, for a theoretical argument it is convenient to consider a small rectangular coil in order to calculate the resultant induction in a given coil. Nevertheless, a coil in rectangular shape is suitable for many practical designs.

Figure 7.7(a) shows a rectangular coil that can be rotated around the axis as indicated. On the four edges currents flow in different directions in the magnetic field B, and hence four different forces, F_1, F_2, F_3, and F_4, determined by (7.9), should be considered together, as shown in the figure. It is noted that F_2 and F_4 are equal in magnitude but go in opposite directions parallel to the axis, thus causing no effect to the rigid coil. In contrast, F_1 and F_3 are equal and opposite, being exerted on different parts of the coil in perpendicular directions to the axis. Consequently, a *torque* or a *couple of forces* acts on the coil. As $|F_1| = |F_3| = IbB$ in this case, the torque can be expressed as $\tau = IbB \times a \sin\theta$, where b and a are edge lengths, and θ is the angle between B and the normal $\boldsymbol{n}$ to the rectangular area of the coil, as indicated in Figures 7.7(a) and (b). Considering the area $ab = S$ as a vector $\boldsymbol{S} = S\boldsymbol{n}$, where the direction of $\boldsymbol{n}$ and the current direction are related by the right-hand rule, the torque is written as a vector

$$\tau = I\boldsymbol{S} \times \boldsymbol{B} = \boldsymbol{p}_{\text{m}} \times \boldsymbol{H}, \quad \text{where} \quad \boldsymbol{p}_{\text{m}} = \mu_{\text{o}} I\boldsymbol{S}. \tag{7.10}$$

Here, the vector $\boldsymbol{p}_{\text{m}}$ is called the *magnetic moment of the coil*. However, the

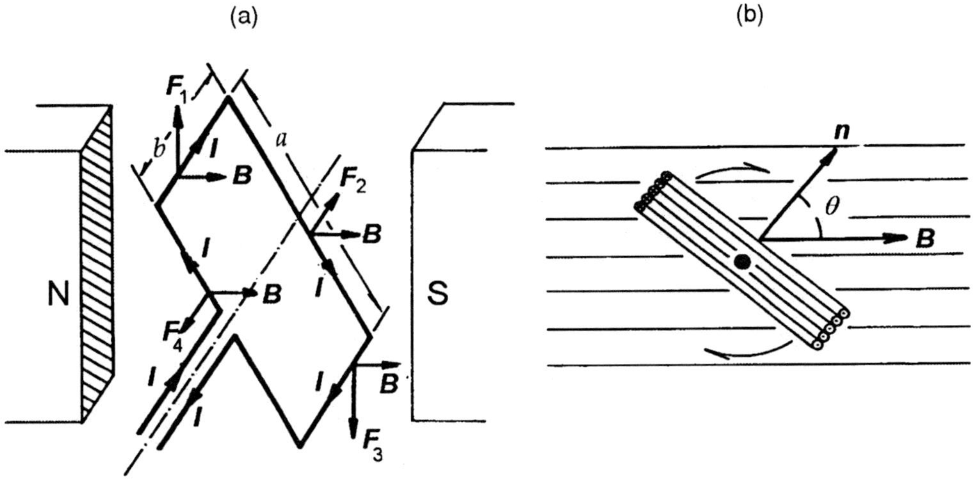

FIGURE 7.7. (a) A magnetic torque on a rectangular current I in a uniform magnetic field B, where the axis is perpendicular to B. (b) A current-carrying short coil in a uniform field B, where the magnetic torque can be expressed as $\tau \propto \boldsymbol{n} \times \boldsymbol{B}$.$\boldsymbol{n}$ is the unit vector normal to the area of the coil.

definition of $\boldsymbol{p}_m$ may often be a matter of preference, e.g., if writing $p_m = I\mathbf{S}$ to exclude μ_o, the torque of (7.10) can be expressed as $\tau = \boldsymbol{p}_m \times \boldsymbol{B}$. As remarked, for a coil in arbitrary shape, $\boldsymbol{S}$ in (7.10) can be considered to represent the whole area, whose direction can be related to the rim current by the right-hand screw rule.

The torque τ is responsible for coil in rotational motion. Therefore, the angle θ between $\boldsymbol{p}_m = IS\boldsymbol{n}$ and $\boldsymbol{B}$, shown in Figure 7.7(b), depends on the energy U for $\boldsymbol{p}_m$ to be rotated by the torque $\tau = -\partial U/\partial\theta$. For a given τ by the field as in (7.10), U can be expressed as

$$U = -\int_{\theta_o}^{\theta} \tau \mathrm{d}\theta = -p_m B(\theta - \cos\theta_o) = -\boldsymbol{p}_m . \boldsymbol{B} + \text{const} \tag{7.11}$$

where const $= p_m B \cos\theta_o$. Assuming $\theta_o = \frac{1}{2}\pi$, the minimum of U is determined at $\theta = 0$ corresponding to $\boldsymbol{p}_m || \boldsymbol{B}$. In this case, $U = p_m B(1 - \cos\theta)$ represents the potential energy of $\boldsymbol{p}_m$ in the field of $\boldsymbol{B}$.

Example 1. An AC Generator.

When such a coil as called the *armature* is rotated mechanically in a magnetic field, an induction voltage is generated, which can then be utilized for practical applications. In this case the angle θ of rotation can be expressed as $\theta = \omega(t - t_o)$, where ω is the angular frequency of rotation, and t_o the initial time. The flux in the armature is given by $\Phi = BS \sin\omega(t - t_o)$, and hence the induced emf is expressed by

$$V_{\mathrm{emf}} = -\frac{\mathrm{d}\Phi}{\mathrm{d}t} = BS\omega \cos \omega (t - t_o) = V_o \cos \omega (t - t_o),$$

where $V_o = BS\omega$ is the amplitude of the *alternating voltage*, varying in sinusoidal form with time. Therefore, when connected to a resistor R, a sinusoidal current $I(t - t_o)$ at the frequency ω flows through the circuit of R and the armature coil. In terms of energy, the required input power for rotating the armature depends on its moment of inertia, whereas the electrical output power is given by the product $[V_{\mathrm{emf}}(t - t_o)][I(t - t_o)]$. For such an AC circuit, the variable $\phi(t) = \omega(t - t_o)$, called the *phase*, plays an essential role, as will be discussed in detail in Chapter 9.

Example 2. An Electric Motor.

Instead of rotating mechanically, the armature can also be turned by a DC current that flows through the coil, and the rotational energy can be utilized for mechanical work. In such a device, an electrical power is converted into mechanical output, constituting the principle for a *motor*. However, we note from (7.10) that the armature can rotate in both directions depending on the phase, for which rotation an additional mechanism must be devised to keep the rotation in one direction. A *commutator* with brush contacts is usually installed with a DC motor for reversing the phase, so that the current flows in phase with the rotating armature. In a DC motor the frequency of rotation can be made variable in a limited range by adjusting the current.

In almost the same device, but one without a commutator, an armature can be rotated in one direction by using an AC current with a fixed phase. In such an AC motor the speed of rotation is determined by the frequency ω of the AC source.

Example 3. A Velocity Selector.

Electrons and ions can usually be evaporated from a high-temperature filament or oven placed in vacuum, in which their kinetic energies are distributed in a thermal range determined by the temperature. From these charged particles, those of a specific velocity v can be selected for a given experimental task. In a uniform magnetic field B, charged particles travelling at a velocity $\boldsymbol{v}$ in the straight path are defected into a circular orbit by the magnetic force $q(\boldsymbol{v} \times \boldsymbol{B})$. On the other hand, in an electric field $\boldsymbol{E}$ they are deflected into a parabolic path by the static force $q\boldsymbol{E}$. In a space where $\boldsymbol{E}$ and $\boldsymbol{B}$ are *crossed*, as shown in Figure 7.8, these deflections of particles at a speed $\boldsymbol{v}$ can be compensated if $\boldsymbol{E}$ and $\boldsymbol{B}$ are arranged in such a way that $q\boldsymbol{E} = -q(\boldsymbol{v} \times \boldsymbol{B})$. In other words, only those particles of a speed $v = |\boldsymbol{E}|/|\boldsymbol{B}|$

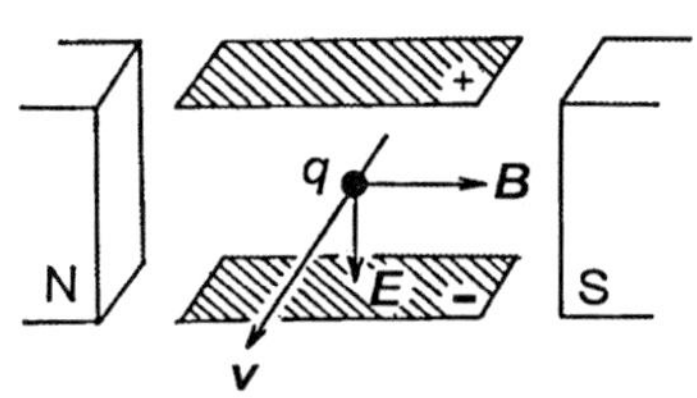

FIGURE 7.8. A velocity selector of crossed magnetic and electric fields. The magnitudes can be selected as $v = E/B$.

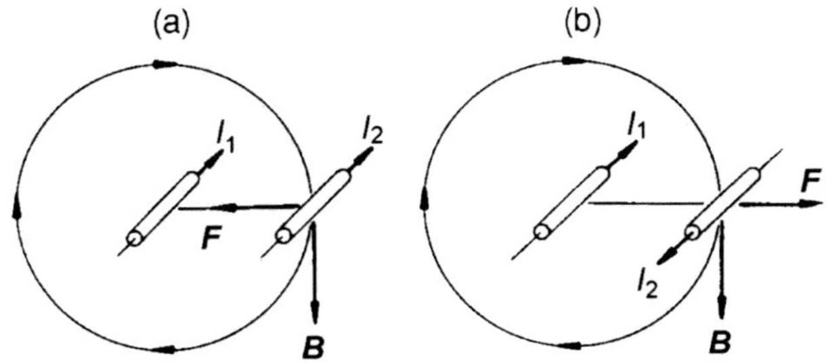

FIGURE 7.9. Magnetic force between (a) parallel currents, and (b) anti-parallel currents. ***B*** shows the magnetic field of one of the currents at the other.

can pass through the crossed fields with no defection. A desired speed $v$can thus be selected by choosing suitable magnitudes of ***E*** and ***B*** , and such a device can be used to control charged particles for many applications.

Example 4. Magnetic Forces Between Parallel Wires.

It can be demonstrated that two long parallel wires carrying currents I_1 and I_2 attract or repel with each other when flowing in the same or opposite directions, respectively. To obtain such an interaction force between them, we can consider that one of these currents generates a magnetic field, as specified by the Ampère law (6.1), at the position of the other, at which the magnetic force can be calculated by (7.9). In Figures 7.9(a) and (b), a circular field-line arising from I_1 is shown at the position of I_2 that is either parallel or anti-parallel to I_1, respectively. In this case, from (6.1), we can write $B = \mu_o I_1/2\pi r$, where r is the distance between the wires. Then, using (7.8) the magnetic forces are given by $F = \pm\ I_2 lB$, where l is the length of interacting wires that are sufficiently longer than r. The magnitudes of these forces are equal, i.e., $|F|/l = \mu_o I_1 I_2/2\pi r$, and directions are opposite, as shown in the figure.

8
Scalar and Vector Potentials

8.1. Magnets

Natural magnets played the central role in magnetic studies in early days, leading to the Ampère law of a magnetic field, which was discovered by using iron magnets. Characterized as N- and S-poles, magnetic charges were postulated by analogy to electric charges. However, it was only after Heisenberg's quantum theory (1928) that the origin of magnetism was attributed to correlated electronic spin angular momenta in materials.

With iron magnets and by induction experiments, it was recognized a current carrying coil generates such magnetic poles upon which field-lines converge, just like in natural magnets. Figure 8.1 illustrates a bar magnet and a uniform coil carrying an electric current, where voltages $V(t)$ induced by moving a search coil L_o along their axes are plotted as a function of time t. Positive and negative voltage pulses can be observed near the end regions as shown, indicating high flux densities of magnetic field-lines corresponding to the magnetic poles. In the middle part of the coil no appreciable induction takes place because of nearly uniform flux densities. On the other hand, to interpret the observed results from the magnet, we have to consider the *internal* magnetic flux $\boldsymbol{M}$, which is virtually constant due to saturated spontaneous *magnetization*. Related to these poles, magnetic fields exist inside and outside of the magnetic material, where the flux of field-lines should be described by the respective densities; that is,

$$\boldsymbol{B}_{\text{in}} = \mu_o \boldsymbol{H}_{\text{in}} + \boldsymbol{M} \quad \text{and} \quad \boldsymbol{B}_{\text{out}} = \mu_o \boldsymbol{H}_{\text{out}}. \tag{8.1}$$

A significant feature of the magnetism is that a *single magnetic pole* cannot be isolated in nature. If such a single pole could exist at all, it was believed that magnetic field-lines would emerge from it, similar to what is seen with electric charges. However magnetic poles appear always in pairs, and no attempt to separate them has been successful. Therefore, accepting this phenomenon as a fact in nature, we consider it a law of magnetism. In fact, ionic and molecular magnetic moments are responsible for such a pair of poles, so that macroscopically the absence of single magnetic charges can be expressed by zero density per volume, i.e., $\rho_m = 0$. Noting that the Gauss theorem expresses the general relation between flux densities

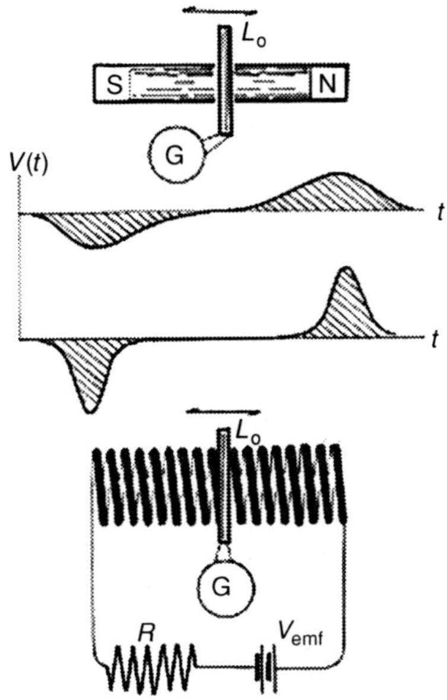

FIGURE 8.1. Magnetic induction pulses as the search coil L_o is moved through a bar magnet (top) and a coil (bottom). The voltage pulses $V(t)$ vs. t in these cases are compared in the center graphs. These pulses represent magnetic poles.

and the responsible source, for the magnetic flux we can write a specific relation

$$\operatorname{div}\boldsymbol{B} = 0. \tag{8.2}$$

Therefore, magnetic fields $\boldsymbol{H}_{\text{in}}$ and $\boldsymbol{H}_{\text{out}}$ inside and outside of a magnet can be expressed, respectively, by

$$\operatorname{div}\boldsymbol{H}_{\text{in}} = -\frac{1}{\mu_o}\operatorname{div}\boldsymbol{M} \quad \text{and} \quad \operatorname{div}\boldsymbol{H}_{\text{out}} = 0. \tag{8.3}$$

Assuming $\boldsymbol{M}$ is constant over an idealized magnet, div $\boldsymbol{H}_{\text{in}} = 0$ inside, with which we can obtain the boundary condition on the magnet surface. Considering a cylindrical bar magnet with flat ends, we can apply the integrated form of (8.2) with (8.3) to a short Gaussian surface of flat areas ΔS across the boundaries, and we obtain $B_{\text{out}}\Delta S = B_{\text{in}}\Delta S$. Therefore,

$$H_{\text{out}} = H_{\text{in}} + \frac{1}{\mu_o}\left(\frac{\mathrm{d}M}{\mathrm{d}n}\Delta n\right)_{\Delta S} = H_{\text{in}} + \frac{\sigma'_{\text{m}}}{\mu_o}, \tag{8.4}$$

where $\sigma'_{\text{m}} = \left(\frac{\mathrm{d}M}{\mathrm{d}n}\Delta n\right)_{\Delta S}$ expresses the apparent magnetic charge density on the boundary surface. In Figure 8.1, the voltage $V(t)$ corresponds to distributed σ'_{m} in the pole regions, and consequently field-lines of $\boldsymbol{H}_{\text{out}}$ emerge from $+\sigma'_{\text{m}}/\mu_o$, ending at $-\sigma'_{\text{m}}/\mu_o$. However, it is noted that such an interpretation arises from curl $\boldsymbol{H}_{\text{out}} = 0$ due to $\boldsymbol{j} = 0$, and for such $\boldsymbol{H}_{\text{out}}$ we always have the relation div $\boldsymbol{H}_{\text{out}} = 0$, as deduced from (8.2).

It is noted that div $\boldsymbol{H} = 0$ should also be significant at a point where $\boldsymbol{j} =$ curl $\boldsymbol{H} = 0$ in a stationary magnetic field that originates from a distant current. As

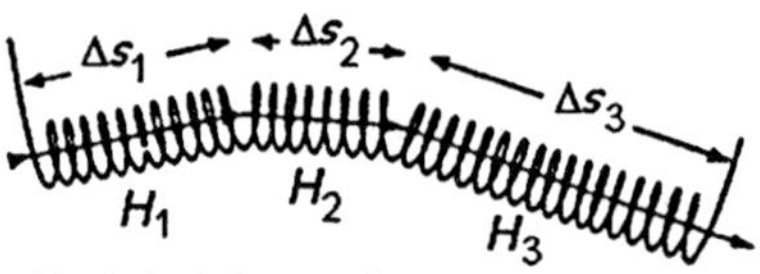

Abb. 6.13. Schema eines magnetischen Spannungsmessers (A. P. Chattock, 1887; W. Rogowski, 1912)

FIGURE 8.2. A chain of differential coils of lengths Δs_1, Δs_2, $\Delta s_3, \ldots,$ monitoring magnitudes of magnetic fields $\boldsymbol{H}_1$, $\boldsymbol{H}_2$, $\boldsymbol{H}_3, \ldots.$ (R.W. Pohl, *Electrizitätslehre*)

discussed in Section 6.6, to obtain the solution at such a point, it is not sufficient to solve curl $\boldsymbol{H} = 0$, but necessary for $\boldsymbol{H}$ to satisfy div $\boldsymbol{H} = 0$. In any case, the equation div $\boldsymbol{B} = 0$ holds in both magnetic and a steady current fields, and (8.2) is regarded as a fundamental equation for any magnetic field.

8.2. Pohl's Magnetic Potentiometer

Using a search coil, it is possible to map a magnetic field in terms of the induced voltage pulse U. As implied by the relation $\boldsymbol{B} = \mu_o \boldsymbol{H}$ in empty space, a small solenoid can be used for investigating the vector character of a distributed field. For a long flexible solenoid of a cross-sectional area $\boldsymbol{S} = S\boldsymbol{n}$, we consider a short part of length $\mathrm{d}\boldsymbol{s} = \mathrm{d}s\boldsymbol{n}$, whose orientation is represented by $\boldsymbol{n}$. A magnetic induction should occur as related to the rate $\mathrm{d}(\boldsymbol{B}.\boldsymbol{S})/\mathrm{d}t$, which is proportional to $\boldsymbol{H}.\mathrm{d}\boldsymbol{s}$, and both are specified by $\boldsymbol{n}$. In such a device, plus and minus voltage pulses in pair can be recorded in L_o, as illustrated in Figure 8.1. Pohl constructed a long flexible solenoid attached to a galvanometer, which was, in fact, displayed for demonstration in his lecture theater at Göttigen. Figures (8.2) and (8.3) show sketches of his device and related concept, which is quoted in the following from his textbook "Elektrizitätslehre."

With such a device the experimental purpose is two-fold: to identify locations for high and low flux densities in the field, and to verify that $\int_A^B \boldsymbol{H}.\mathrm{d}\boldsymbol{s}$ is independent of the path of integration terminated at points A and B. Figure 8.2 illustrates that such a long solenoid can be considered to consist of small segments of lengths $\mathrm{d}s_i$, so that the integral $\int_A^B \boldsymbol{H}.\mathrm{d}\boldsymbol{s}$ can be replaced approximately by the sum $\sum_i \boldsymbol{H}_i.\mathrm{d}\boldsymbol{s}_i$ where voltage pulses at each joint of adjacent segments are cancelled. Figure 8.3 illustrates some practical operations of Pohl's device. Here, Figure 8.3(a) shows measurements of voltage pulses at the ends a and b of a solenoidal magnet, and in Figure 8.3(b) is shown an instrument that measures the N and S poles of a horseshoe magnet. In these experiments it is significant that all the results are independent of the device's "shape," confirming that the integral is determined only by the end positions. Figures 8.3(c) and 8.3(d) illustrate experiments for such voltage pulses at the ends of a solenoidal coil, giving the same results $\int_a^b \boldsymbol{H}.\mathrm{d}\boldsymbol{s} = nI$, where I is current and n is number of turns per length, confirming the Ampère law.

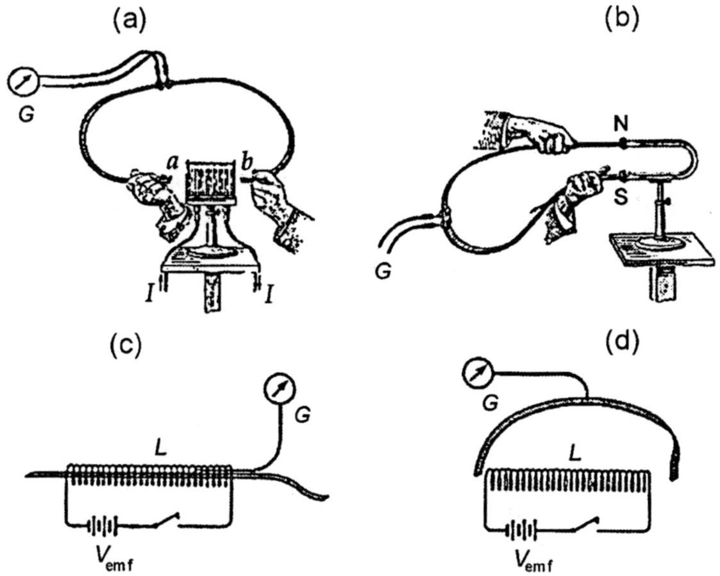

FIGURE 8.3. Demonstrations of Pohl's magnetic potentiometers. (a) Obtaining magnetic induction pulses from a coil, and (b) from a horseshoe magnet. (c) and (d) are for measurements with a long air-core solenoid. (R.W. Pohl, *Electrizitätslehre*)

As signified by the Ampère law, the field of $\boldsymbol{H}(\boldsymbol{r})$ is not conservative, however in the outside space of a magnet, $\boldsymbol{H}.\mathrm{d}\boldsymbol{s}$ can be expressed by a perfect differential of a scalar function $V_{\mathrm{m}}(\boldsymbol{r})$, i.e., $-\boldsymbol{H}.\mathrm{d}\boldsymbol{s} = \mathrm{d}V_{\mathrm{m}}$ between the limits. Such a *scalar* function $V_{\mathrm{m}}(\boldsymbol{r})$, called *pseudo-potential*, is useful for calculation of magnetostatic problems. If a magnetic field signified by a pair of apparent charges, the pseudo-potential is particularly useful. Writing that

$$\boldsymbol{H}_{\mathrm{out}} = -\mathrm{grad}\ V_{\mathrm{m}} \tag{8.5}$$

and the relation

$$\mathrm{div}\ \boldsymbol{H}_{\mathrm{in}} = -\frac{\sigma'_{\mathrm{m}}}{\mu_{\mathrm{o}}}$$

for connecting $\boldsymbol{H}_{\mathrm{in}}$ and $\boldsymbol{H}_{\mathrm{out}}$ at the boundaries, we have to satisfy the Laplace–Poisson equation for V_{m}, i.e.,

$$\nabla^2 V_{\mathrm{m}} = 0 \quad \text{and} \quad \nabla^2 V_{\mathrm{m}} = -\frac{\sigma'_{\mathrm{m}}}{\mu_{\mathrm{o}}}. \tag{8.6}$$

Magnetostatic problems can be solved with such a pseudo-potential V_{m} analogous to electrostatic cases, for which Pohl's potentiometer is designed to determine the magnitude of magnetic poles at A and B, as discussed for a practical laboratory magnet in the next section.

8.3. Scalar Potentials of Magnets

8.3.1. A Laboratory Magnet

For practical laboratory use the magnet yoke is manufactured in closed-ring shape with flat ends facing parallel to each other, as sketched in Fig. 8.4(b). When such a device is magnetized, the flux of field-lines is almost entirely enclosed inside the yoke, although continuous across the gap, apart from small "leaks" in the vicinity of pole edges. Practical magnets are designed to produce a uniform field between the poles with minimum leak; however, leak is ignored in discussions of idealized magnets.

Denoting the cross-sectional area, length of the yoke, and gap by S, l, and d, respectively, the magnet volume is expressed as $v = S(l + d)$, as shown in the figure. As given by equation (8.4), we consider the field along the central line, whose intensities are H_{in} and H_{gap}; inside the yoke material we consider that $H_{in} = \text{constant} - \sigma'_m/\mu_o = -M/\mu_o$. Since there is no electric current, we can apply the Ampère law $\oint_C \boldsymbol{H}.\mathrm{d}\boldsymbol{s} = 0$ to this magnet, where the closed curve C is the path of integration, as shown in the figure. Accordingly, we write the relation

$$H_{in}l + H_{gap}d = 0, \tag{8.7a}$$

and

$$H_{gap} = -\frac{l}{d}H_{in} = \sigma'_m, \tag{8.7b}$$

indicating that H_{gap} in the gap can be very large if $d \ll l$.

Between the pole faces the magnitude of H_{gap} at a given point P, shown in Figure 8.4(a), can be calculated with the Laplace equation $\nabla^2 V_m = 0$. Assuming

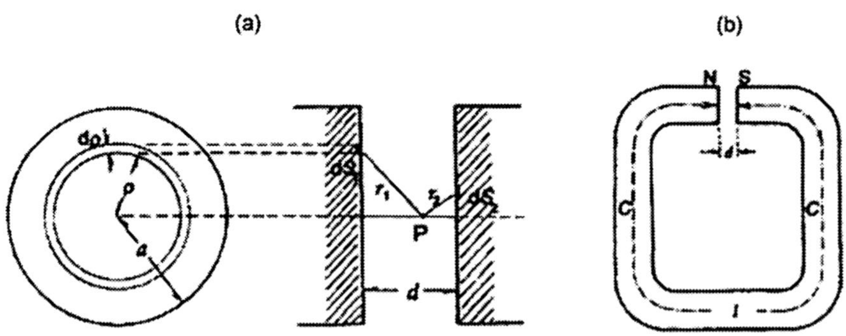

FIGURE 8.4. (a) Calculating the magnetic potential at P between poles of a magnet. (b) A typical ring magnet with an air gap.

circular pole-faces, we set cylindrical coordinates ρ, θ, z with respect to the center O in the gap and express magnetic potentials due to charge elements $\sigma'_\mathrm{m}(\boldsymbol{r}_1)\mathrm{d}S_1$ and $\sigma'_\mathrm{m}(\boldsymbol{r}_2)\mathrm{d}S_2$ on the N- and S-surfaces, respectively, where their positions are specified by vectors $\boldsymbol{r}_1$ and $\boldsymbol{r}_2$ from O. At the point P, where $\boldsymbol{r} = (0, 0, z)$, the potential is generally expressed by

$$V_\mathrm{m}(\boldsymbol{r}) = \frac{1}{4\pi\mu_\mathrm{o}} \int_S \frac{\sigma'_\mathrm{m}(\boldsymbol{r}')\,\mathrm{d}S}{|\boldsymbol{r} - \boldsymbol{r}'|}.$$

Assuming that σ'_m is constant over a circular region $0 \le \rho \le a'$ on the pole faces, this potential is written for the present case as

$$V_\mathrm{m}(z) = \frac{1}{4\pi\mu_\mathrm{o}} \left(\int_{S_1} \frac{\mathrm{d}S_1}{r_1} - \int_{S_2} \frac{\mathrm{d}S_2}{r_2} \right),$$

where

$$\int_{S_1} \frac{\mathrm{d}S_1}{r_1} = \int_0^{2\pi} \mathrm{d}\theta \int_0^{a'} \frac{\rho\,\mathrm{d}\rho}{\sqrt{\rho^2 + z^2}} = 2\pi \left(\sqrt{a'^2 + z^2} - z \right) \approx 2\pi \left(a' - z \right)$$

and

$$\int_{S_2} \frac{\mathrm{d}S_2}{r_2} = 2\pi \left\{ a' - (d - z) \right\}.$$

Hence, we obtain

$$V_\mathrm{m}(z) = \frac{\sigma'_\mathrm{m}}{2\mu_\mathrm{o}}(d - 2z) \quad \text{and} \quad H_\mathrm{gap} = -\frac{\partial V_\mathrm{m}(z)}{\partial z} = \frac{\sigma'_\mathrm{m}}{\mu_\mathrm{o}}.$$

In the above argument, in the field across the narrow gap, the lines of H_gap start from σ'_m and end at $-\sigma'_\mathrm{m}$. The magnitude of σ'_m determined by the magnetization M is a function of temperature, but is usually adjustable in adjustable electromagnets. In any case, a certain amount of energy u is required for keeping σ'_m on the pole surfaces, which can be calculated as the amount to increase charges from 0 to $\pm\sigma'_\mathrm{m}$, i.e.,

$$u = \int_0^{\sigma'_\mathrm{m}} H_\mathrm{gap}\mathrm{d}\sigma'_\mathrm{m} = \frac{1}{\mu_\mathrm{o}} \int_0^{\sigma'_\mathrm{m}} \sigma'_\mathrm{m}\mathrm{d}\sigma'_\mathrm{m} = \frac{\sigma'^2_\mathrm{m}}{2\mu_\mathrm{o}}$$

per unit area.

Therefore, we can consider that a magnetic energy $U_\mathrm{m} = uSd$ is stored in the gap space, which is responsible for a strong attractive force F_m between the poles. That is,

$$F_\mathrm{m} = -\frac{\partial U_\mathrm{m}}{\partial d} = -uS.$$

Exercise. A small current-carrying coil (cross-sectional area S, turn number N, length l) is suspended in perpendicular to a uniform horizontal magnetic field $B = \mu_o H$. Express the magnetic moment, angular momentum of this coil and discuss the concept of magnetic charges as related to its rotational behavior.

8.3.2. A Uniformly Magnetized Sphere

A ferromagnet is spontaneously magnetized under ambient conditions, where the magnetization appears often along a unique axis of the crystalline state. An iron block can be fabricated in spherical shape showing axial magnetization related to apparent magnetic charges in high densities on opposite surfaces. Such a magnetized sphere is characterized by internal flux of field-lines $\boldsymbol{M}$ in near parallel with the axis. On the other hand, the vector $\boldsymbol{M}$ of the sphere appears to be a dipolar moment, when observed at a distant point.

For such a spherical magnet we have a basic relation curl $\boldsymbol{H} = 0$ in the outside space, where the field $\boldsymbol{H}$ can be expressed by a scalar potential V_m, i.e., $\boldsymbol{H} = -\text{grad}\, V_m$. On the other hand, from the basic law div $\boldsymbol{B} = 0$ we can write div $\boldsymbol{H} = -\dfrac{1}{\mu_o}\text{div}\, M$, and hence the potential V_m should be determined by the Poisson equation $\nabla^2 V_m = \dfrac{1}{\mu_o}\text{div}\, \boldsymbol{M}$. Applying this idea to outside of the sphere, because $\boldsymbol{M} = 0$ we are to solve the Laplace equation $\nabla^2 V_m = 0$, whereas inside the sphere we also have the Laplace equation because $\boldsymbol{M} =$ constant. Solutions of the Laplace equation can therefore be connected on the spherical surface, as required by the boundary conditions

$$(\boldsymbol{H}_{out} - \boldsymbol{H}_{in})_{tangent} = 0 \quad \text{and} \quad (\boldsymbol{H}_{out} - \boldsymbol{H}_{in})_{normal} = \sigma'_m. \tag{8.8}$$

Here, the constant magnetization $\boldsymbol{M}$ is responsible for $\boldsymbol{H}_{in}$, as given by $H_{in} = -M/\mu_o$, whereas $\boldsymbol{H}_{out}$ should be represented approximately by a dipolar field at a distant point.

In this problem, the potential function in axial symmetry can be obtained from the Laplace equation $\nabla^2 V_m = 0$, which can be solved by analogy with an equivalent electrostatic problem. As discussed for the electrostatic case, the magnetic potential in axial symmetry can be expressed as

$$V_m(r, \theta) = \sum_n \left(A_n r^n + \frac{B_n}{r^{n+1}} \right) P_n(\cos\theta),$$

where the constants A_n and B_n should be determined by boundary conditions on the sphere.

Inside the sphere $r < R$, the potential should be given exclusively by terms of A_n, because terms of B_n are singular at $r = 0$. On the other hand, for $r > R$ the field is dominated by B_n terms, because A_n terms become infinite for $r \to \infty$. Further, only the term for $n = 1$ is necessary for the spherical problem because

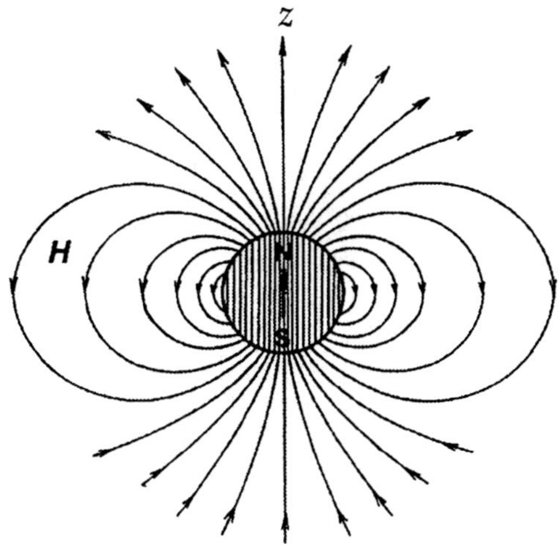

FIGURE 8.5. Magnetic field lines of a uniformly magnetized sphere.

$V_m = -H_{in}\, r \cos\theta$ for $r < R$, and hence $A_1 = -H_{in}$. Thus, we can write

$$V_{in}(r, \theta) = -H_{in} \cos\theta \qquad \text{and} \quad V_{out}(r, \theta) = \frac{B_1}{r^2} \cos\theta,$$

with which the conditions (8.8) can be expressed as

$$V_{in}(R, \theta) = V_{out}(R, \theta) \quad \text{and} \quad \left(\frac{\partial V_{out}}{\partial r}\right)_{r=R} - \left(\frac{\partial V_{in}}{\partial r}\right)_{r=R} = \sigma'_m.$$

From the first condition, we obtain $-H_{in} R \cos\theta = \dfrac{B_1 \cos\theta}{R^2}$, and hence $B_1 = -H_1 R^3$, regardless of θ. The second condition gives rise to the expression for the distributed charge density

$$\sigma'_m = 3\mu_o H_{in} \cos\theta.$$

Field-lines of a magnetized sphere are sketched in Figure 8.5.

8.4. Vector Potentials

In the absence of magnetic materials, electric currents are solely responsible for magnetic field, where the flux density $\boldsymbol{B}$ and the field intensity $\boldsymbol{H}$ are related as $\boldsymbol{B} = \mu_o \boldsymbol{H}$. If currents are steady the magnetic field is stationary and described by the equations

$$\text{div}\, \boldsymbol{B} = 0 \tag{8.2}$$

and

$$\text{curl}\, \mathbf{H} = \boldsymbol{j}. \tag{6.12}$$

Such a magnetic field is not *conservative* in character, and hence cannot be described by a scalar potential. In this case, it is noted that (8.2) can be satisfied by a vector $\boldsymbol{A}$ in a manner that

$$\boldsymbol{B} = \text{curl}\ \boldsymbol{A} \tag{8.9}$$

because of the identity div(curl $\boldsymbol{A}$) $\equiv 0$, where the vector $\boldsymbol{A}$ is called a *vector potential*. In this definition there is an ambiguity because another vector $\boldsymbol{A}'$ defined as $\boldsymbol{A}' = \boldsymbol{A} + \text{grad}\ \chi$, where χ is a scalar function, can also satisfy the equation div $\boldsymbol{B} = 0$, and hence curl $\boldsymbol{A}' =$ curl $\boldsymbol{A}$. Accordingly, using this ambiguity, we can define the vector potential $\boldsymbol{A}$ with additional conditions

$$\text{div}\boldsymbol{A} = 0 \quad \text{and} \quad \nabla^2\chi = 0. \tag{8.10}$$

With (8.10) and the identity relation

$$\text{curl}(\text{curl}\boldsymbol{A}) = \text{grad}(\text{div}\boldsymbol{A}) - \nabla^2\boldsymbol{A} = -\nabla^2\boldsymbol{A},$$

the Ampère law, curl $\boldsymbol{H} = \boldsymbol{j}$ can be written in the form of Poisson's equation, i.e.,

$$-\nabla^2\boldsymbol{A} = \mu_o\boldsymbol{j}, \tag{8.11}$$

indicating that the vector potential $\boldsymbol{A}$ at a given point $\boldsymbol{r}$ can be determined by a current density $\boldsymbol{j}$. However, if $\boldsymbol{j} = 0$ at this point $\boldsymbol{r}$, (8.11) is the Laplace equation

$$\nabla^2\boldsymbol{A} = 0. \tag{8.12}$$

Similar to electrostatic cases, we need to obtain the vector potential $\boldsymbol{A}(\boldsymbol{r})$ at a given point $\boldsymbol{r}$ when a current distribution $\boldsymbol{j}(\boldsymbol{r}')$ is specified at another point $\boldsymbol{r}'$. The potential $\boldsymbol{A}(\boldsymbol{r})$ is determined by the Laplace equation, $\nabla^2\boldsymbol{A}(\boldsymbol{r}) = 0$, if $\boldsymbol{j}(\boldsymbol{r}) = 0$, whereas the current density $\boldsymbol{j}(\boldsymbol{r}')$ at $\boldsymbol{r}'$ is related with $\boldsymbol{A}(\boldsymbol{r}')$ by the Poisson equation $\nabla^2\boldsymbol{A}(\boldsymbol{r}') = -\mu_o\boldsymbol{j}(\boldsymbol{r}')$. Therefore, we are to look for $\boldsymbol{A}(\boldsymbol{r})$ as related to $\boldsymbol{A}(\boldsymbol{r}')$, and the problem can be analyzed by the Greens function method. Here, we simply transfer the result to the present problem for stationary currents, as already discussed the detail in Chapter 4.

For a given current density $\boldsymbol{j}(\boldsymbol{r}')$ at $\boldsymbol{r}'$, the vector potential $\boldsymbol{A}(\boldsymbol{r})$ at $\boldsymbol{r}$ can be expressed by

$$\boldsymbol{A}(\boldsymbol{r}) = \frac{\mu_o}{4\pi}\int_{v(r')}\frac{\boldsymbol{j}(\boldsymbol{r}')}{|\boldsymbol{r}-\boldsymbol{r}'|}\text{d}^3\boldsymbol{r}', \tag{8.13}$$

where the integral is performed over the volume $v(\boldsymbol{r}')$ of distributed currents. The flux density at r can then be calculated by $\boldsymbol{B}(\boldsymbol{r}) = \text{curl}_r\boldsymbol{A}(\boldsymbol{r})$. To calculate $\boldsymbol{A}(\boldsymbol{r})$, we replace curl_r of the integrand in (8.13) by

$$\text{curl}_r\frac{\boldsymbol{j}(\boldsymbol{r}')}{|\boldsymbol{r}-\boldsymbol{r}'|} = -\frac{\boldsymbol{j}(\boldsymbol{r}')\times(\boldsymbol{r}-\boldsymbol{r}')}{|\boldsymbol{r}-\boldsymbol{r}'|^3},$$

FIGURE 8.6. A differential current element for the Biot-Savart formula.

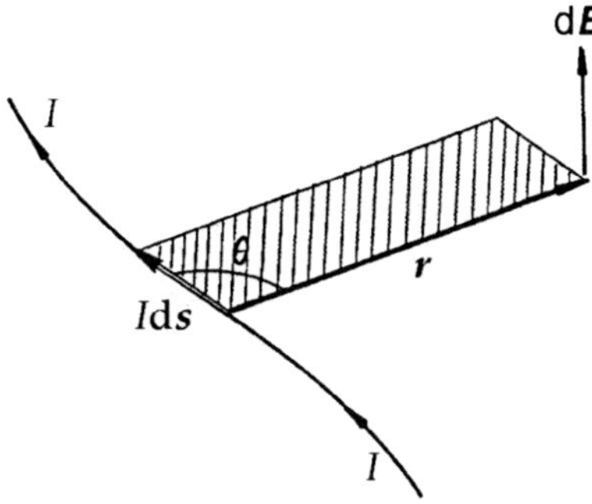

and hence

$$\boldsymbol{B}(\boldsymbol{r}) = \frac{\mu_o}{4\pi} \int_{v(\boldsymbol{r}')} \left\{ \frac{-\boldsymbol{j}(\boldsymbol{r}') \times (\boldsymbol{r} - \boldsymbol{r}')}{|\boldsymbol{r} - \boldsymbol{r}'|^3} \right\} \mathrm{d}^3\boldsymbol{r}'. \tag{8.14}$$

In the following, we modify (8.13) specifically for a line current. Writing the volume element as $\mathrm{d}^3\boldsymbol{r}' = \boldsymbol{S}.\mathrm{d}\boldsymbol{s}$, where $|\boldsymbol{S}| = S$ is the cross-sectional area, the current element along the line element $\mathrm{d}\boldsymbol{s}$ can be expressed as $\boldsymbol{j}(\boldsymbol{r}')\mathrm{d}^3\boldsymbol{r}' = I\mathrm{d}\boldsymbol{s}$. Further, replacing $\boldsymbol{r} - \boldsymbol{r}'$ by $\boldsymbol{r}$ for brevity, we obtain the formula

$$\boldsymbol{B}(\boldsymbol{r}) = \frac{\mu_o I}{4\pi} \oint_C \frac{\mathrm{d}\boldsymbol{s} \times \boldsymbol{r}}{r^3}, \tag{8.15}$$

where C represents the curve of the current. Equation (8.15) is known as the *Biot–Savart law*, for which the relation among $\mathrm{d}\boldsymbol{B}(r)$, $I\mathrm{d}\boldsymbol{s}$, and $\boldsymbol{r}$ is illustrated in Figure 8.6. Among various distributed currents a curved line current is most practical, for which current (8.15) is a useful formula for calculating the magnetic field.

8.5. Examples of Steady Magnetic Fields

Problems of steady magnetic fields can be solved with (8.2) and (6.12) in general, and the Biot–Savart formula (8.15) is particularly useful for a line current. Some simple examples are discussed in this section.

Example 1. A Straight Current.

Consider a straight long wire of radius a that carries a steady current $I = j_z(\pi a^2)$ along the z direction. Setting cylindrical coordinates on the wire with z along the axis, the current density vector is expressed as $\boldsymbol{j} = (0, 0, j_z)$, which is equal to curl $\boldsymbol{H}$. Equation (6.11) can be written in the cylindrical coordinates (ρ, θ, z) as

$$\frac{1}{\rho}\frac{\partial H_z}{\partial \theta} - \frac{\partial H_\theta}{\partial z} = 0, \tag{i}$$

$$\frac{\partial H_\rho}{\partial z} - \frac{\partial H_z}{\partial \rho} = 0, \tag{ii}$$

and

$$\frac{1}{\rho}\frac{\partial(\rho H_\theta)}{\partial\rho} - \frac{1}{\rho}\frac{\partial H_\rho}{\partial\theta} = j_z. \tag{iii}$$

The equation div $\boldsymbol{H} = 0$ can also be expressed as

$$\frac{1}{\rho}\frac{\partial(\rho H_\rho)}{\partial\rho} + \frac{1}{\rho}\frac{\partial H_\theta}{\partial\theta} + \frac{\partial H_z}{\partial z} = 0. \tag{iv}$$

By symmetry, H_z and H_θ should be constant, and hence (i) is not necessary. From (ii), H_z is constant and independent of ρ with no relation to j_z, and therefore insignificant. Then from (iv), we have

$$\frac{\partial(\rho H_\rho)}{\partial\rho} = 0, \quad \text{hence} \quad H_\rho = \frac{c}{\rho}$$

where the constant c should be zero in order for H_ρ to be finite at $\rho = 0$, and hence $H_\rho = 0$. Thus, only H_θ is significant, for which (iii) should be solved against a given j_z. Namely,

$$\frac{1}{\rho}\frac{\mathrm{d}(\rho H_\theta)}{\mathrm{d}\rho} = j_z, \tag{v}$$

from which we obtain $j_z = I/\pi a^2$ for $0 < \rho < a$, whereas $j_z = 0$ for $\rho > a$. Integrating (v), we obtain the solutions

$$H_\theta = \frac{1}{2}j_z\rho + \frac{c'}{\rho} \quad \text{and} \quad H_\theta = \frac{c''}{\rho},$$

in these regions, respectively. Here, the constant c' should be zero in order for H_θ to be definite at $\rho = 0$. Further, on the surface of the wire $\rho = a$, these two H_θ inside and outside should be connected continuously because $\boldsymbol{j}$ has only a z component, and therefore $c'' = I/2\pi$. Thus, we obtain the expressions $(H_\theta)_{\text{in}} = \frac{1}{2}j_z\rho$ and $(H_\theta)_{\text{out}} = I/2\pi\rho$, describing circular field-lines in these regions.

Example 2. A Uniform Solenoid.

We discussed a long solenoid in Section 6.3, where the magnetic field inside was considered uniform, although unverified from the first principle. Here, we discuss the problem with the basic equation (6.11) for a steady current $\boldsymbol{j}$. Setting the z direction on the solenoid axis, we use cylindrical coordinates, ρ, θ and z. Assume that the coiled wire occupies a narrow radial region between $\rho = a$ and $\rho = b$, in which a and b are considered to be almost equal, but $a < b$, and that the number of turns per length of the coil is n. Then, the steady current is expressed by $-I = j_\theta(b-a)/n$, where j_θ can be assumed to be constant, and $a \approx b$. In such a long solenoid, the field vector $\boldsymbol{H}$ should only be a function of ρ, independent of θ and z, hence, using equations (i) to (iv) in Example 8.1, we can derive the following equations for the components H_ρ, H_θ and H_z. That is,

$$\frac{1}{\rho}\frac{\mathrm{d}(\rho H_\rho)}{\mathrm{d}\rho} = 0, \quad \frac{1}{\rho}\frac{\mathrm{d}(\rho H_\theta)}{\mathrm{d}\rho} = 0, \quad \text{and} \quad -\frac{\mathrm{d}H_z}{\mathrm{d}\rho} = j_\theta.$$

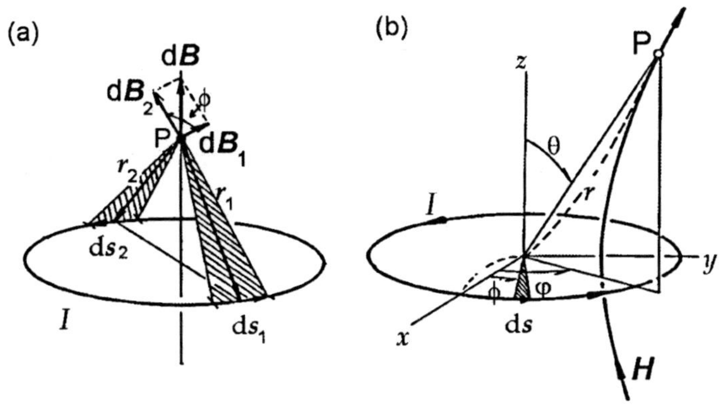

FIGURE 8.7. (a) An application of the Biot-Savart formula to a ring current I. (b) Calculating the magnetic field H at an off-axis point P due to a ring current I.

Integrating the first and second relations, we obtain that $H_\rho = c/\rho$ and $H_\theta = c'/\rho$, where the constants c and c' should both be zero in order to avoid singularities at $\rho = 0$. Hence $H_\rho = H_\theta = 0$. From the third equation in the above, we can easily obtain that

$$H_z = -\int_a^b j_\theta \mathrm{d}\rho = -j_\theta(b-a) = nI,$$

describing a uniform field for a constant I.

Example 3. Magnetic Field on the Axis of a Ring Current.

A stationary current on a conducting ring is hypothetical, unless the current source, e.g., a battery, is specified on the ring. However, the problem is discussed, serving as a model for calculating the field in a coil or of an orbiting electron. We can apply the Biot–Savart formula (8.15) to calculate the magnetic field at a point P on the axis perpendicular to the plane of the ring current, as shown in Figure 8.7(a).

Consider two current elements $I\mathrm{d}\boldsymbol{s}_1$ and $I\mathrm{d}\boldsymbol{s}_2$ that are located on the ring symmetrically with respect to the ring center. Related differential fields $\mathrm{d}\boldsymbol{B}_1$ and $\mathrm{d}\boldsymbol{B}_2$ at P are perpendicular to the planes of $\mathrm{d}\boldsymbol{s}_1 \times \boldsymbol{r}_1$ and $\mathrm{d}\boldsymbol{s}_2 \times \boldsymbol{r}_2$, where $\boldsymbol{r}_1$ and $\boldsymbol{r}_2$ are the distances between these elements and P, respectively. These vector fields can be added as shown in the figure, i.e.,

$$\mathrm{d}\boldsymbol{B} = \mathrm{d}\boldsymbol{B}_1 + \mathrm{d}\boldsymbol{B}_2 = \frac{\mu_o I}{4\pi}\left(\frac{\mathrm{d}\boldsymbol{s}_1 \times \boldsymbol{r}_1}{r_1^3} + \frac{\mathrm{d}\boldsymbol{s}_2 \times \boldsymbol{r}_2}{r_2^3}\right),$$

which is parallel to the z-axis at P where $|\boldsymbol{r}_1| = |\boldsymbol{r}_2| = r$. Therefore,

$$B = \oint |\mathrm{d}\boldsymbol{B}|\cos\phi = \frac{\mu_o I}{4\pi}\cos\phi \oint \frac{\sin\theta \mathrm{d}s}{r^2}$$

$$= \frac{\mu_o I\cos\phi}{4\pi r^2}\left(2 \times \int_0^{\pi r} \mathrm{d}s\right) = \frac{\mu_o I\cos\phi}{2r}, \qquad (8.16)$$

where $\theta = \frac{1}{2}\pi$. If P is at the center of the ring, $\phi = 0$ and $r = a$ (the radius), and hence $B = \mu_o I/2a$.

Example 4. Off-Axis Fields of a Ring Current.

In this example, we calculate the magnetic field at an off-axis point P of a ring current, as shown in Fig. 8.7(b).

The vector potential at the point P($\boldsymbol{r}$), where $\boldsymbol{r} = (r, \theta, \varphi)$, is given by

$$\boldsymbol{A}(\boldsymbol{r}) = \frac{\mu_o}{4\pi}\oint \frac{\mathrm{d}\boldsymbol{s}'}{|\boldsymbol{r} - \boldsymbol{r}'|},$$

where $\boldsymbol{r}' = (a, \phi, \frac{1}{2}\pi)$ and

$$\begin{aligned} \mathrm{d}\boldsymbol{s}' &= (-a\sin\phi\,\mathrm{d}\phi, a\cos\phi\,\mathrm{d}\phi, 0),\\ (\boldsymbol{r} - \boldsymbol{r}')^2 &= (r\sin\theta\cos\varphi - a\cos\phi)^2 + (r\sin\theta\sin\varphi - a\sin\phi)^2 + (r\cos\varphi)^2\\ &= r^2 + a^2 - 2ra\sin\theta\cos(\varphi - \phi). \end{aligned}$$

Since the differential ring current $I\mathrm{d}\boldsymbol{s}$ is on the xy-plane, the vector potential is signified only by A_φ component, which is expressed as

$$A_\varphi = -A_x\sin\varphi + A_y\sin\varphi = \frac{\mu_o I a}{4\pi}\int_0^{2\pi}\frac{\cos(\varphi - \phi)\,\mathrm{d}\phi}{\sqrt{r^2 + a^2 - 2ra\sin\theta\cos(\varphi - \phi)}}$$

Writing $\varphi - \phi = \psi$ and $2\,ar\sin\theta/(r^2 + a^2) = \alpha$, this expression can be simplified as

$$A_\varphi = \frac{\mu_o I a}{4\pi\sqrt{r^2 + a^2}}\int_0^{2\pi}\frac{\cos\psi}{\sqrt{1 - \alpha\cos\psi}}\mathrm{d}\psi.$$

It is noted that for all points of (r, θ, ψ), the parameter α is in the range $0 \le \alpha \le 1$, except for $\theta = \frac{1}{2}\pi$. Therefore, we can expand the above integrand into a Taylor series with respect to a small α; that is,

$$\frac{1}{\sqrt{1 - \alpha\cos\psi}} = 1 + \frac{1}{2}\alpha\cos\psi + \frac{3}{8}\alpha^2\cos^3\psi + \cdots.$$

Using the formula

$$\int_0^{2\pi}\cos^{2n}\psi\,\mathrm{d}\psi = \frac{2\pi\,(2n!)}{2^{2n}\,(n!)^2} \quad \text{and} \quad \int_0^{2\pi}\cos^{2n+1}\psi\,\mathrm{d}\psi = 0,$$

we obtain

$$A_\varphi = \frac{\mu_o I a^2}{4}\frac{r\sin\theta}{\left(r^2 + a^2\right)^{3/2}}\left(1 + \frac{15}{8}\frac{a^2 r^2}{\left(r^2 + a^2\right)^2}\sin^2\theta + \cdots\right).$$

The corresponding flux components calculated from A_φ are

$$B_r = \frac{\mu_o}{r^2 \sin\theta}\frac{\partial\left(r\sin\theta A_\varphi\right)}{\partial\theta},\ B_\theta = -\frac{\mu_o}{r}\frac{\partial\left(rA_\varphi\right)}{\partial r}\quad\text{and}\quad B_\varphi = 0.$$

For a small α, only the first term in the expansion may be considered, hence,

$$B_r = \frac{\mu_o I}{4\pi}\frac{2\pi a^2\cos\theta}{\left(r^2+a^2\right)^{3/2}}\quad\text{and}\quad B_\theta = \frac{\mu_o I}{4\pi}\frac{\pi a^2\left(r^2-2a^2\right)\sin\theta}{\left(r^2+a^2\right)^{5/2}}.$$

For a distant point $r \gg a$, these can be expressed approximately as

$$B_r = \frac{\mu_o}{4\pi}\frac{2p_m\cos\theta}{r^3}\quad\text{and}\quad B_\theta = \frac{\mu_o}{4\pi}\frac{p_m\sin\theta}{r^3},$$

where $p_m = I\pi a^2$ is the magnetic moment of the ring current.

On the other hand, for $r < a$, we have

$$B_r = \frac{\mu_o}{4\pi}\frac{I\cos\theta}{a}\quad\text{and}\quad B_\theta = -\frac{\mu_o}{4\pi}\frac{I\sin\theta}{a},$$

which can be combined for B_z and B_{xy} as

$$B_z = B_r\cos\theta - B_\theta\sin\theta = \frac{\mu_o I}{2a}\quad\text{and}\quad B_{xy} = B_r\sin\theta + B_\theta\cos\theta = 0.$$

This B_z expresses the flux density at the center of the ring.

Exercise. Derive the results of Example 4 directly by the Biot-Savart formula (8.15).

Example 5. The Helmholtz Coils.

Two identical coils wound in the same direction placed on the same axis is shown schematically in Figure 8.8, where magnetic field between the coils can be nearly uniform, depending on the separation $2h$ and the radius a. Such a device designed for maximum uniformity in the central region is referred to as the *Helmholtz coils* and used for a variety of laboratory experiments. In this section, we discuss the field's uniformity with the Biot–Savart formula.

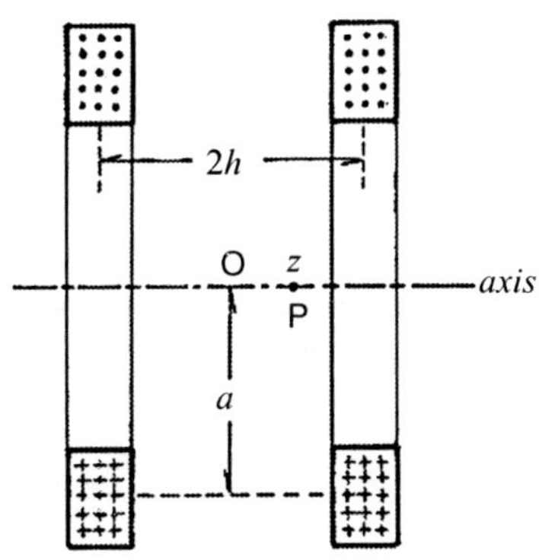

FIGURE 8.8. Helmholtz coils.

We consider a point $P2$ at a distance z from the center O on the axis. In the magnetic field of a coil of N turns, the flux density can be described by the formula (8.15) modified for the present problem. The point P can be specified by the distances $h + z$ and $h - z$ from the centers of coils, therefore the total flux density at P is expressed by

$$B_z = \frac{N\mu_o I}{2}\left[\frac{1}{\left\{1+\frac{(h-a)^2}{a^2}\right\}^{3/2}} + \frac{1}{\left\{1+\frac{(h+a)^2}{a^2}\right\}^{3/2}}\right].$$

Writing $h/a = m$, $z/a = \zeta$, and further defining $p = m/(1 + m^2)$ and $q = 1/(1 + m^2)$, this expression can be changed to

$$B_z = \frac{\mu_o I q^{3/2}}{2a}\left\{\frac{1}{(1-2p\zeta+q\zeta^2)^{3/2}} + \frac{1}{(1+2p\zeta+q\zeta^2)^{3/2}}\right\}.$$

Considering $z < a$ for the Helmholtz coils, this equation can be expanded into a power series of a small parameter ζ, where all terms of odd power vanish by symmetry. Therefore, we keep terms proportional to ζ^2 and ζ^4, ignoring higher terms as insignificant. After some calculation, we obtain

$$B_z = \frac{\mu_o I q^{3/2}}{a}\left\{1 + \left(-\frac{3q}{2} + \frac{15p^2}{2}\right)\zeta^2 + \left(\frac{15q^2}{8} - \frac{105p^2q}{4} + 315p^4\right)\zeta^4\right\}.$$

If the coefficient of ζ^2 is taken as zero, the density B_z is homogeneous in the vicinity of $\zeta = 0$, i.e., $z = 0$, at least approximately, for which we have the relation $q = 5p^2$, or $m = 1$. Hence, the uniform field is approximately obtained at the center O if $2h = a$. In this case, the uniformity is given by the coefficient of ζ^4 in the last expression, which is calculated as –144/125. The *Helmholtz field* is then given by

$$B_z = \frac{\mu_o I}{a}\left(\frac{4}{5}\right)^{3/2}\left(1 - \frac{144z^4}{125a^4}\right).$$

8.6. Vector and Scalar Potentials of a Magnetic Moment

In Section 7.3 a magnetic moment was defined for a rectangular wire of area $\boldsymbol{S}$, carrying a current I, to rotate in a uniform magnetic field, which was expressed as $\boldsymbol{p}_m = I\boldsymbol{S}$ in (7.9). In general, such a magnetic moment can be defined for stationary current distribution $\boldsymbol{j}(\boldsymbol{r}')$ from its vector potential $\boldsymbol{A}(\boldsymbol{r})$. We can expand $\boldsymbol{A}(\boldsymbol{r})$ given by (8.13) into power series of r'/r, assuming $r > r'$. Namely,

$$\frac{1}{|\boldsymbol{r}-\boldsymbol{r}'|} = \frac{1}{r} - \boldsymbol{r}'\cdot\nabla\left(\frac{1}{r}\right) + \frac{1}{2!}\sum_{ij} x_i' x_j' \frac{\partial^2}{\partial x_i \partial x_j}\left(\frac{1}{r}\right) + \ldots,$$

and

$$A(r) = A_o(r) + A_1(r) + A_2(r) + \dots,$$

where

$$A_o(r) = \frac{\mu_o}{4\pi r} \int_{v(r')} j(r')\, d^3 r',$$

$$A_1(r) = -\frac{\mu_o}{4\pi} \int_{v(r')} j(r') \left\{ r' \times \nabla \left(\frac{1}{r} \right) \right\} d^3 r',$$

$$A_2(r) = \frac{\mu_o}{8\pi} \sum \left\{ \int_{v(r')} j(r')\, x'_i x'_j d^3 r' \right\} \frac{\partial^2}{\partial x_i \partial x_j} \left(\frac{1}{r} \right),$$

$$\vdots$$

First, we can prove that $A_o(r) = 0$ for a circular closed current. For a ring current, we have $d^3(r) = S.ds'$ and $j(r').S = I$, where S is the cross-sectional area of the current path C. We can therefore write

$$\int_{v(r')} j(r')d^3 r' = S \oint_C j(r')ds' = I \oint_C ds' = 0,$$

and $A_o\,(r) = 0$.

For the second term, $A_1(r)$, the integrand can be manipulated as

$$j(r') \left(r'.\nabla \frac{1}{r} \right) = \frac{1}{2} \left[j(r') \left(r'.\nabla \frac{1}{r} \right) - r' \left\{ j(r').\nabla \frac{1}{r} \right\} \right] + \frac{1}{2} \left[j(r') \left(r'.\nabla \frac{1}{r} \right) + r' \left\{ j(r').\nabla \frac{1}{r} \right\} \right].$$

For a circular line current along C, $j(r')d^3\,r' = I ds$, so that the second square-bracketed term times $d^3 r'$ can be written as proportional to

$$ds' \left(r'.\nabla \frac{1}{r} \right) + r' \left(ds'.\nabla \frac{1}{r} \right) = d \left\{ r' \left(s'.\nabla \frac{1}{r} \right) \right\},$$

which is a perfect differential, giving rise to zero when integrated along C. Therefore, contributed by the first term in [ ], we have

$$A_1(r) = -\frac{\mu_o}{4\pi} \left\{ \int_{v(r')} \frac{r' \times j(r')}{2} d^3 r' \right\} \times \nabla \frac{1}{r},$$

from which we can define the magnetic moment as

$$p_m = \frac{1}{2} \int_{v(r')} r' \times j(r')\, d^3 r'. \tag{8.17}$$

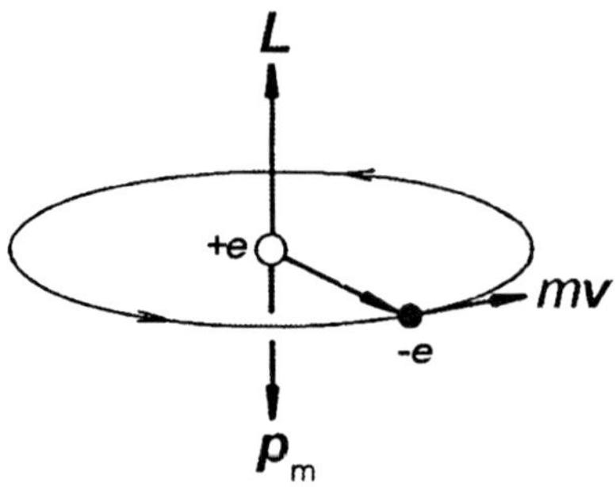

FIGURE 8.9. The Bohr model of a hydrogen atom. The angular momentum $\boldsymbol{L}$ and the magnetic moment $\boldsymbol{p}_{\mathrm{m}}$.

Replacing $\boldsymbol{r}$ by a relative distance between points of observation and of the current source, i.e., $\boldsymbol{R} = \boldsymbol{r} - \boldsymbol{r}'$, and converting ∇ to ∇_R, we can write

$$\boldsymbol{A}_1(\boldsymbol{R}) = -\frac{\mu_o \boldsymbol{p}_{\mathrm{m}}}{4\pi} \times \nabla_R \frac{1}{|\boldsymbol{R}|} = \frac{\mu_o}{4\pi} \nabla_R \times \frac{\boldsymbol{p}_{\mathrm{m}}}{|\boldsymbol{R}|}. \tag{8.18}$$

The corresponding flux density is then expressed as

$$\boldsymbol{B}_1(\boldsymbol{R}) = \mathrm{curl}_R \boldsymbol{A}_1(\boldsymbol{R}) = \frac{\mu_o}{4\pi} \mathrm{curl}_R (\mathrm{curl}_R \frac{\boldsymbol{p}_{\mathrm{m}}}{|\boldsymbol{R}|})$$
$$= \frac{\mu_o}{4\pi} \nabla_R \times \nabla_R \times \frac{\boldsymbol{p}_{\mathrm{m}}}{|\boldsymbol{R}|} = \frac{\mu_o}{4\pi} \nabla_R \left(\nabla_R \cdot \frac{\boldsymbol{p}_{\mathrm{m}}}{|\boldsymbol{R}|} \right) - \frac{\mu_o}{4\pi} \boldsymbol{p}_{\mathrm{m}} \nabla_R^2 \frac{1}{|\boldsymbol{R}|},$$

where the last term should be zero for $|\boldsymbol{R}| \neq 0$. Therefore,

$$\boldsymbol{B}_1(\boldsymbol{R}) = -\nabla_R \left(-\frac{\mu_o}{4\pi} \boldsymbol{p}_{\mathrm{m}} \cdot \nabla_R \frac{1}{|\boldsymbol{R}|} \right) = -\nabla_R V_{\mathrm{m}}(\boldsymbol{R}),$$

where

$$V_{\mathrm{m}}(\boldsymbol{R}) = -\frac{\mu_o}{4\pi} \boldsymbol{p}_{\mathrm{m}} \cdot \nabla_R \frac{1}{|\boldsymbol{R}|} \tag{8.19}$$

is the pseudo-scalar potential, which is analogous to the electrostatic dipolar potential.

The magnetic moment $\boldsymbol{p}_{\mathrm{m}}$ in (8.17) can be written for a circular line current by replacing $\boldsymbol{j}(\boldsymbol{r}')\mathrm{d}^3\boldsymbol{r}' = I\,\mathrm{d}\boldsymbol{s}'$. In this case, $\frac{1}{2}(\boldsymbol{r}' \times \mathrm{d}\boldsymbol{s}') = \mathrm{d}\boldsymbol{S}$ is the differential area swept by the current, and hence $\boldsymbol{p}_{\mathrm{m}} = I\boldsymbol{S}$, being consistent with (7.9).

8.7. Magnetism of a Bohr's Atom

In the Bohr model of a hydrogen atom, an electron is orbiting around the proton, displaying a feature of a ring current discussed in Section 8.4. The model represents the nature of orbiting charges in atoms, offering an elementary interpretation of magnetism in matter.

The orbiting electronic charge $-e$ in a period of time T for a revolution around the proton is equivalent to a circular current $I = e/T$. Therefore, for a circular orbit of radius r the magnetic moment can be expressed as

$$\boldsymbol{p}_{\mathrm{m}} = \left(-\frac{e}{T}\right)\pi r^2 \boldsymbol{n},$$

where the unit vector $\boldsymbol{n}$ is perpendicular to the plane of orbit. Denoting the mass and speed of the electron by m and v, respectively, we have $T = (2\pi r)/v$, and $mvr\boldsymbol{n} = \boldsymbol{L}$ is the angular momentum vector, as illustrated in Figure 8.9. Therefore, we can relate the magnetic moment with the angular momentum as

$$\boldsymbol{p}_{\mathrm{m}} = -\frac{e}{2m}\boldsymbol{L} = \gamma \boldsymbol{L}, \tag{8.20}$$

where $\gamma = -e/2m$ is called the *gyromagnetic ratio*. Noting that $\boldsymbol{L}$ is constant of time for an isolated atom, such a magnetic moment, due to electronic motion, can be detected only in a magnetic field applied externally. Therefore, such an atomic magnetic moment can be observed as a response to an applied field.

Figure 8.10 illustrates the circular motion of an electron in a hydrogen atom, where (a) is the isolated case, and (b) and (c) show the atom's orbit in a uniform magnetic field $\boldsymbol{B}$ applied parallel to $\boldsymbol{n}$ and $-\boldsymbol{n}$, respectively. The Coulomb force $\boldsymbol{F}$ between the electron and the proton is primarily responsible for orbiting motion, as shown in Figure 8.10(a), whereas additional magnetic forces $\boldsymbol{F}_{\pm} = \pm e(\boldsymbol{v} \times \boldsymbol{B})$ modify the motion, as illustrated in Figures 8.10(b) and (c), respectively. However, in such a model, the result is oversimplified, as inferred from the induction effect calculated for the circular orbit, where the normal vector $\boldsymbol{n}$ is not exactly parallel to B.

Magnetic forces arise from the dynamic electric field $\boldsymbol{E}$, which can be derived from basic equations: curl $\boldsymbol{E} = -\partial\boldsymbol{B}/\partial t$ and div $\boldsymbol{B} = 0$. As seen from the relation curl $\boldsymbol{E} =$ curl $(-\partial\boldsymbol{A}/\partial t)$, the quantity $e\boldsymbol{A}$ can be considered an additional *momentum* due to the applied field $\boldsymbol{B}$. For a circular orbit in a uniform field, the Faraday–Lenz

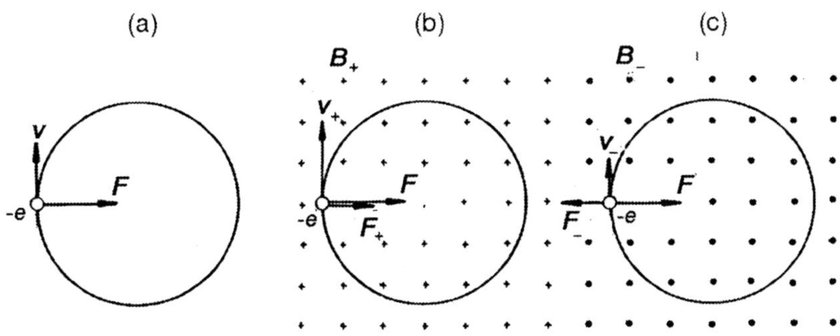

FIGURE 8.10. (a) A Bohr atom, where the force $\boldsymbol{F}$ is the Coulomb attractive force. (b) A Bohr atom on a uniform magnetic field $\boldsymbol{B}$, where the magnetic force $\boldsymbol{F}_{+}$ is additive to $\boldsymbol{F}$, i.e., $\boldsymbol{F} + \boldsymbol{F}_{+}$. (c) With the magnetic force $\boldsymbol{F}_{-}$ the net force on the electron is $\boldsymbol{F} - \boldsymbol{F}_{-}$. The magnetic forces in (b) and (c) are equal in magnitude but opposite in sign.

law can be expressed as

$$-\frac{\mathrm{d}}{\mathrm{d}t}\boldsymbol{B}\cdot\left(\oint_C \frac{\boldsymbol{r}\times \mathrm{d}\boldsymbol{s}}{2}\right) = \oint_C \boldsymbol{E}\cdot \mathrm{d}\boldsymbol{s} = -\oint_C \frac{\partial \boldsymbol{A}}{\partial t}\cdot \mathrm{d}\boldsymbol{s},$$

from which we can derive an expression for the vector potential $\boldsymbol{A}$ of a uniform $\boldsymbol{B}$, that is,

$$\boldsymbol{A} = \frac{1}{2}(\boldsymbol{B}\times\boldsymbol{r}). \tag{8.21}$$

In a uniform field $\boldsymbol{B}$, the *Hamiltonian* of the orbiting electron can be written as

$$\begin{aligned}\mathcal{H} &= \frac{1}{2m}(\boldsymbol{p}+e\boldsymbol{A})^2 - \frac{e^2}{4\pi\varepsilon_o r}\\ &= \mathcal{H}_o + \frac{e}{m}\boldsymbol{p}.\boldsymbol{A} + \frac{e^2}{2m}\boldsymbol{A}^2,\end{aligned} \tag{8.22}$$

where $\mathcal{H}_o = \dfrac{\boldsymbol{p}^2}{2m} - \dfrac{e^2}{4\pi\varepsilon_o r}$ is the Hamiltonian when $\boldsymbol{B} = 0$, and $\boldsymbol{A}$ is given by (8.22). The second term in (8.22) can be modified as

$$\frac{e}{m}\boldsymbol{p}.\boldsymbol{A} = \frac{e}{2m}\{\boldsymbol{p}.(\boldsymbol{B}\times\boldsymbol{r})\} = \frac{e}{2m}\{(\boldsymbol{r}\times\boldsymbol{p}).\boldsymbol{B}\} = -\boldsymbol{p}_{\mathrm{m}}\cdot\boldsymbol{B}, \tag{8.23}$$

representing the magnetic dipolar energy $U_{\mathrm{m}} = -\boldsymbol{p}_{\mathrm{m}}\cdot\boldsymbol{B}$ in a magnetic field, making the dipolar direction stable if $\boldsymbol{p}_{\mathrm{m}}||\boldsymbol{B}$, similar to the electric dipolar energy (4.26a). For a system of N atoms, the corresponding macroscopic energy W_1 should be related to the time average of $\boldsymbol{p}_{\mathrm{m}}$ over many revolutions, and expressed by $W_1 = -\mathrm{N}\langle\boldsymbol{p}_{\mathrm{m}}\rangle.\boldsymbol{B}$.

The third term in (8.22) can be expressed for $\boldsymbol{B} = (0, 0, B)$ as

$$\frac{e^2}{2m}\boldsymbol{A}^2 = \frac{e^2}{8m}\left(x^2+y^2\right)B^2,$$

which should also be averaged over many revolutions of the orbiting electron to represent a macroscopic energy. Therefore, we consider that $\langle x^2+y^2\rangle = \frac{1}{2}\langle r^2\rangle$, where $r^2 = x^2+y^2+z^2$, resulting in a macroscopic energy W_2

$$W_2 = \frac{Ne^2\langle r^2\rangle B^2}{12m}.$$

On applying an external field B, a system of N atoms gains a macroscopic energy $W = W_1 + W_2$, for which we can define the *magnetic susceptibility* by $\chi = -\partial W/\partial B$. Hence, we have

$$\chi = \chi_1 + \chi_2 \quad \text{where} \quad \chi_1 = \mathrm{N} < \boldsymbol{p}_{\mathrm{m}} \gg 0 \quad \text{and} \quad \chi_2 = -\frac{\mathrm{N}e^2\langle r^2\rangle}{6m}B < 0.$$

In this context, for such a system of independent atoms, the magnetic susceptibility can be positive or negative, corresponding to *paramagnetic* or *diamagnetic* cases, respectively.

Any materials other than spontaneously magnetized materials can be magnetized by an applied field. Therefore, we can consider the *induced magnetization* $\boldsymbol{M}$, which is expressed by

$$\boldsymbol{M} = \chi \boldsymbol{B} = \chi \mu_o \boldsymbol{H}, \tag{8.24}$$

excluding *spontaneous magnetization*.

The spontaneous polarization occurs in certain groups of materials, such as the iron group, rare-earth compounds, and others, whose mechanism is attributed to specific internal interactions in these materials, and the relation $\boldsymbol{M} = \boldsymbol{M}(\boldsymbol{H})$ cannot be expressed by a simple function of $\boldsymbol{H}$. In such materials, the magnetizing process is usually characterized by "hysteresis" due to *domain* movement, constituting a subject of specific investigation.

For materials other than such *ferromagnetic* cases, the magnetic flux density is generally expressed as

$$\boldsymbol{B} = \mu_o \boldsymbol{H} + \boldsymbol{M} = (1 + \chi)\mu_o \boldsymbol{H} = \mu \boldsymbol{H}, \tag{8.25}$$

where the constant $\mu = (1 + \chi)\mu_o$ represents the *magnetic permeability*.

9 Inductances and Magnetic Energies

9.1. Inductances

In a uniform solenoid of the cross-sectional area S, number of turns N, and length l, the number of field-lines is given by the flux $\Phi = \mathrm{N}BS$, where $B = \mu_o H$ and $H = \mathrm{N}I/l$ if a current I is flowing through it. Hence,

$$\Phi = \frac{\mu_o \mathrm{N}^2 S}{l} I = L_o I, \tag{9.1}$$

where the constant $L_o = \mu_o \mathrm{N}^2 S/l$ is referred to as the *inductance* of an idealized coil, where the inside field is uniform. If the empty space in the coil is filled with a magnetic material of a permeability μ, the inductance is given by $L = \mu \mathrm{N}^2 S/l$, which is larger than L_o because of the factor $\mu > \mu_o$. If the filling is ferroelectric we obtain a substantially large flux $\Phi = LI$ in the solenoid. With these parameters N, S, l, and μ, *inductors* of various L can be designed for practical use.

According to the Faraday–Lenz law, such a flux Φ changes with varying current as a function of time t, and an induced electromotive force occurs as described by

$$V_{\mathrm{emf}} = -\frac{\mathrm{d}\Phi}{\mathrm{d}t} = -L\frac{\mathrm{d}I}{\mathrm{d}t}, \tag{9.2}$$

where the unit of $L$is V-sec/A in the MKSA system and is called "henry."

Figure 9.1(a) shows a circuit to "activate" the inductor L, where the current can be on or off by switching S to positions "1" or "2," respectively. It is noted that during the process for setting currents, the magnetic induction takes place in L, as specified by (9.2). Therefore, when S is turned to "1" in the circuit including the resistor R, the current Ican be determined by the Ohm law, that is,

$$V_{\mathrm{emf}} + V_{\mathrm{i}} = RI.$$

Hence, using (9.2) we have a differential equation

$$L\frac{\mathrm{d}I}{\mathrm{d}t} + RI = V_{\mathrm{emf}}. \tag{9.3a}$$

First, we discuss a case where the current I starts to flow through L and R after the switch S was turned to "1" at $t = 0$. In this case the transient current $I = I(t)$

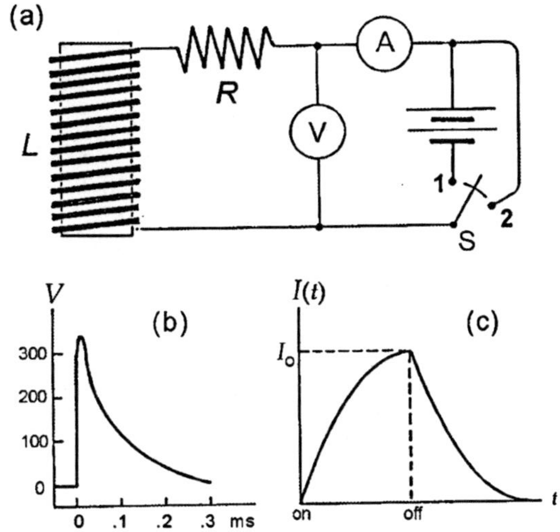

FIGURE 9.1. Transient currents in an inductor L, when the current I is turned on or off. (a) A typical circuit for observing transient effects by a voltmeter V or by an ammeter A. Switching S to 1, the current is "on", while to 2, the current is "off". (b) A voltage pulse. (c) Transient currents vs. time t, when S is on and off in sequence.

flows from the battery of V_{emf}, for which (9.3a) is solved with the initial condition is $I = 0$ at $t = 0$.

Letting $V_{emf}/R = I_o$ and $L/R = \tau$, (9.3a) can be modified as

$$\frac{\mathrm{d}I}{I_o - I} = \frac{\mathrm{d}t}{\tau},$$

and hence

$$-\ln(I_o - I) = \frac{t}{\tau} + \text{const},$$

where the constant can be set equal to $-\ln I_o$ by using the initial condition. Therefore, the transient current is expressed as

$$I = I_o\left[1 - \exp\left(-\frac{t}{\tau}\right)\right]. \tag{9.4a}$$

On the other hand, when S is switched to "2," the current starts to flow, bypassing V_{emf}, for which the equation

$$L\frac{\mathrm{d}I}{\mathrm{d}t} + RI = 0 \tag{9.3b}$$

is solved with the initial condition $I = I_o$ at $t = 0$. The transient current is expressed by the solution of (9.3b), i.e.,

$$I = I_o \exp\left(-\frac{t}{\tau}\right) \tag{9.4b}$$

showing an exponential decay. Figure 9.1(c) shows such build-up and decay curves of the transient current I, as described by (9.4a) and (9.4b), respectively. Such transient currents can be displayed on a CRO screen when square-wave voltages stimulate the on-off switching. We note that the parameter τ is in units of sec and hence called the *time constant* of a given circuit.

If a high inductance L is connected with no resistance or a small R in the circuit, a high V_i surges when it is switched off, due to a large dI/dt and small time constant τ, as shown in Figure 9.1(a). To avoid a dangerously high voltage surge in such a case, as a usual precaution one should include a sufficiently large R in series with the circuit to maintain the current at a low level.

Next, we discuss such transient processes in terms of electrical energy. Equation (9.3a) multiplied by I is written as

$$LI\frac{dI}{dt} + RI^2 = V_{emf}I,$$

where $V_{emf}I$ expresses the power delivered by the battery. Integrating this equation for a process of increasing current during the time interval between $t = 0$ to $t = \infty$, we obtain

$$\frac{1}{2}LI_o^2 + \int_0^\infty RI(t)^2\,dt = V_{emf}\int_0^\infty I(t)\,dt,$$

signifying the energy relation among L, R and V_{emf}. Clearly, the term $1LI_o^2$ in this relationrepresents an energy stored in the inductor $L$when it is activated by the current I_o, and the second term is the heat produced during the charging process. The magnetic energy can further be modified to express it by the field quantities B and H. Namely,

$$\frac{1}{2}LI_o^2 = \frac{1}{2}\frac{N^2\Phi^2}{L} = \frac{N^2B^2S^2l}{2\mu N^2S} = \frac{B^2}{2\mu}Sl = u_mV, \tag{9.5a}$$

where

$$u_m = \frac{B^2}{2\mu} \tag{9.5b}$$

is the energy density in the solenoid of volume V. For a general magnetic field, (9.5b) can be written as

$$u_m = \frac{1}{2\mu}\boldsymbol{B}^2 = \frac{\mu}{2}\boldsymbol{H}^2 = \frac{1}{2}\boldsymbol{B}.\boldsymbol{H}. \tag{9.5c}$$

Example 1. Some numerical examples are instructive for understanding formal arguments. Referring to Figure 9.1, we assume that $L = 3$ henry and $R = 20$ ohm. If a battery of $V_{emf} = 100$ V is turned on at $t = 0$, the transient current for activating L is characterized by the time constant $\tau = L/R = 0.5$ sec. The current is then given by $I = I_o[1 - \exp(-t/0.5)]$, where $I_o = V_{emf}/R = 5$ A. When $t = \tau$, then $I = I_o(1 - e^{-1}) = 3.16$ A, and $dI/dt = 12.3$ A/sec. Therefore, the magnetic energy increases at a rate given by $P_L = d(1LI^2)/dt = LI(dI/dt) = 3$ henry $\times$ 3.16 A $\times$ 12.3 A/sec $=$ 17 watt. The heat dissipation in R occurs at a rate $P_R = I^2R = (3.16\,\text{A})^2 \times$

FIGURE 9.2. Calculating the mutual inductance of a concentric cable.

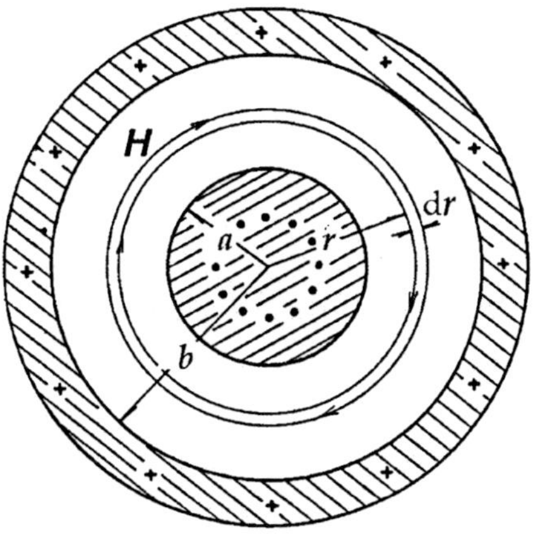

20 ohm = 200 watt. These P_L and P_R should be related to the power delivered by the battery, i.e., $P = 100\ \text{V} \times 3.16\ \text{A} = 317$ watt. That is, $P \approx P_L + P_R$.

Example 2. Inductance of a Coaxial Cable.

Figure 9.2 illustrates a cross-sectional view of a long coaxial cylinders of radii a and $b(a < b)$, carrying steady currents I and $-I$ in opposite directions. In the space between these cylinders, the magnetic flux density $\boldsymbol{B}$ is radial and the magnitude on a coaxial surface of radius r is given by $B\,(r) = \mu_o I/2\pi r$. The magnetic energy stored in this field can be calculated by (9.5a), that is,

$$U = \frac{1}{2}L_o I^2 = \int_a^b u_m(r) 2\pi r \mathrm{d}r = \frac{1}{2\mu_o}\left(\frac{\mu_o I}{2\pi}\right)^2 2\pi l \int_a^b \frac{r\mathrm{d}r}{r^2} = \frac{\mu_o I^2 l}{4\pi}\ln\left(\frac{b}{a}\right),$$

where

$$L_o = \frac{\mu_o l}{2\pi}\ln\left(\frac{b}{a}\right).$$

9.2. Self- and Mutual Inductances

The inductance Lrepresents the magnetic induction effect in a coil when the inside field changes with time. Such a change can take place not only because of a changing current in the coil, but also by changing currents in other coils located nearby, which are called, respectively, *self-* and *mutual inductions*. Although (9.1) is written for an idealized solenoid, the inductance can be expressed in a general form that is convenient for calculation.

The flux Φ defined for the cross-sectional area $\boldsymbol{S}$ at a position $\boldsymbol{r}$ along a coil can be expressed in terms of the vector potential $\boldsymbol{A}(\boldsymbol{r})$, that is,

$$\Phi = \int_S \boldsymbol{B}.\mathrm{d}\boldsymbol{S} = \int_S \text{curl}\boldsymbol{A}.\mathrm{d}\boldsymbol{S} = \oint_C \boldsymbol{A}.\mathrm{d}\boldsymbol{s}, \tag{9.6}$$

where Stokes's theorem was used for conversion from the surface to line integral, and $d\boldsymbol{s}$ is the line element on C. Considering a current $\boldsymbol{j}(\boldsymbol{r}')$ at $\boldsymbol{r}'$, the potential $\boldsymbol{A}(\boldsymbol{r})$ at $\boldsymbol{r}$ is given by

$$\boldsymbol{A}(\boldsymbol{r}) = \frac{\mu_o}{4\pi} \int_{v(\boldsymbol{r}')} \frac{\boldsymbol{j}(\boldsymbol{r}')}{|\boldsymbol{r} - \boldsymbol{r}'|} d^3\boldsymbol{r}'.$$

For the ring current along the coiled wire we can write $\boldsymbol{j}(\boldsymbol{r}')d^3\boldsymbol{r}' = I d\boldsymbol{s}'$, and hence

$$\boldsymbol{A}(\boldsymbol{r}) = \frac{\mu_o I}{4\pi} \oint_C \frac{d\boldsymbol{s}'}{|\boldsymbol{r} - \boldsymbol{r}'|}.$$

Thus, from (9.6) we obtain the expression

$$L = \frac{\mu_o}{4\pi} \oint_C \oint_C \frac{d\boldsymbol{s}.d\boldsymbol{s}'}{|\boldsymbol{r} - \boldsymbol{r}'|}. \tag{9.7}$$

For two interacting coils 1 and 2 fluxes are written as

$$\Phi_1 = \oint_{C_1} (\boldsymbol{A}_{11} + \boldsymbol{A}_{12}) \,.\, d\boldsymbol{s}_1 \quad \text{and} \quad \Phi_2 = \oint_{C_2} (\boldsymbol{A}_{21} + \boldsymbol{A}_{22}) \,.\, d\boldsymbol{s}_2$$

where

$$\boldsymbol{A}_{11} = \frac{\mu_o I}{4\pi} \oint_{C_1} \frac{d\boldsymbol{s}'_1}{|\boldsymbol{r}_1 - \boldsymbol{r}'_1|} \quad \text{and} \quad \boldsymbol{A}_{22} = \frac{\mu_o I}{4\pi} \oint_{C_2} \frac{d\boldsymbol{s}'_2}{|\boldsymbol{r}_2 - \boldsymbol{r}'_2|}$$

are vector potentials due to inductions in these coils 1 and 2 by their own currents I_1 and I_2, and

$$\boldsymbol{A}_{12} = \frac{\mu_o I}{4\pi} \oint_{C_2} \frac{d\boldsymbol{s}'_2}{|\boldsymbol{r}_1 - \boldsymbol{r}'_2|} \quad \text{and} \quad \boldsymbol{A}_{21} = \frac{\mu_o I}{4\pi} \oint_{C_1} \frac{d\boldsymbol{s}'_1}{|\boldsymbol{r}_2 - \boldsymbol{r}'_1|}$$

represent inductions mutually caused by I_2 and I_1, respectively.

Therefore we can write

$$\Phi_1 = L_{11} I_1 + M_{12} I_2 \quad \text{and} \quad \Phi_2 = M_{21} I_1 + L_{22} I_2, \tag{9.8}$$

where

$$L_{11} = \frac{\mu_o}{4\pi} \oint_{C_1} \oint_{C_1} \frac{d\boldsymbol{s}_1.d\boldsymbol{s}'_1}{|\boldsymbol{r}_1 - \boldsymbol{r}'_1|} \quad \text{and} \quad L_{22} = \frac{\mu_o}{4\pi} \oint_{C_2} \oint_{C_2} \frac{d\boldsymbol{s}_2.d\boldsymbol{s}'_2}{|\boldsymbol{r}_2 - \boldsymbol{r}'_2|} \tag{9.9}$$

are called *self-inductances*, and

$$M_{12} = \frac{\mu_o}{4\pi} \oint_{C_1} \oint_{C_2} \frac{d\boldsymbol{s}_1.d\boldsymbol{s}'_2}{|\boldsymbol{r}_1 - \boldsymbol{r}'_2|} \quad \text{and} \quad M_{21} = \frac{\mu_o}{4\pi} \oint_{C_2} \oint_{C_1} \frac{d\boldsymbol{s}_2.d\boldsymbol{s}'_1}{|\boldsymbol{r}_2 - \boldsymbol{r}'_1|} \tag{9.10}$$

the *mutual inductances*. We note that in (9.10) these mutual inductances are symmetrical between 1 and 2, that is,

$$M_{12} = M_{21}. \tag{9.11}$$

Activated independently by currents I_1 and I_2, the two inductors interact magnetically at a close distance. As discussed for a single inductor in Section 9.1, energies from two emfs are dissipated in resistances in part and stored, also, in inductors as magnetic energies. For two conductors the induced voltages are determined by $V_{in}(1) = -d\Phi_1/dt$ and $V_{in}(2) = -d\Phi_2/dt$, respectively, where Φ_1 and Φ_2 are given by (9.8). Therefore, the stored energy is expressed by

$$U(t) = -\int_0^t \frac{d\Phi_1}{dt} I_1 dt - \int_0^t \frac{d\Phi_2}{dt} I_2 dt.$$

Using (9.8), this can be written as

$$U(t) = \frac{1}{2} L_{11} I_1^2 + \frac{1}{2} L_{22} I_2^2 + M_{12} I_1 I_2 + M_{21} I_2 I_1, \tag{9.12a}$$

varying generally as a function of time, as I_1 and I_2 change with different time constants. However, both currents become steady as $t \to \infty$. With stationary currents on, both inductors can store energies in the form $1 L I_o^2$, whereas symmetrical mutual inductances ($M_{12} = M_{21}$) give rise to the same interaction energy. Therefore, the total magnetic energy in the two inductors with steady currents I_1 and I_2 can be expressed as

$$U(t) = \frac{1}{2} L_1 I_1^2 + \frac{1}{2} L_2 I_2^2 + M I_1 I_2, \tag{9.12b}$$

which is the same as (9.12a) where L_{11}, L_{22}, and $M_{12} = M_{21}$ are replaced by L_1, L_2, and M.

Self-inductance characterizes an inductor as a circuit element, while mutual inductance signifies a *coupling* between inductors, depending on their distance and arrangement. In the above we discussed two inductors activated individually by currents $I_1$and I_2; however, in many cases these currents may originate from the same source. A coaxial inductor is the example wherein the magnetic field exists only inside the coaxial cable as related to currents Iandg $- I$on the inside and outside conductors. Therefore, the inductance L calculated in Example 9.2 is the mutual inductance between the currents. Nevertheless, the mutual parameter M refers to an interaction between two well-defined inductors with a fixed configuration.

We can show that at any point in a magnetic field the energy density u_m is expressed by (9.5c). In the above, $\boldsymbol{A}_{11} + \boldsymbol{A}_{12}$ and $\boldsymbol{A}_{21} + \boldsymbol{A}_{22}$ represent in parts the vector potential A in the entire field at the positions of currents I_1 and I_2, respectively. Therefore, the rate at which magnetic energy changes is given by

$$\frac{dU(1,2)}{dt} = I_1 \frac{d\Phi_1}{dt} + I_2 \frac{d\Phi_2}{dt} = I_1 \frac{d}{dt} \oint_{C_1} \mathbf{A.ds} + I_2 \frac{d}{dt} \oint_{C_2} \mathbf{A.d}s;$$

however, the mutual interactions are duplicated in this expression. Writing $I_1 \mathrm{d}\boldsymbol{s} = \boldsymbol{j}_1 \mathrm{d}v$, $I_2 \mathrm{d}\boldsymbol{s} = \boldsymbol{j}_2 \mathrm{d}v$, and $\boldsymbol{j}_1 + \boldsymbol{j}_2 = \boldsymbol{j}$, for a symmetric energy $U = \frac{1}{2}[U(1,2) + U(2,1)]$ we have

$$2\frac{\mathrm{d}U}{\mathrm{d}t} = \frac{\mathrm{d}}{\mathrm{d}t}\int_v \boldsymbol{A}.\boldsymbol{j}\mathrm{d}v = \frac{\mathrm{d}}{\mathrm{d}t}\int_v \boldsymbol{A}.\mathrm{curl}\boldsymbol{H}\mathrm{d}v.$$

Using the identity $\mathrm{div}(\boldsymbol{A} \times \boldsymbol{H}) = (\mathrm{curl}\boldsymbol{A}).\,\boldsymbol{H} - \boldsymbol{A}.\,(\mathrm{curl}\boldsymbol{H})$, the last integral can be expressed as

$$\int_v \boldsymbol{A}.\,(\mathrm{curl}\boldsymbol{H})\,\mathrm{d}v = \int_v (\mathrm{curl}\boldsymbol{A})\,.\,\boldsymbol{H}\mathrm{d}v - \int_v \mathrm{div}\,(\boldsymbol{A} \times \boldsymbol{H})\,\mathrm{d}v.$$

Here, with the Gauss theorem, we have $\int_v \mathrm{div}\,(\boldsymbol{A} \times \boldsymbol{H})\,\mathrm{d}v = \oint_S (\boldsymbol{A} \times \boldsymbol{H})_n\,\mathrm{d}S \to 0$ as $S \to \infty$, and therefore obtain the formula $U = \frac{1}{2}\int_v \boldsymbol{B}.\,\boldsymbol{H}\mathrm{d}v$ for a stationary field with the energy density $u_\mathrm{m} = \frac{1}{2}\boldsymbol{B}.\,\boldsymbol{H}$.

9.3. Mutual Interaction Force Between Currents

We discussed the magnetic field of two current-activated inductors, where the stored energy is not only in individual coils, but also in the space in between. The energy of the field spread between inductors was calculated as in Section 9.2 for a rigid configuration.

On the other hand, in a non-rigid configuration the interaction energy W between two inductors is responsible for a mechanical *force*. Such an interaction force can be considered as arising from a variation in W due to a small displacement $\delta\boldsymbol{r}$ in the mutual distance. Namely,

$$-\boldsymbol{F}.\,\delta\boldsymbol{r} = \delta W. \tag{9.13}$$

Such a small variation δWin the interaction energy can occur with a variation in magnetic flux, so that we can write

$$\delta W = I_1\delta\Phi_{12} = I_2\delta\Phi_{21},$$

where

$$\delta\Phi_{12} = \int_{S_1} \boldsymbol{B}_2(\boldsymbol{r}_1).\delta\boldsymbol{S}_1 = \int_{S_1} \boldsymbol{B}_2(\boldsymbol{r}_1).\,(\mathrm{d}\boldsymbol{s}_1 \times \delta\boldsymbol{r}) = \left\{\oint_{C_1} \boldsymbol{B}_2(\boldsymbol{r}) \times \mathrm{d}\boldsymbol{s}_1\right\}.\delta\boldsymbol{r}.$$

and hence the force $\boldsymbol{F}_{12}$ can be determined by

$$\boldsymbol{F}_{12} = I_1\left(\oint_{C_1} \boldsymbol{B}_2\,(\boldsymbol{r}_1) \times \mathrm{d}\boldsymbol{s}_1\right).$$

In this case, the Biot–Savart formula is written as

$$\boldsymbol{B}_2\,(\boldsymbol{r}_1) = \frac{\mu_\mathrm{o} I_2}{4\pi}\oint_{C_2} \frac{\mathrm{d}\boldsymbol{s}_2 \times (\boldsymbol{r}_1 - \boldsymbol{r}_2)}{|\boldsymbol{r}_1 - \boldsymbol{r}_2|^3},$$

and therefore the interaction force can be expressed as

$$\boldsymbol{F}_{12} = -\boldsymbol{F}_{21} = \frac{\mu_o I_1 I_2}{4\pi} \oint_{C_1} \mathrm{d}\boldsymbol{s}_1 \times \oint_{C_2} \frac{\mathrm{d}\boldsymbol{s}_2 \times (\boldsymbol{r}_1 - \boldsymbol{r}_2)}{|\boldsymbol{r}_1 - \boldsymbol{r}_2|^3}. \tag{9.14}$$

9.4. Examples of Mutual Induction

9.4.1. Parallel Currents

In Section 7.3, we discussed a magnetic force between parallel currents. Here we calculate the mutual inductance and force with formulas (9.10) and (9.14).

Figure 9.3 shows two parallel conductors of finite length l and width d. Taking current elements $I_1 \mathrm{d}z_1$ and $I_2 \mathrm{d}z_2$ at z_1 and z_2 on these conductors, as indicated in the figure, we calculate the mutual inductance with (9.10). That is,

$$M = \frac{\mu_o}{4\pi} \int_0^l \mathrm{d}z_1 \int_0^l \frac{\mathrm{d}z_2}{\sqrt{d^2 + (z_1 - z_2)^2}}.$$

Replacing z_2 by $x = (z_1 - z_2)/d$, we have

$$\int_0^l \frac{\mathrm{d}z_2}{\sqrt{d^2 + (z_1 - z_2)^2}} = \int_{-z_1/d}^{(l-z_1)/d} \frac{\mathrm{d}x}{\sqrt{1 + x^2}} = \sinh^{-1}\left(\frac{l - z_1}{d}\right) + \sinh^{-1}\left(\frac{z_1}{d}\right).$$

These terms on the right give rise to basically the same results as when the expression is integrated once again, namely,

$$M = \frac{\mu_o}{4\pi} 2d \int_0^{l/d} \left(\sinh^{-1} x\right) \mathrm{d}x = x \sinh^{-1} x - \sqrt{1 + x^2}.$$

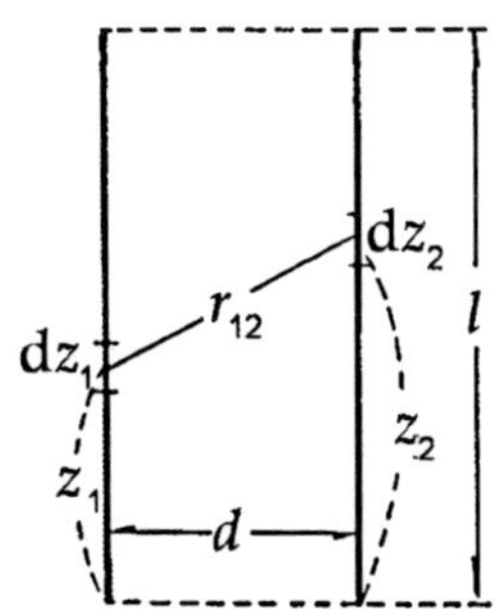

FIGURE 9.3. Calculating the mutual inductance of twin parallel currents.

Using the identity relation $\sinh^{-1} x = \ln\left(x + \sqrt{1 + x^2}\right)$, we obtain

$$M = \frac{\mu_o l}{2\pi d}\left[\ln\left(\frac{l}{d} + 1 + \frac{l^2}{d^2}\right) + 1 - \sqrt{1 + \frac{l^2}{d^2}}\right].$$

If $l \gg d$, this expression can be simplified as

$$M = \frac{\mu_o l}{2\pi d}\left(\ln\frac{2l}{d} - 1\right).$$

For the interaction force, using $\mathrm{d}\boldsymbol{s}_1 = (0, 0, \mathrm{d}z_1)$, $\mathrm{d}\boldsymbol{s}_2 = (0, 0, \mathrm{d}z_2)$, $\boldsymbol{r}_1 - \boldsymbol{r}_2 = (d, 0, z_1 - z_2)$ in (9.14), we can write $\mathrm{d}\boldsymbol{s}_1 \times (\mathrm{d}\boldsymbol{s}_2 \times (\boldsymbol{r}_1 - \boldsymbol{r}_2)) = (d\mathrm{d}z_1\mathrm{d}z_2, 0, 0)$, and therefore,

$$\begin{aligned}F_x &= \frac{\mu_o I_1 I_2 d}{4\pi}\int_0^l \mathrm{d}z_1 \int_0^l \frac{\mathrm{d}z_2}{\left\{d^2 + (z_1 - z_2)^2\right\}^{3/2}}\\ &= \frac{\mu_o I_1 I_2}{2\pi d}\left(\sqrt{d^2 + l^2} - d\right) \sim \frac{\mu_o l}{2\pi d} I_1 I_2.\end{aligned}$$

9.4.2. *Two Ring Currents*

Figure 9.4 shows two parallel ring currents whose centers are located at $z = \pm b$ on the z-axis, flowing along circular paths parallel to the xy-plane. Considering current elements $I_1\mathrm{d}\boldsymbol{s}_1$ and $I_2\mathrm{d}\boldsymbol{s}_2$ on each current at positions $(a\cos\varphi_1, a\sin\varphi_1, b)$ and $(a\cos\varphi_2, a\sin\varphi_2, -b)$, the distance between them can be calculated from

$$\begin{aligned}|\boldsymbol{r}_1 - \boldsymbol{r}_2|^2 &= a^2(\cos\varphi_1 - \cos\varphi_2)^2 + a^2(\sin\varphi_1 - \sin\varphi_2)^2 + 4b^2\\ &= 4b^2 + 2a^2\{1 - \cos(\varphi_1 - \varphi_2)\},\end{aligned}$$

and

$$\mathrm{d}\boldsymbol{s}_1 . \mathrm{d}\boldsymbol{s}_2 = a^2\cos(\varphi_1 - \varphi_2)\mathrm{d}\varphi_1\mathrm{d}\varphi_2.$$

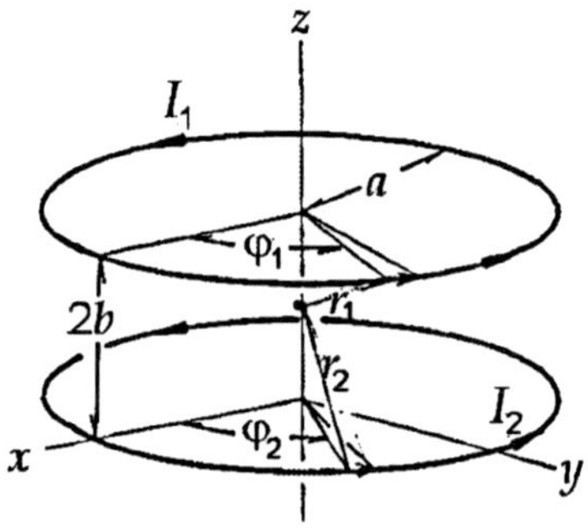

FIGURE 9.4. Calculating the mutual inductance of twin parallel ring currents.

Writing $\varphi = \varphi_1 - \varphi_2$, the mutual inductance is a function of φ and φ_2, but can be expressed as

$$M = \frac{\mu_o}{4\pi}\left(2\pi a^2\right)\int_0^{2\pi}\frac{\cos\varphi d\varphi}{\sqrt{4b^2 + 2a^2(1-\cos\varphi)}},$$

after integrating with respect to φ_2. Further, defining an angle $\theta = \frac{1}{2}(\varphi - \pi)$, M can be written in terms of *elliptic complete integrals*, as derived in the following:

$$\cos\varphi = 2\sin^2\theta - 1, \quad 4b^2 + 2a^2(1-\cos\varphi) = 4\{b^2 + a^2(1-\sin^2\theta)\},$$

and the range of integration becomes $-\frac{1}{2}\pi \le \theta \le \frac{1}{2}\pi$. Thus, we obtain

$$M = \mu_o a^2\int_0^{\pi/2}\frac{\left(2\sin^2\theta - 1\right)d\theta}{\sqrt{a^2 + b^2 - a^2\sin^2\theta}}.$$

Define a parameter $\kappa^2 = \dfrac{a^2}{a^2+b^2}$, and splitting $2\sin^2\theta - 1$ into two parts as

$$2\sin^2\theta - 1 = \frac{2-\kappa^2}{\kappa^2} - \frac{2\left(1-\kappa^2\sin^2\theta\right)}{\kappa^2},$$

the last expression for M is divided into two terms of complete elliptic integrals, i.e.,

$$M = \frac{\mu_o a^2}{\kappa}\left[\left(2-\kappa^2\right)K(\kappa) - 2E(\kappa)\right],$$

where

$$K(\kappa) = \int_0^{\pi/2}\frac{d\theta}{\sqrt{1-\kappa^2\sin^2\theta}} \quad \text{and} \quad E(\kappa) = \int_0^{\pi/2}\sqrt{1-\kappa^2\sin^2\theta}\,d\theta$$

are known as *complete elliptic integrals of the first and second kinds, respectively.*

10
Time-Dependent Currents

10.1. Continuity of Charge and Current

Static charges and steady currents are responsible for electrostatic and steady magnetic fields, respectively. A static field arises from a charge Q at rest, whereas, for a steady field, a constant current $I = \mathrm{d}Q/\mathrm{d}t$ is responsible; both represent states of the charge carriers relative to the observer. However, charge and current are not exclusive, and both electric and magnetic fields can be observed as related with time-dependent Q and I, as discussed for magnetic induction.

Referring to Figure 1.2, if a current I is not steady, not all charges in the volume $S\Delta x$ are necessarily moving out during the time-interval Δt, depending on the position x along the conductor. For such a non-steady flow of charges, we can write $I(x) - I(x - \Delta x) = \dfrac{\Delta Q}{\Delta t}$, which can be expressed as

$$-\frac{\Delta(jS)}{\Delta x}\Delta x = \frac{\Delta\rho}{\Delta t}S\Delta x, \text{ and hence } -\frac{\Delta j}{\Delta x} = \frac{\Delta\rho}{\Delta t}. \tag{10.1a}$$

In general, the charge density ρ and the current vector $\boldsymbol{j}$ are functions of the position $\boldsymbol{r}$ and time t along the current passage for which (10.1a) can be expressed as

$$\frac{\partial\rho}{\partial t} + \operatorname{div}\boldsymbol{j} = 0, \tag{10.1b}$$

which is known as the *equation for continuity*. Obviously, (10.1b) signifies that ρ and $\boldsymbol{j}$ are not independent, so that these quantities, as combined, represent the *source* of electromagnetic fields.

Assuming that all particles move at the same speed v in one-dimensional flow, we have $j = \rho v$, and the equation (10.1b) can be expressed as

$$\frac{\partial\rho}{\partial t} + v\frac{\partial\rho}{\partial x} = 0, \tag{10.1c}$$

where ρ is a function of x and t. We assume that such a function $\rho(x, t)$ is separable into factor functions of x and t only, as given by a product $\rho(x, t) = \rho_1(x)\rho_2(t)$.

Then, from (10.1c) we obtain

$$\frac{v}{\rho_1}\frac{d\rho_1}{dx} = -\frac{1}{\rho_2}\frac{d\rho_2}{dt} = i\omega,$$

where ω is a constant independent of x and t, ignoring frictional effects. Solving for ρ_1 and ρ_2, we can write

$$\rho(x,t) = \rho_o \exp\left\{-i\omega\left(t - \frac{x}{v}\right)\right\},$$

representing the propagating density ρ that can be specified by the phase $\omega\left(t - \frac{x}{v}\right)$, where the constant ω is interpreted as the *frequency* of periodic variation. At a low frequency ω_- the phase is dominated by ωt if $\frac{\omega}{v}x$ is negligible in macroscopic scale. In this case, the specific solution of (10.1c) is given by $j = \rho v = \rho_o v\exp(-i\omega t)$, and the corresponding current can be expressed as $I = I_o\exp(-i\omega t)$, almost independent of x. In contrast, at a higher ω, as signified by the *wavelength* $\lambda = \frac{2\pi v}{\omega}$ the term $\frac{\omega}{v}x$ becomes significant for the density to propagate along x, and the current performance is seriously affected by the size and shape of a circuit. Leaving high-frequency problems to a later discussion, in this chapter we discuss time-dependent currents at low frequencies, occurring almost simultaneously in the whole circuit.

10.2. Alternating Currents

The principle of generating alternating current (AC) was briefly discussed in Section 7.3 as an example of a magnetic force. By rotating a rectangular coil, as shown in Figure 7.7, around the axis placed in perpendicular to the direction of a static magnetic field, an induced voltage V_{in} occurs as described by the Faraday–Lenz law, i.e., $V_{in} = -\frac{d\Phi}{dt}$, where the magnetic flux is $\Phi = \Phi_o \sin\phi$, here ϕ is the rotational angle, i.e., $\phi = \omega t + \phi_o$. Here ϕ_o is the initial angle at $t = 0$, and ω is the *angular frequency* of rotation. Therefore, the induced voltage changes sinusoidally with t, as given by

$$V_{in} = V_o\cos\phi, \quad \text{where} \quad V_o = -\omega\Phi_o,$$

indicating that the amplitude V_o is proportional to ω, and that the variation of V_{in} is signified by the angle ϕ, which is given by the *phase* of the *alternating voltage*.

If such a voltage source is connected with a resistor R, an induction current I flows as determined by the Ohm law, $V_{in} = RI$, and hence

$$I = \frac{V_o}{R}\cos(\omega t + \phi_o), \tag{i}$$

which is *in phase* with V_{in}.

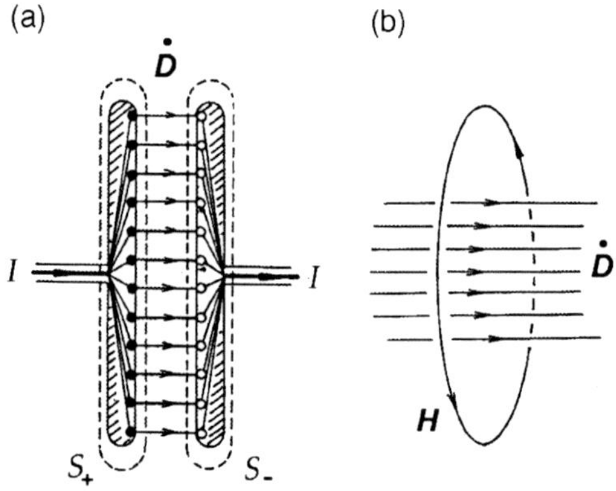

FIGURE 10.1. The displacement current density $\boldsymbol{D}$ in an empty capacitor. (B) The relation between a displacement current of density $\boldsymbol{D}$ and the $\boldsymbol{H}$ vector.

If connected with an inductor L, the induction current I is determined from the relation $V_{\text{in}} - L(\mathrm{d}I/\mathrm{d}t) = 0$, and hence

$$\frac{\mathrm{d}I}{\mathrm{d}t} = \frac{V_o}{L}\cos(\omega t + \phi_o),$$

and

$$I = \frac{V_o}{L}\int_0^t \cos(\omega t + \phi_o)\,\mathrm{d}t = \frac{V_o}{\omega L}\sin(\omega t + \phi_o).$$

Such a sinusoidal current can be written as

$$I = I_o\cos(\omega t + \phi_o - 1\pi) \quad \text{where} \quad I_o = \frac{V_o}{\omega L}, \tag{ii}$$

indicating that the phase of I is behind the voltage V_{in} by 1π, and ωL behaves as if it were the effective resistance, which is called the *impedance* of the coil.

Further, it is interesting when a capacitor C is connected with an AC source. In fact, a current appears to flow across the gap in the capacitor, despite the open space that is not conductive. Considering that capacitor plates can always be charged to equal amounts, $+Q$ and $-Q$, on the connecting wire the current $+I$ flows into one plate, and $-I$ goes out from the other, so that the current appears to flow through the gap continuously. Although contradictory to non-conductive empty space, Maxwell proposed his idea of a *displacement current* to dissolve the problem, and his idea was in fact verified experimentally.

Figure 10.1(a) illustrates that charges $\pm Q$ on the plates are distributed at a uniform density $\pm\sigma(t)$ over the area A of an idealized capacitor. To be consistent with a static capacitor where $\sigma = D$, Maxwell proposed a current given by the density $j_D = \dfrac{\mathrm{d}\sigma}{\mathrm{d}t} = \dfrac{\mathrm{d}D}{\mathrm{d}t}$ across the gap, by considering that these σ and D in the space are functions of time, and called the displacement current. With this proposal, the current becomes continuous on the capacitor plate, i.e., $(jS)_{\text{wire}} = (j_D A)_{\text{gap}}$. From the continuity relation (10.1b), we can write that

$$\frac{\mathrm{d}}{\mathrm{d}x}\left(j + \frac{\mathrm{d}D}{\mathrm{d}t}\right) = 0,$$

which can be generalized as $\mathrm{div}\left(\boldsymbol{j}+\frac{\partial \boldsymbol{D}}{\partial t}\right)=0$. Maxwell considered that the net current density $\boldsymbol{j}+\frac{\partial \boldsymbol{D}}{\partial t}$ is responsible for the magnetic field $\boldsymbol{H}$, and wrote the Ampère law as

$$\boldsymbol{j}+\frac{\partial \boldsymbol{D}}{\partial t}=\mathrm{curl}\,\boldsymbol{H}. \tag{10.2}$$

This is one of the Maxwell equations governing entire electromagnetic phenomena. Figure 10.1(b) shows the right-hand relation between $\partial \boldsymbol{D}/\partial t$ and $\boldsymbol{H}$ vectors.

Returning to a capacitor connected with an AC generator, we have a voltage relation

$$V_{\mathrm{in}}=\frac{Q}{C}=\frac{1}{C}\int_0^t I\,\mathrm{d}t=V_{\mathrm{o}}\cos(\omega t+\phi_{\mathrm{o}}).$$

Differentiating this, we obtain

$$I=-I_{\mathrm{o}}\sin(\omega t+\phi_{\mathrm{o}})=I_{\mathrm{o}}\cos\left(\omega t+\phi_{\mathrm{o}}+\frac{1}{2}\pi\right),\quad \text{where}\quad I_{\mathrm{o}}=\omega C V_{\mathrm{o}}. \tag{iii}$$

Therefore, the phase of AC current is advanced from that of V_{in} by $\frac{1}{2}\pi$, and the impedance of a capacitor is given by $1/\omega C$.

10.3. Impedances

A practical AC circuit consists of resistances, inductances, and capacitances connected in various ways with an AC source. Figures 10.2(a), (b), and (c) show

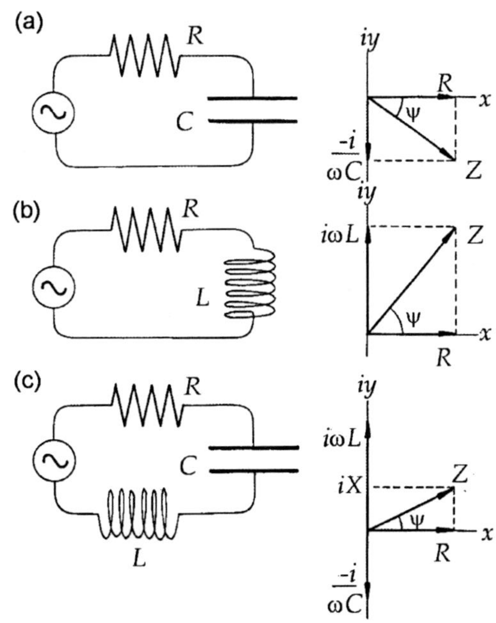

FIGURE 10.2. Impedances of AC circuits in a complex plane. The phase ψ of AC current can delay or advance from the applied voltage. (a) *RC*, (b) *RL*, and (c) *RCL* combinations.

simple cases of *RC, RL*, and *RCL* connected in series with V_{in}, respectively. In the first case, the voltage relation can be written as

$$V_{\text{in}} = RI + \frac{Q_{\text{plate}}}{C} = RI + \frac{1}{C}\int_0^t I_D \mathrm{d}t, \quad \text{where} \quad I = I_D,$$

Hence, the equation for I is given by

$$-\omega V_{\text{o}} \sin(\omega t + \phi_{\text{o}}) = R\frac{\mathrm{d}I}{\mathrm{d}t} + \frac{I}{C}.$$

Letting $I = I_{\text{o}}\sin(\omega t + \phi_{\text{o}})$, the right side can be expressed by

$$R\frac{\mathrm{d}I}{\mathrm{d}t} + \frac{I}{C} = \omega R I_{\text{o}} \cos(\omega t + \phi_{\text{o}}) + \frac{I_{\text{o}}}{C}\sin(\omega t + \phi_{\text{o}});$$

therefore we obtain the relation

$$\begin{aligned} -V_{\text{o}}\cos(\omega t + \phi_{\text{o}}) &= R I_{\text{o}}\cos(\omega t + \phi_{\text{o}}) + \frac{I_{\text{o}}}{\omega C}\sin(\omega t + \phi_{\text{o}}) \\ &= -\sqrt{R^2 + \left(\frac{1}{\omega C}\right)^2}\, I_{\text{o}}\cos(\omega t + \phi_{\text{o}} - \psi), \end{aligned} \tag{iv}$$

where $\tan\psi = 1/\omega CR$. The result can be expressed as

$$V_{\text{in}} = Z(\omega)I, \quad I = I_{\text{o}}\cos(\omega\tau + \phi_{\text{o}} - \psi)$$

and

$$Z(\omega) = \sqrt{R^2 + \frac{1}{\omega^2 C^2}} = \frac{V_{\text{o}}}{I_{\text{o}}}. \tag{10.3a}$$

Here, the phase of I shows delay by ψ, and $Z(\omega)$ is the impedance.

In the case of *RL* shown in Figure 10.2(b), we can carry out a calculation similar to the one above. The equation for the current is

$$V_{\text{o}}\cos(\omega t + \phi_{\text{o}}) - L\frac{\mathrm{d}I}{\mathrm{d}t} = RI,$$

which can be solved by letting $I = I_{\text{o}}\cos(\omega t + \phi_{\text{o}})$. We obtain the relation

$$\begin{aligned} V_{\text{o}}\cos(\omega t + \phi_{\text{o}}) &= \omega L I_{\text{o}}\sin(\omega t + \phi_{\text{o}}) + R I_{\text{o}}\cos(\omega t + \phi_{\text{o}}) \\ &= \sqrt{R^2 + (\omega L)^2}\, I_{\text{o}}\cos(\omega t_{\text{o}} + \phi_{\text{o}} + \psi), \end{aligned} \tag{v}$$

where $\tan\psi = R/\omega L$. Therefore, we can define the impedance as

$$Z(\omega) = \sqrt{R^2 + \omega^2 L^2}, \tag{10.3b}$$

and the angle ψ represents the current phase, which advances the AC voltage.

For the *LCR* case shown in Figure 10.2(c), we have the equation

$$V_{\text{o}}\cos(\omega t + \phi_{\text{o}}) - L\frac{\mathrm{d}I}{\mathrm{d}t} = \frac{Q}{C} + RI,$$

which can be solved in similar manner, and the result is

$$V_{\text{o}}\cos(\omega t + \phi_{\text{o}}) = Z(\omega)\, I_{\text{o}}\cos(\omega t + \phi_{\text{o}} - \psi),$$

where

$$Z(\omega) = \sqrt{R^2 + \left(\omega L - \frac{1}{\omega C}\right)^2} \quad \text{and} \quad \tan\psi = \frac{\omega L - \dfrac{1}{\omega C}}{R}. \tag{10.3c}$$

The power delivered instantaneously by an AC source is expressed by

$$P(t) = V(t)\,I(t) = V_o I_o \cos\phi \cos(\phi - \psi),$$

where $\phi = \omega t + \phi_o$ is the voltage phase. However, for usual measurements with meters it is convenient to define its *average* over a period of the AC cycle $T_o = 2\pi/\omega$; that is,

$$\langle P\rangle = \frac{1}{T_o}\int_0^{T_o} P\mathrm{d}t = V_o I_o \frac{\omega}{2\pi}\int_0^{T_o} \cos\phi\cos(\phi-\psi)\,\mathrm{d}t = \frac{1}{2}V_o I_o \cos\psi. \tag{10.4}$$

Accordingly, the AC voltage and current are defined as square roots of the *mean-square averages* of $V(t)$ and $I(t)$, i.e. $\bar{V} = \sqrt{\langle V^2\rangle} = \dfrac{V_o}{\sqrt{2}}$ and $\bar{I} = \dfrac{I_o}{\sqrt{2}}$, respectively. With these mean-square roots, (10.4) can be expressed as $\langle P\rangle = \bar{V}\ \bar{I}\cos\psi$. It is noted that such averages are referred to conventional AC measurements over repeated periods of oscillation, which are characterized by a long timescale of observation.

10.4. Complex Vector Diagrams

Equations (iv) and (v) indicate that AC voltages across R, L, and C can be given effectively as if these elements are connected individually with V_{in}, namely, RI, ωLI, and $I/\omega C$, respectively. Further, (10.3a) and (10.3b) imply that the impedances can be geometrically determined in x, y-planes by points $(R, \omega L)$ and $(R, -1/\omega C)$, where ψ is clearly the angle between the impedance Z and the x-axis, as illustrated in Figures 10.2(a) and (b). Equation (10.2c) can be interpreted that Z is represented by a point $\left(R, \omega L - \dfrac{1}{\omega C}\right)$ and an angle ψ, as in Figure 10.2(c).

It is convenient to express such impedances on a complex plane, considering the y-axis to be the imaginary axis iy. Actually, the diagrams shown in Figures 10.2(a), (b), and (c) are complex planes. Thus, the impedance can be expressed as a complex function of ω, that is

$$Z(\omega) = R + iX(\omega), \tag{10.5}$$

where the imaginary part $X(\omega)$ is called the *reactance*, e.g., $X(\omega) = \omega L - \dfrac{1}{\omega C}$ in Figure 10.2(c).

The vector-diagram method can also be obtained by considering the AC voltage and current as complex variables, whose *real* parts represent actual variables. For example, in an *RLC* circuit, the voltage relation

$$V_{\text{in}} = L\frac{dI}{dt} + \frac{1}{C}\int_0^t I dt + RI$$

can be solved by assuming $V_{\text{in}} = V_o \exp(i\omega t)$. Then, writing $I = I_o \exp(i\omega t)$, we obtain

$$V_o \exp(i\omega t) = \omega L I_o \exp(i\omega t) + \frac{1}{i\omega C} I_o \exp(i\omega t) + R I_o \exp(i\omega t).$$

That is,

$$V_o = \left(i\omega L + \frac{1}{i\omega C} + R\right) I_o = \mathbf{Z}(\omega) I_o, \tag{10.6a}$$

where

$$\mathbf{Z}(\omega) = R + i\left(\omega L - \frac{1}{\omega C}\right) \tag{10.6b}$$

is the *complex impedance*. This expression can be written in the form that

$$\mathbf{Z}(\omega) = Z(\omega)\exp(-i\psi), \tag{10.6c}$$

where

$$Z(\omega) = \sqrt{R^2 + \left(\omega L - \frac{1}{\omega C}\right)^2} \quad \text{and} \quad \tan\psi = \frac{\omega L - \frac{1}{\omega C}}{R},$$

and from (10.4a) we obtain

$$V_o = Z(\omega) I_o \exp(-i\psi).$$

Hence, $I = I_o \exp i(\phi - \psi)$ when $V = V_o \exp(i\phi)$, where the phase of the AC voltage is written as $\phi = \omega t + \phi_o$. Figure 10.3 illustrates the relation between complex AC voltage and current, characterized by a phase difference ψ.

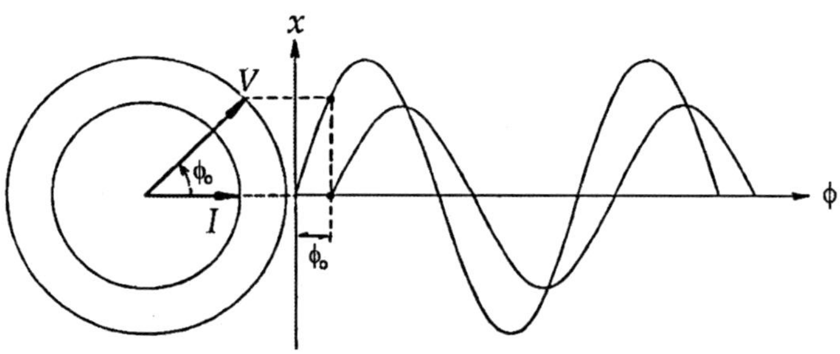

FIGURE 10.3. A typical relation between AC voltage vs. AC current.

10.5. Resonances

10.5.1. A Free LC Oscillation

In a circuit shown in Figure 10.4(a), when the switch S is turned on "1," the inductor L becomes active with a current I. Switching S to "2" disconnects the battery, and the capacitor C is charged to $\pm Q$, which is then discharged through L, then charges C, and continues repeatedly. As illustrated in Figure 10.4(b), such an oscillation takes place in a cycle of 1-2-3-4 for a long time if the resistance of the coil is negligibly small.

A free oscillation between an idealized inductor and capacitor can be described by

$$V_{\text{in}} = V_C \quad \text{or} \quad -L\frac{\mathrm{d}I}{\mathrm{d}t} = \frac{Q}{C}.$$

Multiplying by $I = \mathrm{d}Q/\mathrm{d}t$, we have $-LI\dfrac{\mathrm{d}I}{\mathrm{d}t} = \dfrac{Q}{C}\dfrac{\mathrm{d}Q}{\mathrm{d}t}$, which can then be integrated with respect to time, that is,

$$\frac{1}{2}LI^2 + \frac{1}{2}\frac{Q^2}{C} = \text{const.}, \tag{10.7a}$$

implying that the magnetic energy in the inductor is added to the electrostatic energy in the capacitor in a given time t. Equation (10.7a) represents the energy of a harmonic oscillation of (Q, I), analogous to a mass particle, in which $\frac{1}{2}LI^2$ and $Q^2/2C$ signify kinetic and potential energies of the moving charge, respectively.

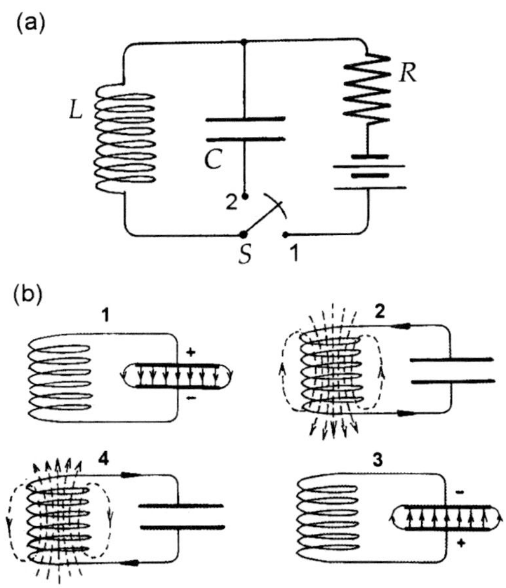

FIGURE 10.4. Free oscillation in an $L - C$ circuit. (a) The inductor L is activated when the switch S is turned onto 1, and then to 2 for oscillation. (b) Repeated charging of the capacitor in sequence of 1-2-3-4-1.

The characteristic *frequency* of oscillation can be expressed as

$$\omega_o = \frac{1}{\sqrt{LC}}, \tag{10.7b}$$

and $Q(t) = Q_o \exp(i\omega_o t)$, $I(t) = i\omega_o Q_o \exp(i\omega_o t)$, where Q_o is the amplitude, shifting the current phase by $\frac{1}{2}\pi$.

10.5.2. Series Resonance

Consider a series of *RCL* connected with an AC source, as shown in Figure 10.2(c). In this case, the voltage-current relation is given by (10.6a), (b), and (c). It is noted that if the frequency of AC voltage is equal to ω_o given by (10.7b), the reactance $X(\omega)$ vanishes at $\omega = \omega_o$, i.e. $X(\omega_o) = 0$, signifying that $Z(\omega_o) = R$ is minimum. The AC current exhibits maximum amplitude at this condition, which is therefore called the *resonance*.

At resonance the current and charge in the inductor and capacitor are in oscillatory motion with maximum amplitude at ω_o, while their energy is being dissipated in the resistor. Hence, the *resonance quality* can be expressed by the ratio between the energies of oscillation and dissipation averaged over a cycle of oscillation, i.e.,

$$\omega_o \frac{L\langle I^2\rangle}{R\langle I^2\rangle} = \frac{\omega_o L}{R} = \frac{1}{Q_o}, \tag{10.8}$$

where Q_o is called the *quality factor* or *Q-factor* of resonance.

The reactance $X(\omega)$ can be modified by using the definitions of ω_o and Q_o given by (10.7b) and (10.8) as

$$X(\omega) = RQ_o\left(\frac{\omega}{\omega_o} - \frac{\omega_o}{\omega}\right),$$

and, hence, the impedance is expressed as

$$Z(\omega)^2 = R^2 + X(\omega)^2 = R^2\left\{1 + Q_o^2\left(\frac{\omega}{\omega_o} - \frac{\omega_o}{\omega}\right)^2\right\}.$$

Therefore, currents at off- and on resonance can be calculated as $I_o(\omega) = V_o/Z(\omega)$ and $I_o(\omega_o) = V_o/R$, respectively, for which we have the formula,

$$\frac{I_o(\omega)^2}{I_o(\omega_o)^2} = \frac{1}{1 + Q_o^2\left(\frac{\omega}{\omega_o} - \frac{\omega_o}{\omega}\right)^2}. \tag{10.9a}$$

Figure 10.5 sketches the curve of (10.9a) against $\Delta\omega/\omega_o$, where $\pm\Delta\omega = \omega - \omega_o$. In such a curve, frequencies $\omega_\pm$ at *half-power points*, defined by $I_o(\omega_\pm)^2/I_o(\omega_o)^2 = \frac{1}{2}$, are of particular interest. Writing $\omega_\pm = \omega_o \pm \Delta\omega$, $\frac{\omega_\pm}{\omega_o} - \frac{\omega_o}{\omega_\pm} \cong \pm\frac{2\Delta\omega_\pm}{\omega_o}$, and

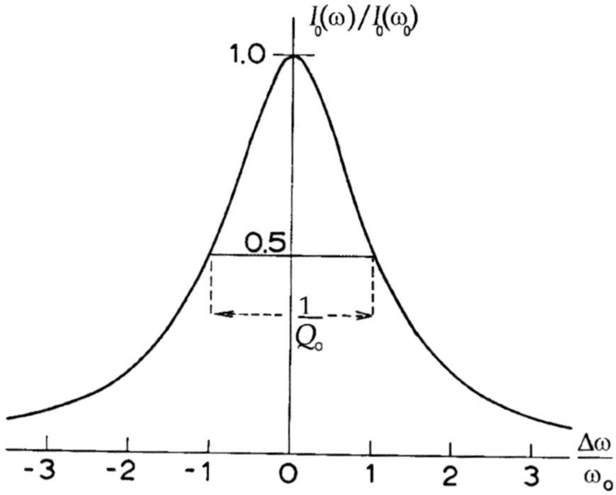

FIGURE 10.5. A typical L-C resonance. The half-width is indicated as $1/Q_o$, where the quality factor Q_o depends on the internal resistance R in the inductor L.

from the relation $1 + 4Q_o^2\left(\frac{\Delta\omega_\pm}{\omega_o}\right)^2 = 2$, we obtain

$$\frac{2\Delta\omega_\pm}{\omega_o} = \frac{1}{Q_o}. \tag{10.9b}$$

The quality factor Q_o can be easily estimated from half-power points, as shown in the figure.

10.5.3. Parallel Resonance

Figure 10.6 shows a parallel connection of L and C, where the resistance R can represent an internal resistance in L as well as added resistances, for which the Kirchhoff rules may be applied as in a DC circuit. With regard to junctions, marked a and b in the figure, the impedances $R + i\omega L$ and $1/i\omega C$ are connected in parallel, for which the voltage difference V_{ab} is expressed as

$$V_{ab} = (R + i\omega L)\, I_1 = \frac{1}{i\omega C} I_2,$$

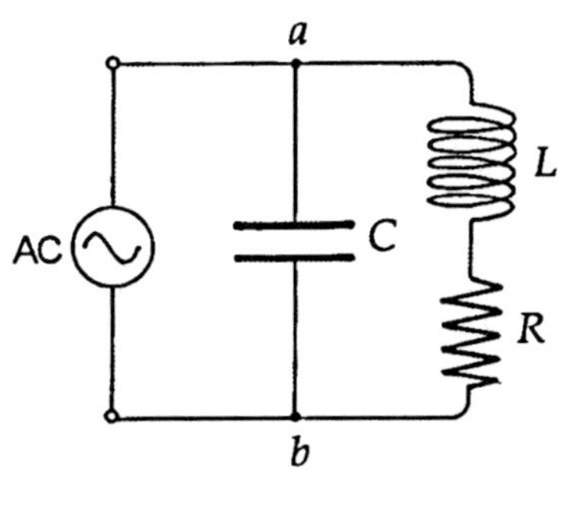

FIGURE 10.6. A circuit for L-C resonance. Normally, R represents the internal resistance due to a finite conductivity of the coil.

where $I_1 + I_2 = I$ is the net current to satisfy $V_{\text{in}} = Z(\omega)I$. Therefore,

$$I = \frac{V_{ab}}{R + i\omega L} + i\omega C V_{ab} = Y(\omega)V_{ab},$$

where $Y(\omega)$ is the *complex admittance*

$$Y(\omega) = \frac{1}{R + i\omega L} + i\omega C = \frac{R - i\omega L}{R^2 + \omega^2 L^2} + i\omega C.$$

In this case, the resonance is signified by a particular frequency ω_o for the imaginary part to become zero, that is, $\dfrac{L}{R^2 + \omega_o^2 L^2} - C = 0$, in which case the resonance frequency is determined by

$$\omega_o^2 = \frac{1}{LC} - \frac{R^2}{L^2}, \tag{10.10}$$

giving the characteristic frequency identical to (10.7b) in a series resonance, provided R is negligible. However, the impedance at resonance can be written as

$$Z(\omega_o) = \frac{V(\omega_o)}{I(\omega_o)} = \frac{R^2 + \omega_o^2 L^2}{R} = \frac{L}{CR},$$

for which an internal resistance R cannot be considered zero. Nevertheless, if so assumed, $I(\omega_o) = 0$ and $I_1(\omega_o) = -I_2(\omega_o)$, so that the parallel resonance is often referred to as *anti-resonance*.

In practice, for a parallel resonator characterized by L and C, $1/\sqrt{LC}$ is a useful parameter for approximate resonance. We can therefore write an approximate relation between the exact resonance frequency ϖ_o and $\omega_o = 1/\sqrt{LC}$, with which (10.8) can be re-expressed as

$$\varpi_o^2 = \omega_o^2\left(1 - \frac{1}{Q_o^2}\right) \quad \text{where} \quad \frac{1}{Q_o} = \frac{\omega_o L}{R}.$$

Therefore,

$$\varpi_o \approx \omega_o\left(1 - \frac{1}{2Q_o^2}\right) \quad \text{and} \quad Z(\varpi_o) \approx Z(\omega_o)\left(1 + \frac{1}{Q_o^2}\right).$$

10.6. Four-Terminal Networks

In this section, we discuss circuits with two *input* and two *output* terminals, known generally as a *four-terminal networks*. When an AC voltage is applied to an input, the output voltage is not necessarily sinusoidal, as caused by relaxation effects with resistive elements. A guiding principle for solving network problems is the Kirchhoff rules at low frequencies, thereby considering a unique time-dependent potential at each point in a given circuit as related to the input phase. As will be discussed in Chapter 11 for high frequencies, this assumption is valid only at sufficiently low frequencies, where the physical size of a circuit is smaller than a wavelength of radiation.

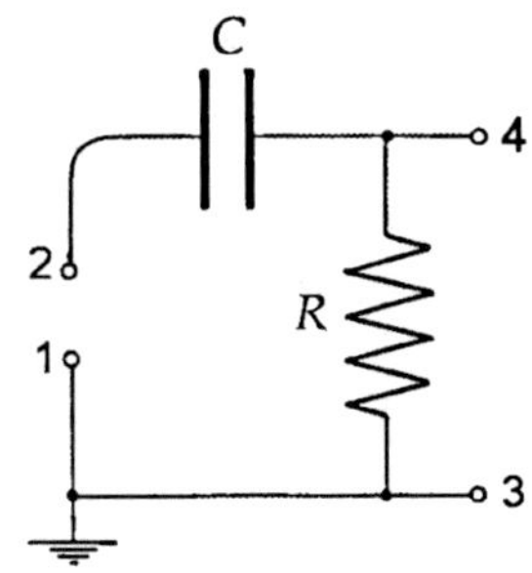

FIGURE 10.7. A simple four-terminal network. The input and output are 1-2 and 3-4, respectively.

10.6.1. RC Network

Figure 10.7 shows a circuit of C and R connected in series, where terminals 1 and 2 constitute the input, to which a varying voltage can be applied. The output terminals 3 and 4 are "open," meaning *unloaded*, while we assume that terminals 1 and 3 are connected to the "ground," i. e., $V_1 = V_3 = 0$.

We first calculate the output voltage V_4 at the open terminal 4, when an AC voltage at a frequency ω is applied to the input. The voltage relation can be written as

$$V_2 - V_4 = \frac{Q}{C} \quad \text{and} \quad V_4 = RI.$$

Writing the input voltage as $V_2 = V(t)$, we have

$$I = C\frac{\mathrm{d}(V_2 - V_4)}{\mathrm{d}t} = \frac{V_4}{R}.$$

Therefore, the differential equation for V_4 is given by

$$\frac{\mathrm{d}V_4}{\mathrm{d}t} + \frac{V_4}{\tau} = \frac{\mathrm{d}V_2}{\mathrm{d}t}, \tag{10.11}$$

where $\tau = RC$ is the time constant. For a sinusoidal input $V_2 = V_o\exp(i\omega t)$, we can set $V_4 = V_{4o}\exp(i\omega t)$ in (10.11) and obtain

$$\left(i\omega + \frac{1}{\tau}\right) V_{4o} = i\omega V_o,$$

and hence

$$\begin{aligned}\frac{V_{4o}}{V_o} &= \frac{i\omega\tau}{1 + i\omega\tau} = \frac{i\omega\tau\,(1 - i\omega\tau)}{1 + \omega^2\tau^2}\\ &= \frac{\omega\tau}{\sqrt{1+\omega^2\tau^2}}\frac{i + \omega\tau}{\sqrt{1+\omega^2\tau^2}} = \frac{\omega\tau}{\sqrt{1+\omega^2\tau^2}}\exp i\delta,\end{aligned}$$

where $\tan\delta = \dfrac{1}{\omega\tau}$. Therefore, the output voltage is given by

$$V_4 = \frac{V_o\omega\tau}{\sqrt{1+\omega^2\tau^2}}\exp i(\omega t + \delta), \tag{10.12}$$

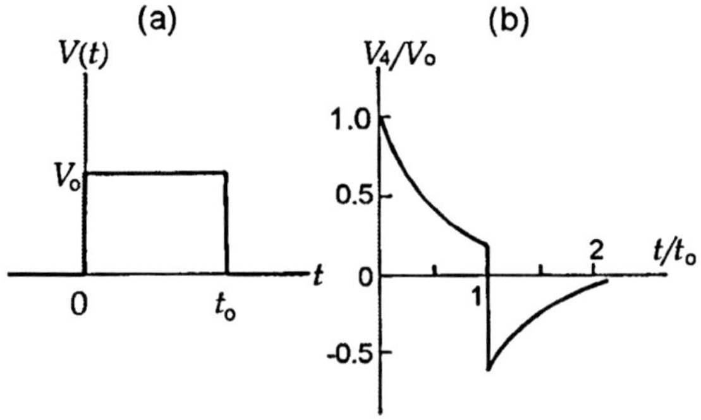

FIGURE 10.8. (a) A square-shaped voltage input; (b) its distorted output.

where δ is a phase shift. Obviously, for a static input, $V_4 = 0$ as $\omega = 0$.

Next, we consider an input voltage pulse V_o in a rectangular shape, applied during an interval $0 \le t \le t_o$, as shown in Figure 10.8(a). In such a case it is customary to express the pulse by a Fourier integral

$$V_2(t) = \frac{1}{\sqrt{2\pi}} \int_{-\infty}^{+\infty} f(\omega) \exp(i\omega t)\, d\omega,$$

where the function $f(\omega)$ is given by

$$f(\omega) = \frac{1}{\sqrt{2\pi}} \int_0^{t_o} V_2(t) \exp(-i\omega t)\, dt.$$

In the present case, $V_2 = V_o$ for $0 \le t \le t_o$, and 0 in other time regions, hence

$$f(\omega) = \frac{V_o}{\sqrt{2\pi}} \int_0^{t_o} \exp(-i\omega t)\, dt = \frac{V_o}{\sqrt{2\pi}} \frac{1 - \exp(-i\omega t_o)}{i\omega}.$$

Using this expression in (10.11), we can write

$$V_4(t) = \frac{V_o}{2\pi i} \int_{-\infty}^{+\infty} \frac{\exp i\omega t - \exp i\omega(t - t_o)}{\omega - \dfrac{i}{\tau}} d\omega. \tag{10.13}$$

When we consider that ω is a complex variable, we can evaluate the integral in (10.13) by the Cauchy theorem from the theory of complex variables. The integrand has a "pole" at $\omega = \dfrac{i}{\tau}$, hence the analytic function of the numerator, $f(\omega) = \exp(i\omega t) - \exp[i\omega(t - t_o)]$, can be determined by the principal value that is determined by

$$f\left(\frac{i}{\tau}\right) = \frac{1}{2\pi i} \mathrm{P}\left[\oint_C \frac{f(\omega)}{\omega - \dfrac{i}{\tau}} d\omega\right],$$

where a closed path C is divided into semi-circular C_1 and C_2, as shown in Figure 10.9.

FIGURE 10.9. Calculating a complex integral with a pole using the Cauchy theorem.

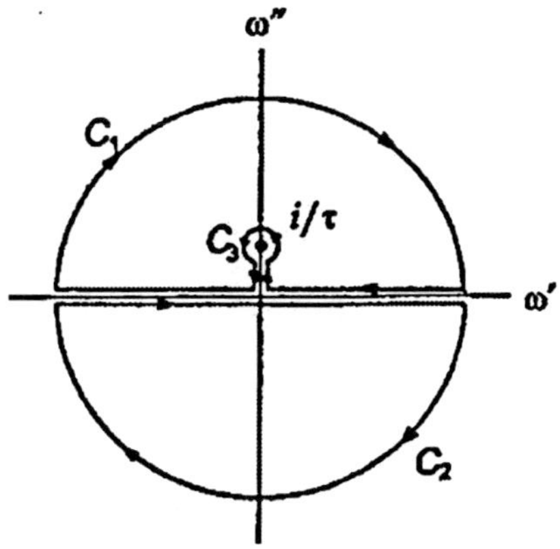

Equation (10.13) indicates that $V_4(t)/V_o$ should be given by the principal value $f(i/\tau)$, and the output voltage can be described by

$$\begin{aligned} V_4(t) &= 0 \qquad \text{for } t < 0, \\ &= V_o \exp(-\frac{t}{\tau}) \quad \text{for } 0 < t < t_o \\ &= V_o \left\{ \exp\left(-\frac{t}{\tau}\right) - \exp\left(-\frac{t - t_o}{\tau}\right) \right\} \text{ for } t > t_o, \end{aligned}$$

which are sketched in Figure 10.8(b).

10.6.2. Loaded Transformer

A transformer for converting voltages is shown in Figure 10.10, where an AC voltage V_1 can be applied to the input terminals 1 and 2 of the primary coil L_1, whereas the secondary L_2 is loaded with an external resistor R. We consider a mutual inductance M between L_1 and L_2. In practice, the internal resistances r_1 and r_2 of these coils are not negligible, so are therefore included in calculation.

We can write voltage relations for these closed primary and secondary circuits; that is,

$$(r_1 + i\omega L_1)\, I_1 + i\omega M I_1 = V_1$$

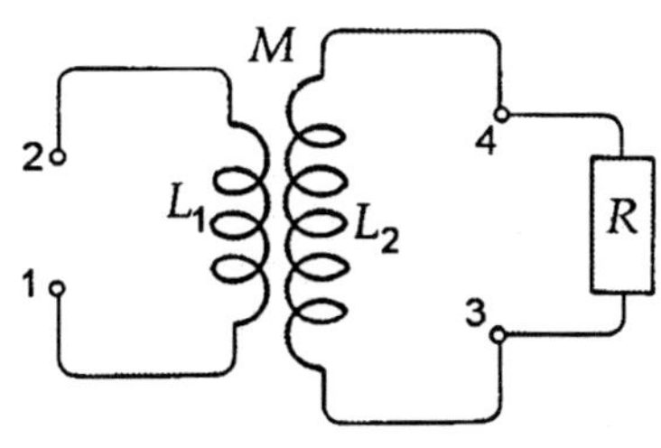

FIGURE 10.10. An AC transformer.

and

$$i\omega M I_1 + (r_2 + R + i\omega L_2) I_2 = 0.$$

Eliminating I_1, the secondary current I_2 can be obtained from these equations, and expressed as

$$I_2 = \frac{-i\omega M}{r_1 (r_2 + R) + i\omega \{L_1 (r_2 + R) + L_2 r_1\} + \omega^2 \left(M^2 - L_1 L_2\right)} V_1.$$

Between the terminal 3 and 4, we have the relation $V_2 = R I_2$, and therefore

$$\frac{V_2}{V_1} = \frac{-i\omega M R}{(r_2 + R)\left(r_1 + i\omega L_1 + \dfrac{i\omega L_2 r_1 + \omega^2 \left(M^2 - L_1 L_2\right)}{r_2 + R}\right)}.$$

In the limit of $R \to \infty$, $R \approx r_2 + R \to \infty$, hence this expression can be reduced to

$$\frac{V_2}{V_1} = \frac{-i\omega M}{r_1 + i\omega L_1}$$

which may further be approximated as $\dfrac{V_2}{V_1} = -\dfrac{M}{L_1}$, if $r_1 \ll \omega L_1$.

10.6.3. An Input-Output Relation in a Series RCL Circuit

For a circuit of *RCL* connected in series, we already discussed the resonance in Section 10.5. In this subsection, the input-output relation is obtained for a *RCL* circuit, shown in Figure 10.11, where an oscillating voltage is applied to the input terminals 1 and 2.

The input voltage in arbitrary shape can be expanded into a Fourier series

$$V_{21}(t) = \frac{1}{\sqrt{2\pi}} \int_{-\infty}^{+\infty} V(\omega) \exp i\omega t \, d\omega, \tag{i}$$

and the current can be expressed as

$$I(t) = \frac{1}{\sqrt{2\pi}} \int_{-\infty}^{+\infty} I(t) \exp i\omega t \, d\omega. \tag{ii}$$

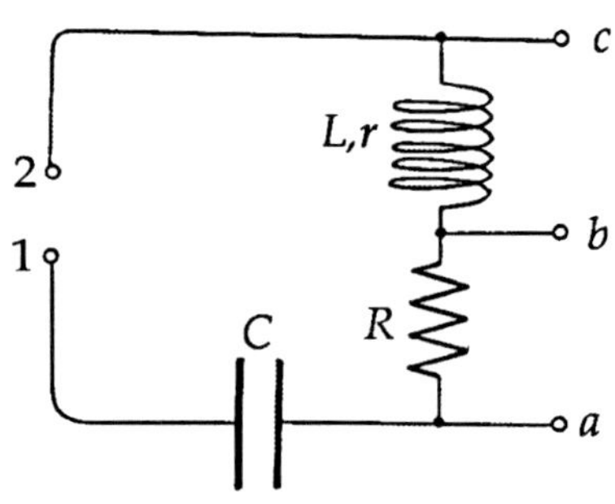

FIGURE 10.11. An example of a four-terminal input-output circuit.

The voltage relation is

$$V_{21} - L\frac{dI}{dt} = (R + r)I + \frac{Q}{C},$$

from which the equation to be solved for $I(t)$ is

$$\frac{d^2 I}{dt^2} + (R + r)\frac{dI}{dt} + \frac{1}{C}I = \frac{dV_{21}}{dt}. \quad \text{(iii)}$$

This is a linear differential equation, and (ii) is a linear combination of sinusoidal functions $\exp(i\omega t)$ indexed by ω. Therefore, we can derive from (iii) solutions with respect to individual ω. Using the series (i) and (ii) in the equation (iii), for each component at ω we obtain

$$\left\{-\omega^2 L + i\omega(R + r) + \frac{1}{\omega C}\right\} I(\omega) = i\omega V(\omega),$$

and hence

$$I(\omega) = \frac{V(\omega)}{R + r + i\left(\omega L - \dfrac{1}{\omega C}\right)}.$$

The output voltage between the terminals c and b in Figure 10.10, is given by

$$V_{cb} = \frac{1}{\sqrt{2\pi}}\int_{-\infty}^{+\infty}(r + i\omega L)I(\omega)\exp(i\omega t)d\omega,$$

whereas those between b-a and a-1 are

$$V_{ba} = \frac{1}{\sqrt{2\pi}}\int_{-\infty}^{+\infty} RI(\omega)\exp(i\omega t)d\omega$$

and

$$V_{a1} = \frac{1}{\sqrt{2\pi}}\int_{-\infty}^{+\infty}\frac{I(\omega)}{C}\exp(i\omega t)d\omega.$$

At resonance, i.e., $\omega = \omega_o = 1/\sqrt{LC}$, these output voltages are signified by the current, i.e., $I(\omega_o)\exp(-i\psi)$, where the phase shift ψ is determined by

$$\tan\psi = \frac{1}{R + r}\left(\omega L - \frac{1}{\omega C}\right).$$

10.6.4. *Free Oscillation in an RCL Circuit*

In an unloaded *RCL* circuit, as shown in Figure10.11, we consider a free oscillation caused by a short voltage pulse applied to 1 and 2.

An input pulse is assumed as a voltage V_o in a short duration of time τ, being described by a delta function, i.e., $V_{in} = V_o\delta(\tau)$. The current in the circuit is given in such a Fourier series as (ii) in the previous subsection, hence,

$$V_{in}(t) = \frac{V_o}{2\pi}\int_{-\infty}^{+\infty}\exp(i\omega t)d\omega,$$

and

$$I(t) = \frac{V_o}{2\pi}\int_{-\infty}^{+\infty} \frac{\exp(i\omega t)}{R + i\left(\omega L - \dfrac{1}{\omega C}\right)} d\omega.$$

This expression can be modified as

$$I(t) = \frac{V_o}{2\pi i L}\int_{-\infty}^{+\infty} \frac{\omega \exp(i\omega t) d\omega}{\omega^2 - i\omega \dfrac{R}{L} - \dfrac{1}{LC}}.$$

Factorizing the denominator as $(\omega - \omega_1)(\omega - \omega_2)$, where $\omega_{1,2} = \frac{iR}{2L} \pm \sqrt{\frac{1}{LC} - \left(\frac{R}{2L}\right)^2}$, if $R < 2\sqrt{\frac{L}{C}}$,

$$\begin{aligned} I(t) &= \frac{V_o}{2\pi i L}\int_{-\infty}^{+\infty} \frac{\exp(i\omega t)\omega d\omega}{(\omega - \omega_1)(\omega - \omega_2)} \\ &= \frac{V_o}{2\pi i(\omega_1 - \omega_2)}\int_{-\infty}^{+\infty}\left(\frac{1}{\omega - \omega_1} - \frac{1}{\omega - \omega_2}\right)\left[\omega \exp(i\omega t)\right] d\omega. \end{aligned}$$

Using the Cauchy theorem, the integrals are given by the principal values at the poles ω_1 and ω_2, and

$$I(t) = \frac{V_o}{L}\left[\frac{\omega_1}{\omega_1 - \omega_2}\exp(i\omega_1 t) - \frac{\omega_2}{\omega_1 - \omega_2}\exp(i\omega_2 t)\right].$$

Writing that $\frac{R}{2L} = \lambda$ and $\frac{1}{LC} - \frac{R^2}{4L^2} = \omega_o^2$,

$$\begin{aligned} I(t) &= \frac{V_o}{L}\frac{\exp(-\lambda t)}{2\omega_o}\left[(i\lambda + \omega_o)\exp(i\omega_o t) - (i\lambda - \omega_o)\exp(-i\omega_o t)\right] \\ &= \frac{V_o}{L}\exp(-\lambda t)\left[\cos(\omega_o t) - \frac{\lambda}{\omega_o}\sin(\omega_o t)\right] \\ &= \frac{V_o}{\sqrt{\omega_o^2 L^2 + \frac{1}{4}R^2}}\exp(-\lambda t)\cos(\omega_o t + \delta), \end{aligned}$$

where

$$\tan\delta = \frac{R}{2L\omega_o}.$$

Here, the constants λ and δ signify an exponential decay and phase shift of the free oscillation.

Part 3
Electromagnetic Waves

11
Transmission Lines

11.1. Self-Sustained Oscillators

Free oscillation in a circuit is generally short-lived because of the presence of a resistance. Normally AC current flows continuously, as the generator keeps supplying energy, but it is not simple to obtain a source of persisting electrical oscillations at a desired frequency. Today, vacuum tubes and semi-conducting devices, oscillators, signal amplifiers, and other electronically controlled circuits are available at almost any frequency, thereby, basic electromagnetic experiments can be performed with precision in modern electronics. A self-sustaining oscillator with a sizable output power is a particularly important device to study high-frequency phenomena.

A sustaining oscillator is basically a forced oscillator whose output power is fed back in part to the input to sustain oscillation. In this section, to understand the *feedback* principle, we take a vacuum tube as an example, although it is almost obsolete in modern electronics. In fact, a feedback mechanism is significant for various growing processes in natural phenomena, and described effectively in terms of *negative resistance* in circuit theory. While a conventional (positive) resistance signifies energy loss, a negative resistance can be interpreted as sustained oscillation.

Figure 11.1(a) shows a circuit of a *triode* tube providing an electronically generated current $I(t)$ at the characteristic frequency ω_o determined by the LC resonator. The triode consists of an electron-emitting *cathode* (K), an *anode* (A), and a *grid* (G) that controls the electronic flow in the tube. The electronic current driven by the anode voltage V_a can be modified by the grid voltage V_g, exhibiting, typically, a sharp rise at a negative value of $V_g = -V_b$, called a *bias* voltage, as illustrated in Figure 11.1(b). In such a four-terminal circuit, the current-voltage relation can normally be described by a linear relation

$$I = Y(\beta V_a + V_g),$$

where Y is the *electronic admittance* and β the coupling between anode- and grid voltages. In this case the output voltage is given by $V = \beta V_a + V_g$, and the current

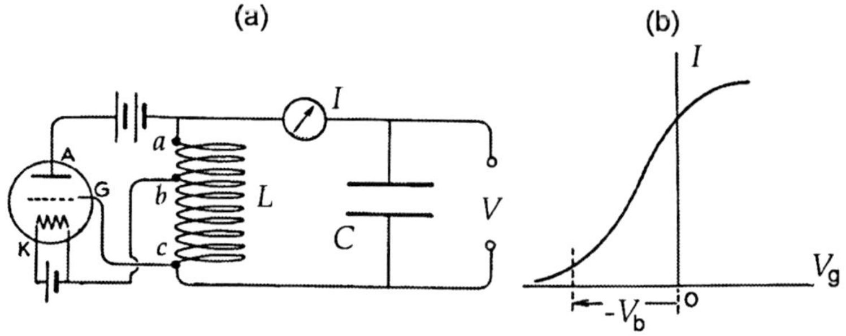

FIGURE 11.1. (a) A triode oscillator. (b) A typical non-linear $I - V$ character.

modulated by the grid voltage ΔV_g, can be expressed by

$$\Delta I = Y\left(\beta \frac{\partial V_a}{\partial V_g} + 1\right) \Delta V_g = (-\beta\alpha_o + 1)\,\Delta V_g,$$

where $\alpha_o = -\partial V_a / \partial V_g$ is called the *amplification factor*. Hence,

$$\frac{\Delta V_g}{\Delta I} = \frac{1}{Y(1 - \beta\alpha_o)},$$

which is positive if $0 < \beta\alpha_o < 1$, but negative if $\beta\alpha_o > 1$. The former case represents an amplification, whereas the latter condition gives rise to a negative resistance for the current to be controlled by the bias potential. In solid state electronics the I-V relation in some semi-conductors is signified by a negative slope in a certain region as shown in Figures 11.2(a) and (b). This condition provides a negative damping condition similar to a triode case.

Electric oscillations in a wide range of frequencies are utilized for a variety of applications, such as telecommunications and other purposes in modern electronics for which oscillators are built on the basis of the principle of negative feedback. Although no details of oscillating mechanisms are given here, we should bear in mind in the following discussions that self-sustaining oscillators are always involved in interpreting theoretical results.

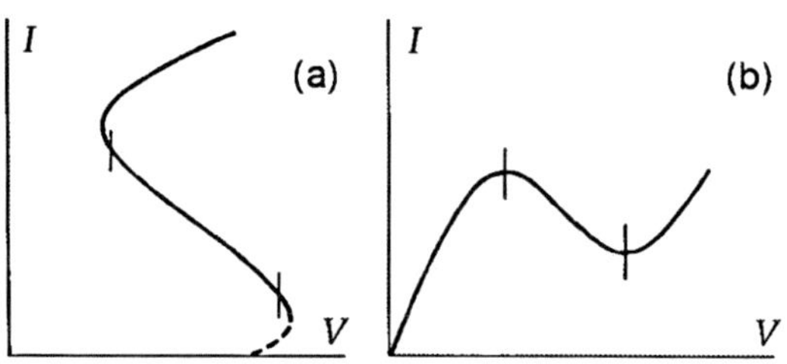

FIGURE 11.2. Examples for negative resistances.

11.2. Transmission Lines

Long parallel wires and concentric conductors are of particular importance for signal transmission. Known as *Lecher's wires* and *coaxial cable*, these transmission lines are mathematically equivalent, sharing similar principles. Physically, however, their performances are not identical: the electromagnetic field is completely confined to the space of a coaxial cable, whereas it is somewhat diffused from Lecher's wires. Both devices are in practical use, but a coaxial cable is an ideal transmission line at high frequencies. Practically, the space between two conductors is filled with insulating material to maintain uniform spacing, although the signal can be modified by the dielectric property. At a high frequency, the oscillator energy is transferred to such a transmission line, thus it is transferred to a terminal load that is expressed by an impedance Z_L, as illustrated in Figure 11.3.

In such a long device required for power transmission or communication purposes, the internal resistance cannot be ignored, and the capacitance between two conductors plays an important role. Besides, their self- and mutual inductances are also significant at high frequencies. Under the circumstances the voltage difference across the conductors and the currents on two conductors should be described as functions of the position x and time t, to which an oscillator and a terminal load are connected.

For a uniform transmission line a potential difference ΔV between x and $x + dx$ along the line should be related to a resistance $R\Delta x$ by the Ohm law $\Delta V = (R\Delta x)\Delta I_{\text{con}}$, where R is defined as the resistance per unit length, and the conduction current is $\Delta I_{\text{con}} = I(x + \Delta x) - I(x)$. Writing for convenience $G\Delta x = 1/R\Delta x$, where G is the *conductance* per unit length, we obtain the relation

$$I_{\text{con}} = (G\Delta x)\Delta V.$$

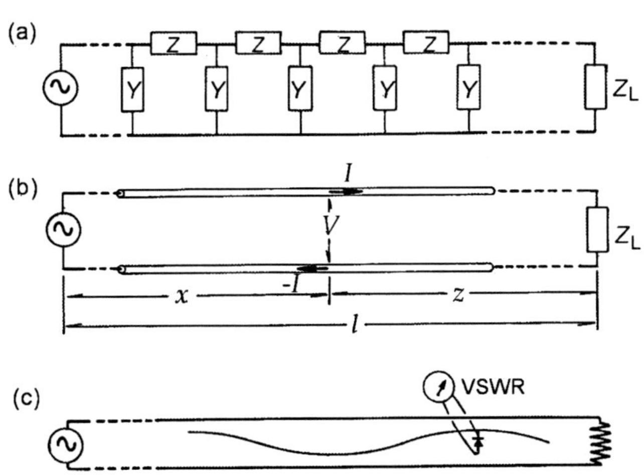

FIGURE 11.3. (a) Distributed impedances of a transmission line and a terminal impedance Z_L. (b) Current and voltage functions. (c) Measuring VSWR in a transmission line.

Next, with regard to capacitive effect, the potential difference δV across the two conductors due to varying time can be expressed as $\delta V = \Delta Q / C\Delta x$, where C represents the capacity per unit length. Hence, writing as $\delta V = \dfrac{\partial(\delta V)}{\partial t}\Delta t$ and $\Delta Q = \Delta I_{\text{cap}}\Delta t$, we have

$$\Delta I_{\text{cap}} = \frac{\Delta Q}{\Delta t} = C\Delta x \frac{\partial(\delta V)}{\partial t}.$$

Here, ΔI_{cap} represents displacement current between the conductors. Therefore, on each conductor of the line we connect a loop of a net current $I = I_{\text{con}} + I_{\text{cap}}$, whose variation along the section Δx is given by $\Delta I_{\text{con}} + \Delta I_{\text{cap}} = \Delta I = \dfrac{\partial I}{\partial x}\Delta x$. If such a line segment is isolated, $\Delta I_{\text{con}} = -\Delta I_{\text{cap}}$, so that $\Delta I = 0$. However, the propagating current is signified by $\Delta I \neq 0$, which is $\pm\Delta I$ or $\mp\Delta I$, due to the continuity theorem, in opposite directions on each conducting segment.

On the other hand, the voltage relation in the loop must be $\Delta V + \delta V = 0$, so that

$$\delta V = -\Delta V. \tag{11.1}$$

Accordingly, we have

$$\frac{\partial I}{\partial x} = -G\delta V - C\frac{\partial(\delta V)}{\partial t},$$

which is the relation between the voltage difference $\delta V(x,t)$ and currents $\pm I(x,t)$. Writing $V(x,t) = \delta V$ for the potential difference between two conductors at x and t, this relation can be expressed as

$$\frac{\partial I}{\partial x} = -GV - C\frac{\partial V}{\partial t}, \tag{11.2}$$

representing the voltage across the conductor $V(x,t)$ and the conducting currents $\pm I(x,t)$ along the transmission line.

For $\dfrac{\partial V}{\partial x}$ on the two conductors we can write

$$\frac{\partial V}{\partial x} = -R(+I) - L\frac{\partial(+I)}{\partial t} - M\frac{\partial(-I)}{\partial t}$$

and

$$-\frac{\partial V}{\partial x} = -R(-I) - L\frac{\partial(-I)}{\partial t} - M\frac{\partial(+I)}{\partial t},$$

where L and M are self- and mutual inductances per unit length. Combining these relations, we can write

$$\frac{\partial V}{\partial x} = -RI - (L-M)\frac{\partial I}{\partial t},$$

where $L - M$ is the effective inductance of coupled conductors per unit length, which we may rewrite as L for simplicity. Hence, we have

$$\frac{\partial V}{\partial x} = -RI - L\frac{\partial I}{\partial t}. \tag{11.3}$$

Differentiating (11.2) and (11.3) with respect to t and x, respectively, we obtain

$$\frac{\partial^2 I}{\partial x \partial t} = -G\frac{\partial V}{\partial x} - C\frac{\partial^2 V}{\partial t^2} \quad \text{and} \quad \frac{\partial^2 V}{\partial x^2} = -R\frac{\partial I}{\partial x} - L\frac{\partial^2 V}{\partial x \partial t}.$$

Eliminating $\frac{\partial^2 I}{\partial x \partial t}$ from these and replacing $\frac{\partial I}{\partial x}$ by (11.2), we obtain a differential equation for $V(x, t)$, namely,

$$\frac{\partial^2 V}{\partial x^2} = LC\frac{\partial^2 V}{\partial t^2} + (LG + RC)\frac{\partial V}{\partial t} + RGV. \tag{11.4}$$

For the function $I(x, t)$, we can write exactly the same equation, and (11.4) is known as the *telegraph equation*, indicating that the two functions $V(x, t)$ and $I(x, t)$ express propagation along the x direction, being proportional at all values of x and t. For these functions $V(x, t)$ and $I(x, t)$, we can write the relation

$$V(x, t) = ZI(x, t), \tag{11.5}$$

and call the factor Z the *impedance* at x and t.

As remarked in Section 10.1, a propagating current $I(x, t)$ is a time-dependent current that also depends on its position in a circuit. Such a propagation is a significant phenomenon at high frequencies, as described in the following discussions.

11.3. Fourier Transforms

In Section 11.2 we defined two functions $V(x, t)$ and $I(x, t)$ for propagation along a transmission line. Such functions can generally be expressed by Fourier series, i.e.,

$$V(x, t) = \int_\omega \int_k V(k, \omega) \exp[i(kx - \omega t)]\, dk\, d\omega \tag{11.6a}$$

and

$$I(x, t) = \int_\omega \int_k I(k, \omega) \exp[i(kx - \omega t)]\, dk\, d\omega, \tag{11.6b}$$

where $V(k, \omega)$ and $I(k, \omega)$ are the Fourier transforms of $V(x, t)$ and $I(x, t)$, respectively. As (11.6a) and (11.6b) are written as linear combinations of elementary waves $\exp[i(kx - \omega t)]$ that is specified by k and ω educes the problem to individual waves. Using Fourier expansions in (11.2) and (11.3), we obtain relations between Fourier transforms at k and ω. Namely,

$$\begin{aligned} ikI(k,\omega)+(G-i\omega C)V(k,\omega)&=0,\\ ikV(k,\omega)+(R-i\omega L)I(k,\omega)&=0. \end{aligned} \tag{11.7}$$

Eliminating $I(k,\omega)$ and $V(k,\omega)$ from (11.6), we obtain a determinant equation

$$\begin{vmatrix} ik & G-i\omega C \\ ik & R-i\omega L \end{vmatrix}=0,$$

from which we have

$$k^2=\left(LC\omega^2-GR\right)+i(RC-GL)\omega,$$

indicating that k is generally a complex function of ω, which is a real number. Therefore, writing

$$k=k'+ik'', \text{ i.e., } k^2=k'^2+k''^2+2ik'k'',$$

and

$$k'^2=LC\omega^2, \quad k''^2=GR \quad \text{and} \quad 2k'k''=(RC+GL)\omega.$$

From the first and second relations, we obtain $k'=\sqrt{LC}\omega$ and $k''=\sqrt{GR}$. Putting these into the third expression, we derive the relation

$$2\sqrt{LC}\sqrt{GR}\omega=(RC+GL)\omega,$$

and hence the relation

$$RC=GL \tag{11.8}$$

should be held, independent of ω

Thus, for the component wave specified by k and ω, we have

$$V(x,t)=V(k,\omega)\exp(-k''x)\exp[i(k'x-\omega t)]$$

and

$$I(x,t)=I(k,\omega)\exp(-k''x)\exp[i(k'x-\omega t)],$$

showing propagaton with a phase $k'x-\omega t$, with amplitudes decreasing at a rate $1/k''$ along the x direction.

For the amplitude ratio, from (11.7) with (11.8) we obtain

$$\frac{V(k,\omega)}{I(k,\omega)}=\frac{-ik}{G-i\omega C}=\sqrt{\frac{L}{C}}\frac{(GR/L)^{1/2}-i\omega C^{1/2}}{GC^{-1/2}-i\omega C^{1/2}}=\sqrt{\frac{L}{C}}=\sqrt{\frac{R}{G}}.$$

Therefore,

$$V(k,\omega)=Z_oI(k,\omega), \quad \text{where} \quad Z_o=\sqrt{\frac{L}{C}} \tag{11.9}$$

is a constant independent of ω and called the *wave impedance* of the transmission line.

For a uniform coaxial cable of radii a and $b\,(a < b)$, we obtained expressions for C and L for static charges $(Q, -Q)$ and stationary currents $(I, -I)$ in Sections 3.2 and 9.1, respectively, namely,

$$C = \frac{2\pi\varepsilon_o}{\ln\dfrac{b}{a}} \quad \text{and} \quad L = \frac{\mu_o}{2\pi}\ln\frac{b}{a}$$

per unit length. However, it is basically incorrect to use these expressions to write

$$Z_o = \frac{1}{2\pi}\sqrt{\frac{\mu_o}{\varepsilon_o}}\ln\frac{b}{a}$$

for a transmission line; at least such a calculation does not give the correct wave impedance of Lecher's wires, although it conforms to a coaxial cable.

In the foregoing argument a uniform transmission line is characterized by four parameters R, L, G, and C, defined per unit length. Among these, R and L represent basic properties of conductors, while G and C are related to the spacing material. Therefore the transmission line can be regarded as many identical four-terminal networks connected in succession, as illustrated in Figure 11.3(a), each of which is characterized at a given ω by a series impedance $\boldsymbol{Z}(\omega) = R + i\omega L$ on conductors and an admittance $\boldsymbol{Y}(\omega) = G + i\omega C$ across the space.

Assuming that R and G are negligible, the telegraph equation (11.4) can be written as

$$\frac{\partial^2(V, I)}{\partial x^2} - LC\frac{\partial^2(V, I)}{\partial t^2} = 0, \tag{11.10}$$

which is the *wave equation.* We can see that the functions (V, I) for $k'' = 0$ of (11.8) satisfy (11.9) when the wavevector $k' = \omega\sqrt{LC}$, and hence the speed of propagation is given by $v = 1/\sqrt{LC}$. For a coaxial cable, using the expressions for C and L quoted in Section 11.3, we obtain

$$v = \frac{1}{\sqrt{\varepsilon_o\mu_o}} = 2.998 \times 10^8\ \text{m-sec}^{-1}, \tag{11.11}$$

which is exactly the same as the speed of light in a vacuum.

It is noted that $k' = -\omega\sqrt{LC}$, or the functions (V, I) with the phase $-k'x - \omega t$, also satisfy the wave equation (11.9). Such a wave is characterized by the speed $-v$, expressing a wave propagating in the $-x$ direction, and the general solution of (11.10) is expressed by a linear combination of these forward and backward waves. Waves of (V, I) are also characterized by decreasing amplitudes $[V(k, \omega), I(k, \omega)]\exp(-k''x)$, where $x_d \sim 1/k''$ specifies a distance over which the amplitude diminishes by the factor $1/e$.

11.4. Reflection and Standing Waves

On two conductors of a transmission line in infinite length, the currents flow in opposite directions $\pm I(\phi)$, where a single voltage function $V(\phi)$ describes the propagation. A combination of $(V, +I)$ and $(V, -I)$ can therefore be considered for practical propagation of electromagnetic waves in other transmission lines of finite length. Figure 11.3(b) illustrates a transmission line connected with an oscillator at one end $x = 0$, and an impedance Z_L at the other end $x = l$. At an arbitrary point x and time t along such a line, we can express the voltage and current functions as linear combinations

$$V(x, t) = V_+\exp(i\phi + V_-\exp(-i\phi)$$
$$I(x, t) = \frac{V_+}{Z_o}\exp(i\phi) - \frac{V_-}{Z_o}\exp(-i\phi),$$

where the amplitudes V_+ and V_- are determined by Z_L and the length l of the transmission line.

The terminal impedance is related with $V(l, t)$ and $I(l, t)$, and written as

$$Z_L = \frac{V(l, t)}{I(l, t)} = Z_o\frac{V_+\exp(ikl) + V_-\exp(-ikl)}{V_+\exp(ikl) - V_-\exp(-ikl)}.$$

Accordingly,

$$Z_L = Z_o\frac{1+\Gamma}{1-\Gamma}, \quad \text{where} \quad \Gamma = \frac{V_-}{V_+}\exp(-2ikl) \tag{11.12a}$$

is called the *reflection coefficient*, which can also be expressed as

$$\Gamma = \frac{Z_L - Z_o}{Z_L + Z_o}. \tag{11.12b}$$

It is clear that $\Gamma = 0$, as given by $Z_L = Z_o$, signifies no reflection from Z_L.

Normally, the terminal impedance Z_L represents an unknown object at $x = l$, which can be detected from the reflected wave. For that purpose, the *standing wave* composed of incident and reflected waves is a significant measure to analyze the impedance Z_L. Denoting the distance from Z_L by z, the voltage function can be specified by z and t for convenience, that is,

$$\begin{aligned}
V(z, t) &= \{V_+\exp[ik(l-z)] + V_-\exp[-ik(l-z)]\}\exp(-i\omega t)\\
&= V_+\exp(ikl)\left[\exp(-ikz) + \Gamma\exp(ikz)\right]\exp(-i\omega t)\\
&= V_+\exp(ikl)\left[\exp(-ikz) + \frac{Z_L - Z_o}{Z_L + Z_o}\exp(ikz)\right]\exp(-i\omega t)\\
&= \frac{2V_+\exp ikl}{Z_L + Z_o}(Z_L\cos kz + iZ_o\sin kz)\exp(-i\omega t).
\end{aligned} \tag{11.13}$$

Therefore, the quantity

$$V(z, t)V^*(z, t) = \frac{4V_+^2}{(Z_L + Z_o)^2}\left(Z_L^2\cos^2 kz + Z_o^2\sin^2 kz\right) \tag{11.14}$$

is independent of time, showing $V(z)^2$ as a standing wave along the line. It is noted that the function $V(z)^2$ is just a constant, if $Z_L = Z_o$, or $\Gamma = 0$, otherwise showing a sinusoidal variation along the line with maximum and minimum occurring between two positions a quarter wavelength apart. Usually, the ratio $V(z_{max})^2/V(z_{min})^2$ can be measured with a device equipped with a sliding detector, as illustrated in Figure 11.3(c), and such a ratio is called the *standing-wave ratio* (SWR). However, using a detector sensitive to $V(z)$, the standing wave can be specified by a ratio of the root mean-square (rms) voltages, i.e., the *voltage standing wave ratio* (VSWR) given by $V_{rms}(z_{max})/V_{rms}(z_{min})$, where $V_{rms}(z) = \sqrt{V(z)^2}$. Using (11.14), we can derive expressions

$$\mathrm{SWR} = \left(\frac{Z_L}{Z_o}\right)^2 \text{ or } \left(\frac{Z_o}{Z_L}\right)^2, \text{ and } \mathrm{VSWR} = \frac{|Z_L|}{Z_o} \text{ or } \frac{Z_o}{|Z_L|}. \tag{11.15}$$

In the above argument, the complex character of an impedance, $\mathbf{Z}_L = Z_L\exp(i\psi)$, was not particularly emphasized, but in practice it is important to determine the angle ψ to determine if $\mathbf{Z}_L$ is *inductive* or *capacitive*. On the other hand, the complex reflection coefficient, $\Gamma = \Gamma\exp(i\varphi)$ can be measured with a standing wave; however the angle φ is not the same as ψ. Nevertheless, ψ is calculable from measured SWR; namely, from (11.11) we have

$$\Gamma = \frac{|Z_L| - Z_o}{|Z_L| + Z_o}\exp(i\varphi) = \frac{\mathrm{VSWR} - 1}{\mathrm{VSWR} + 1}\exp(i\varphi).$$

From a complex load $\mathbf{Z}_L$, the reflection coefficient is complex, and hence (11.12) can be re-expressed as

$$\begin{aligned} V(z,t) &= V_+\{\exp[ik(l-z)] + \Gamma\exp(i\varphi)\exp[-ik(l-z)]\} \\ &= V_+\exp i\left(kl + \frac{\varphi}{2}\right)\left[\exp\left(-ikz - \frac{\varphi}{2}\right) + \Gamma\exp\left(ikz + \frac{\varphi}{2}\right)\right]\exp(i\omega t) \\ &= \frac{2V_+\exp i\left(kl + \frac{\varphi}{2}\right)}{Z_L + Z_o}\left\{Z_L\cos\left(kz + \frac{\varphi}{2}\right) + iZ_o\sin\left(kz + \frac{\varphi}{2}\right)\right\}\exp(i\omega t). \end{aligned}$$

Therefore

$$V(z,t)V^*(z,t) = \frac{4V_o^2}{(Z_L + Z_o)^2}\left[Z_L^2\cos^2\left(kz + \frac{\varphi}{2}\right) + Z_o^2\sin^2\left(kz + \frac{\varphi}{2}\right)\right]. \tag{11.16}$$

If observed with a power-sensitive detector (*square-wave detector*), the standing wave shows power maximum and minimum in succession, where their distance is determined by the phase relation $kz + \frac{1}{2}\varphi = \frac{1}{2}\pi \times \mathrm{n}$, where n is a positive integer. If $Z_L > Z_o$, maxima and minima of SWR are determined by $kz_{max} + \frac{1}{2}\varphi = \frac{1}{2}\pi \times$ odd and even numbers, respectively, and hence the distances between adjacent maxima Δz_{max} and adjacent minima Δz_{min} are given by $\pi/2k = \lambda/4$, where λ is the wavelength defined by $k = 2\pi/\lambda$.

Figure 11.4 illustrates standing waves from such a complex impedance $|Z_L\exp(i\psi)| > Z_o$, where the minima are shifted by $\pm\varphi$. If φ is not zero, the $z = 0$ position does not give an SWR maximum, shifting to $z = \varphi/2k$. If the impedance Z_L cannot be determined by well-defined extrema, the angle φ can usually be

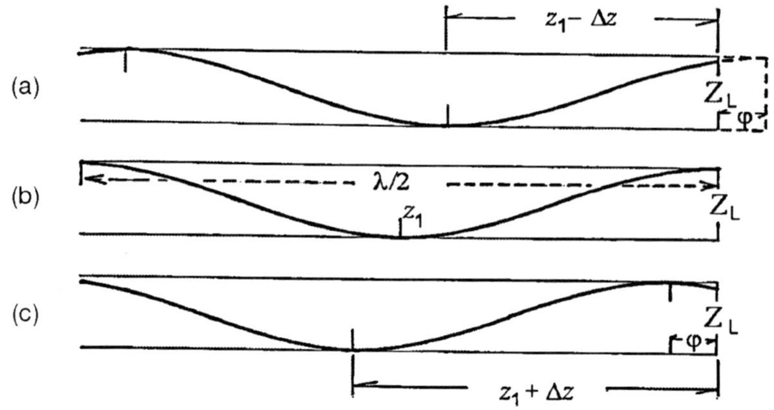

FIGURE 11.4. The terminal impedance. (a) Z_L is complex with a positive phase angle $\varphi > 0$, (b) Z_L is real, and $\varphi = 0$. (c) Z_L is complex and $\varphi < 0$.

estimated from the first minimum position of SWR, i.e., $k(z_{\min})_1 = \pi - \frac{1}{2}\varphi$. As remarked, the phase angle $\pm\psi$ of $\mathbf{Z}_L$ can be calculated from φ, whose signs indicate if the imaginary part of $\mathbf{Z}_L = R_L + iX_L$ is inductive or capacitive.

11.5. The Smith Chart

Measuring a complex impedance is a significant experimental task for which there are various methods to determine its real and imaginary parts. Although practical details are not our present concern, it is worth discussing a graphical method, known as the Smith Chart, as it is useful for many practical applications.

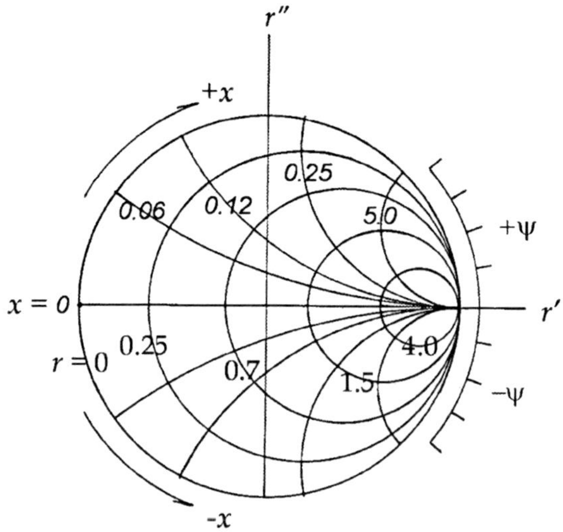

FIGURE 11.5. A Smith chart.

For convenience, we define the reduced impedance $\mathbf{Z}_L/Z_o = z' + iz''$, and write $\Gamma = r' + ir''$. From the relation $z' + iz'' = \dfrac{1 + r' + ir''}{1 - r' - ir''}$, we can obtain the relation

$$\left(r' - \frac{z'}{1+z'}\right)^2 + z''^2 = \frac{1}{(1+z')^2} \tag{i}$$

and

$$(r'-1)^2 + \left(r'' - \frac{r'}{z''}\right)^2 = \left(\frac{r'}{z''}\right)^2. \tag{ii}$$

For given values of z' and z'', (i) and (ii) are both equations of circles in the plane of r' and r''. Figure 11.5 sketches such families of circles (i) and of (ii) for $r' = 1$, showing a graphical idea of the Smith chart to evaluate the values of z' and z'' from a measured impedance. (For details, see "Transmission Line Calculator" by P. H. Smith, *Electronics*, **12**, 29 (1939)).

12 The Maxwell Equations

12.1. The Maxwell Equations

Maxwell formulated the laws of electromagnetism by hydrodynamic analogy, basing them on Faraday's concept of fields. The Maxwell equations constitute fundamental laws of physics, as recognized in Einstein's special theory of relativity. We have, so far, followed these pioneers' footsteps in arriving at Maxwell's equations, the physical implications of which are formally discussed in this chapter.

Except for what is seen in spontaneously magnetized materials, electric charges at rest and in motion are regarded as the source of electromagnetic fields in matter, including vacuum space, and dynamic electric fields are derived from the time variation of magnetic fields. On the other hand, the absence of magnetic charges is accepted as a natural law for the electromagnetic field.

Acting on electric charges and magnetic moments, electric and magnetic field vectors $\boldsymbol{E}$ and $\boldsymbol{H}$ are defined as force and torque, respectively, which are tangential to field-lines, whereas their densities of distribution are signified by another set of vectors, $\boldsymbol{D}$ and $\boldsymbol{B}$. Although considered as theoretically the same in vacuum space, the vectors $\boldsymbol{D}$ and $\boldsymbol{E}$ are not necessarily identical in dielectric media, and neither are the vectors $\boldsymbol{B}$ and $\boldsymbol{H}$ in magnetized materials. And, these field vectors are regarded as originating from charges and currents that are expressed by their densities ρ and $\boldsymbol{j}$.

The Maxwell theory consists of four differential equations:

(I) $$\boldsymbol{j} + \frac{\partial \boldsymbol{D}}{\partial t} = \operatorname{curl} \boldsymbol{H}, \tag{10.2}$$

(II) $$-\frac{\partial \boldsymbol{B}}{\partial t} = \operatorname{curl} \boldsymbol{E}, \tag{7.5}$$

(III) $$\operatorname{div} \boldsymbol{D} = \rho, \tag{4.2c}$$

(IV) $$\operatorname{div} \boldsymbol{B} = 0. \tag{8.5}$$

For the field vectors in a vacuum, we have additional relations:

(V.1) $$\boldsymbol{D} = \varepsilon_o \boldsymbol{E} \quad \text{and} \quad \boldsymbol{B} = \mu_o \boldsymbol{H}; \qquad (2.7), (7.3)$$

whereas, in dielectric and magnetic media, these are written as

(V.2) $$\boldsymbol{D} = \varepsilon_o \boldsymbol{E} + \boldsymbol{P} \quad \text{and} \quad \boldsymbol{B} = \mu_o \boldsymbol{H} + \boldsymbol{M}, \qquad (2.12), (8.23)$$

where

(V.2) $$\boldsymbol{P} = \chi \varepsilon_o \boldsymbol{E} \quad \text{and} \quad \boldsymbol{M} = \chi_m \mu_o \boldsymbol{H}, \qquad (2.13), (8.22)$$

representing electric and magnetic properties of matter, respectively. Combining these, (V.2) can be expressed as

(V.3) $$\boldsymbol{D} = \varepsilon \boldsymbol{E}, \varepsilon = (1 + \chi)\varepsilon_o \quad \text{and} \quad \boldsymbol{B} = \mu \boldsymbol{H}, \mu = (1 + \chi_m)\mu_o. \qquad (2.14a), (8.23)$$

On the other hand, conducting materials are characterized by the electrical conductivity σ, for which the Ohm law

(V.4) $$\boldsymbol{j} = \sigma \boldsymbol{E},$$

describes electric properties of normal conductors.

In the Maxwell equations, all field vectors are functions of the position $\boldsymbol{r}$ and time t in the electromagnetic field. In contrast to the densities ρ and $\boldsymbol{j}$ that are normally restricted in limited regions, the field vectors are functions in much broader region of space. Some regions in the field can be free from ρ and $\boldsymbol{j}$, implying that the electromagnetic field can be extended to a distant point in free space. Hertz (1888) first demonstrated that an electromagnetic signal identified by its frequency travels a distance through an empty space from one location to another, demonstrating the presence of a field in a free space.

Figure 12.1(a) sketches Hertz's experiment, wherein a spark gap attached to an induction coil generates an intense electromagnetic field, which then propagates to a "resonance ring" with a spark gap located at a distant point. The ring was

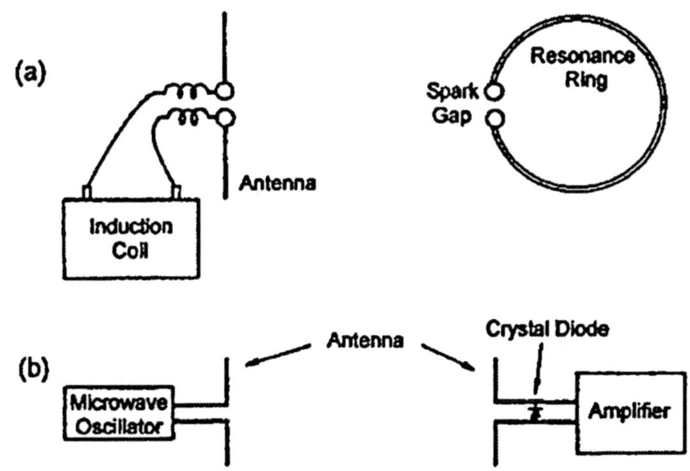

FIGURE 12.1. (a) Hertz's experiment. (b) A modern version of Hertz's experiment for propagating electromagnetic waves.

designed for its circumference to match the wavelength, where the detected signal is evident from a spark in the gap.

Such an apparatus was primitive by today's standards, but sufficiently demonstrated a propagating field through an empty space.

The empty space can be characterized by $\sigma = 0$ and $\boldsymbol{j} = 0$, in which the Maxwell equations can be written as

$$\mathrm{curl}\,\boldsymbol{H} = \varepsilon_o \frac{\partial \boldsymbol{E}}{\partial t}, \tag{i}$$

$$\mathrm{curl}\,\boldsymbol{E} = -\mu_o \frac{\partial \boldsymbol{H}}{\partial t}, \tag{ii}$$

$$\mathrm{div}\,\boldsymbol{E} = 0 \tag{iii}$$

and

$$\mathrm{div}\,\boldsymbol{H} = 0. \tag{iv}$$

Operating "curl" on both sides of (i) and (ii), the operator "curlcurl" can be re-expressed for any arbitrary vector $\boldsymbol{F}$ by using the identity formula as curl curl $\boldsymbol{F} \equiv$ grad div$\boldsymbol{F} - \nabla^2 \boldsymbol{F}$. Thus, from (i) we obtain

$$\mathrm{grad}\,\mathrm{div}\boldsymbol{H} - \nabla^2 \boldsymbol{H} = \varepsilon_o \mathrm{curl} \frac{\partial \boldsymbol{E}}{\partial t} = \frac{\partial}{\partial t} \mathrm{curl}\,\boldsymbol{E}.$$

Here, from (iv) and (ii) div $\boldsymbol{H} = 0$ and curl $\boldsymbol{E} = -\mu_o \dfrac{\partial \boldsymbol{H}}{\partial t}$, and therefore,

$$\nabla^2 \boldsymbol{H} - \varepsilon_o \mu_o \frac{\partial^2 H}{\partial t^2} = 0. \tag{12.1a}$$

Similarly from (ii),

$$\nabla^2 \boldsymbol{E} - \varepsilon_o \mu_o \frac{\partial^2 \boldsymbol{E}}{\partial t^2} = 0, \tag{12.1b}$$

where

$$\frac{1}{\sqrt{\varepsilon_o \mu_o}} = c \tag{12.1c}$$

is found to be equal to the empirical value of the speed of light in a vacuum. Equations (12.1a) and (12.1b) are *wave equations* in free space.

The field vectors $\boldsymbol{E}$ and $\boldsymbol{H}$ determined by solutions of these equations are characterized as proportional to $\exp i(\boldsymbol{k} \cdot \boldsymbol{r} - \omega t)$, representing propagation along the direction of the wave vector $\boldsymbol{k}$ at the speed c. It is a well-established empirical fact that the electromagnetic field propagates at this speed c, identical to the speed of light within experimental accuracy, providing solid evidence that optical rays are electromagnetic waves at shorter wavelengths.

All components of $\boldsymbol{E}$ and $\boldsymbol{H}$ in a free space obey the wave equation, indicating that in a given direction their space-time variations are basically sinusoidal. Factorizing the components into individual functions of space-time, these factors can all be expressed as sinusoidal. Accordingly, it is logical to consider a space-time variation as described by the phase $\boldsymbol{k} \, . \, \boldsymbol{r} - \omega t$. Therefore, assuming $(\boldsymbol{E}, \boldsymbol{H}) \propto \exp i(\boldsymbol{k} \, . \, \boldsymbol{r} - \omega t)$ in (12.1ab), we obtain

$$c^2 \boldsymbol{k}^2 = \omega^2 \quad \text{and hence} \quad \pm |\boldsymbol{k}| = \frac{\omega}{c},$$

for the vector $\boldsymbol{k}$, indicating that the propagation occurs in opposite directions, $+\boldsymbol{k}$ and $-\boldsymbol{k}$.

12.2. Electromagnetic Energy and the Poynting Theorem

The electromagnetic field should have an energy, as described by the volume densities $\frac{1}{2}\boldsymbol{D}.\boldsymbol{E}$ and $\frac{1}{2}\boldsymbol{B}.\boldsymbol{H}$ in, respectively, the static and stationary fields. To obtain the expression for energies in a time-dependent field, we calculate $\boldsymbol{E}.\dfrac{\partial \boldsymbol{D}}{\partial t} + \boldsymbol{H}.\dfrac{\partial \boldsymbol{B}}{\partial t}$ from (I) and (II); that is,

$$\boldsymbol{E}.\frac{\partial \boldsymbol{D}}{\partial t} + \boldsymbol{H}.\frac{\partial \boldsymbol{B}}{\partial t} = \boldsymbol{E}.\text{curl}\,\boldsymbol{H} - \boldsymbol{H}.\text{curl}\,\boldsymbol{E} - \boldsymbol{j}.\boldsymbol{E},$$

which can be written as

$$\frac{\partial}{\partial t}\left(\frac{1}{2}\boldsymbol{D}.\boldsymbol{E} + \frac{1}{2}\boldsymbol{B}.\boldsymbol{H}\right) = -\,\text{div}\,(\boldsymbol{E} \times \boldsymbol{H}) - \boldsymbol{j}.\boldsymbol{E}.$$

Considering a volume v enclosed in a surface $\boldsymbol{S} = S\boldsymbol{n}$, where $\boldsymbol{n}$ is an outward normal, we can define the energy

$$U = \int_v u \, \mathrm{d}v, \quad \text{where} \quad u = \frac{1}{2}(\boldsymbol{D}.\boldsymbol{E} + \boldsymbol{B}.\boldsymbol{H}) \tag{12.2}$$

is the energy density, and hence obtain the expression

$$\frac{\partial U}{\partial t} = -\int_v \text{div}\,(\boldsymbol{E} \times \boldsymbol{H})\,\mathrm{d}v - \int_v \boldsymbol{j}.\boldsymbol{E}\mathrm{d}v = -\oint_S (\boldsymbol{E} \times \boldsymbol{H}).\mathrm{d}\boldsymbol{S} - \int_v \boldsymbol{j}.\boldsymbol{E}\,\mathrm{d}v, \tag{12.3}$$

in which the Gauss theorem was applied to convert the first volume integral in the second expression to the surface integral in the last one, whereas the $\boldsymbol{j}.\boldsymbol{E}$ term represents the heat dissipation in the volume.

Writing

$$\boldsymbol{P} = \boldsymbol{E} \times \boldsymbol{H} \tag{12.4}$$

which is called the *Poynting* vector, the integral $\oint_S \boldsymbol{P}.\mathrm{d}\boldsymbol{S}$ represents an electromagnetic energy flowing out of the volume v per unit time, expressing electromagnetic *radiation*. With definition (12.4), equation (12.3) can be expressed as

$$-\frac{\partial U}{\partial t} = \oint_S \boldsymbol{P}.\mathrm{d}\boldsymbol{S}, \tag{12.5}$$

which is known as the *Poynting theorem*.

12.3. Vector and Scalar Potentials

The static electric field arising from an electric charge can be described by a potential function V that satisfies the Laplace-Poisson equation $\nabla^2 V = -\frac{\rho}{\varepsilon_o}$, and the stationary magnetic field is determined by a vector potential $\boldsymbol{A}$ that is given by $\nabla^2 \boldsymbol{A} = -\mu_o \boldsymbol{j}$. A time-dependent electromagnetic field due to ρ and $\boldsymbol{j}$ changing with time should, therefore, be represented by scalar and vector potentials V and $\boldsymbol{A}$ as combined.

Equation (IV), signifying the absence of magnetic charge, indicates that

$$\boldsymbol{B} = \text{curl}\,\boldsymbol{A}. \tag{8.9}$$

Hence, from (II) we have

$$\text{curl}\,\boldsymbol{E} = -\text{curl}\left(\frac{\partial \boldsymbol{A}}{\partial t}\right) \quad \text{and} \quad \text{curl}\left(\boldsymbol{E} + \frac{\partial \boldsymbol{A}}{\partial t}\right) = 0.$$

The last equation indicates that the vector $\boldsymbol{E} + \frac{\partial \boldsymbol{A}}{\partial t}$ is given by a gradient of a scalar function V, that is,

$$\boldsymbol{E} = -\text{grad}\ V - \frac{\partial \boldsymbol{A}}{\partial t}. \tag{12.6}$$

Using (12.6) and (8.9) in (III) and (I), we obtain

$$\text{div}(\varepsilon_o \boldsymbol{E}) = -\varepsilon_o \text{div}\left(\text{grad}\ V + \frac{\partial \boldsymbol{A}}{\partial t}\right) = \rho, \tag{i}$$

and

$$\text{curl}\left(\frac{\text{curl}\,\boldsymbol{A}}{\mu_o}\right) = \boldsymbol{j} + \varepsilon_o \frac{\partial}{\partial t}\left(-\text{grad}\ V - \frac{\partial \boldsymbol{A}}{\partial t}\right). \tag{ii}$$

Here, in (ii), curl(curl $\boldsymbol{A}$) $\equiv$ grad(div $\boldsymbol{A}$) $-$ $\nabla^2 \boldsymbol{A}$, and, therefore,

$$\nabla^2 \boldsymbol{A} - \varepsilon_o \mu_o \frac{\partial^2 \boldsymbol{A}}{\partial t^2} = -\mu_o \boldsymbol{j} + \text{grad}\left(\varepsilon_o \mu_o \frac{\partial V}{\partial t} + \text{div}\,\boldsymbol{A}\right). \tag{12.7a}$$

As discussed in Section 8.3, the vector potential $\boldsymbol{A}$ cannot be uniquely defined by (8.9), as any other vector

$$\boldsymbol{A}' = \boldsymbol{A} + \text{grad}\,\chi \tag{12.7b}$$

gives $\boldsymbol{B} = \text{curl}\,\boldsymbol{A}'$ if $\nabla^2\chi = 0$. Also, we note that if we utilize this ambiguity, the equation (12.6) is unchanged with such a redefinition of the scalar potential as

$$V' = V - \frac{\partial \chi}{\partial t}, \tag{12.7c}$$

as verified by $\boldsymbol{E} = -\text{grad}\,V - \frac{\partial}{\partial t}\left(\boldsymbol{A}' - \text{grad}\,\chi\right) = -\text{grad}\,V' - \frac{\partial \boldsymbol{A}'}{\partial t}$. Therefore, it is always possible to choose a set of $(V', \boldsymbol{A}')$ with a suitable scalar function χ atisfying $\nabla^2\chi = 0$, such that

$$\varepsilon_o\mu_o\frac{\partial V}{\partial t} + \text{div}\,\boldsymbol{A} = 0. \tag{12.7d}$$

Such a change of $(V, \boldsymbol{A})$ with the scalar χ stablishes the relation between V and $\boldsymbol{A}$, and (12.7b) and (12.7c) are referred to as the *gauge transformation*. The equation (12.7d) unchanged with a gauge transformation is known as the *Lorentz condition*.

Using (12.7d) in (12.7a), we arrive at an inhomogeneous differential equation

$$\nabla^2\boldsymbol{A} - \varepsilon_o\mu_o\frac{\partial^2\boldsymbol{A}}{\partial t^2} = -\mu_o\boldsymbol{j}. \tag{12.8a}$$

Equation (i) can also modified by (12.5) and (12.7d) as a form similar to (12.8a), i.e.,

$$\nabla^2 V - \varepsilon_o\mu_o\frac{\partial^2 V}{\partial t^2} = -\frac{\rho}{\varepsilon_o}. \tag{12.8b}$$

The equations (12.8ab) are the basic equations to be solved for electromagnetic potentials, when a charge-current distribution is specified. Obviously, at space- and time coordinates $(\boldsymbol{r}, t)$ far away from ρ and $\boldsymbol{j}$, equations (12.8a) and (b) become *homogeneous wave equations* similar to (12.1a) and (b) for $\boldsymbol{E}$ and $\boldsymbol{H}$ in free space.

12.4. Retarded Potentials

In general, we look for the potentials $(V, \boldsymbol{A})$ at a given space-time $(\boldsymbol{r}, t)$, when the sources $(\rho, \boldsymbol{j})$ are specified as functions of space-time $(\boldsymbol{r}', t')$. It is significant that an event at a point $\boldsymbol{r}$ is not detectable at a distant point $\boldsymbol{r}'$ at the same time as it occurs at t'. In other words, it takes certain time for the information to arrive at a point of observation.

We consider that a point charge ρ at $\boldsymbol{r}'$ appears during a short time interval between t' and $t' + \delta t$, and that at a point $\boldsymbol{r}$ no information about $\rho(\boldsymbol{r}',t')$ is available until a later time. In this case, the observed charge at a point $\boldsymbol{r}$ at t can be expressed as $\rho(\boldsymbol{r}')\delta(\boldsymbol{r} - \boldsymbol{r}')\delta(t - t')$, where the δ are delta functions. Here $\boldsymbol{r} - \boldsymbol{r}'$ is the distance, whereas $t - t' = \tau$ represents a period of time after ρ starts to vary at $\boldsymbol{r}$; that is, τ is a positive time for $\rho(\boldsymbol{r}',t')$ to be detected at $\boldsymbol{r}$. We define a function $G(\boldsymbol{r} - \boldsymbol{r}', t - t')$ to satisfy the equation

$$\left(\nabla^2 - \frac{1}{c^2}\frac{\partial^2}{\partial\tau^2}\right)G(\boldsymbol{r} - \boldsymbol{r}', t - t') = -\delta(\boldsymbol{r} - \boldsymbol{r}')\delta(t - t'). \tag{12.9}$$

If such a function $G(\boldsymbol{r} - \boldsymbol{r}', \tau)$, namely the Green function, can be obtained, the potential $V(\boldsymbol{r})$ can be calculated as

$$V(\boldsymbol{r}, t) = -\frac{1}{\varepsilon_0} \iint G(\boldsymbol{r} - \boldsymbol{r}', t - t')\rho(\boldsymbol{r}', t')\, \mathrm{d}^3\boldsymbol{r}'\mathrm{d}t'. \tag{12.10}$$

For convenience, if instead of $\boldsymbol{r} - \boldsymbol{r}'$ and $t - t'$, we write $\boldsymbol{r}$ and τ, respectively, the solution of the inhomogeneous equation (12.9) can be expressed by a Fourier expansion

$$G(\boldsymbol{r}, \tau) = \frac{1}{(2\pi)^4} \int_{\boldsymbol{k}\text{-space}} \mathrm{d}^3\boldsymbol{k} \int_{-\infty}^{+\infty} \mathrm{d}\omega \Gamma(\boldsymbol{k}, \omega) \exp i(\boldsymbol{k}.\boldsymbol{r} - \omega\tau)$$

and hence

$$\delta(\boldsymbol{r})\delta(\tau) = \frac{1}{(2\pi)^4} \int_{k\text{-space}} \mathrm{d}^3k \int_{-\infty}^{+\infty} \mathrm{d}\omega \exp i(\boldsymbol{k}.\boldsymbol{r} - \omega\tau).$$

Therefore, the Fourier transform $\Gamma(\boldsymbol{k}, \omega)$ can be written as

$$\Gamma(\boldsymbol{k}, \tau) = \frac{1}{\left(\frac{\omega}{c}\right)^2 - |\boldsymbol{k}|^2},$$

and hence

$$G(\boldsymbol{r}, \tau) = \frac{1}{(2\pi)^4} \int_0^\infty \boldsymbol{k}^2\, \mathrm{d}\boldsymbol{k} \int_0^\pi \exp(i\boldsymbol{k}\cdot\boldsymbol{r}) \sin\theta\, \mathrm{d}\theta \int_0^{2\pi} \mathrm{d}\varphi \int_{-\infty}^{+\infty} \frac{\exp(-i\omega\tau)}{\frac{\omega^2}{c^2} - \boldsymbol{k}^2}\, \mathrm{d}\omega,$$

where the volume element, i.e., $\mathrm{d}^3\boldsymbol{k} = k^2\mathrm{d}\boldsymbol{k} \sin\theta \mathrm{d}\theta \mathrm{d}\varphi$, is expressed in polar coordinates k, θ, and φ with respect to a fixed reference system in the $\boldsymbol{k}$ space.

Here, in the last integral, the integrand has two poles, at $\omega = \pm ck$ on the real axis in the Reω–Imω plane, assuming ω to be a complex variable. Modifying the last integral as shown below, we can apply the Cauchy theorem to integrate it; that is,

$$\int_{-\infty}^{+\infty} \frac{\exp(-i\omega\tau)}{\left(\frac{\omega}{c}\right)^2 - k^2} \mathrm{d}\omega = \frac{1}{2k} \lim_{|\omega|\to\infty} \oint_C \left(\frac{1}{\omega - ck} - \frac{1}{\omega + ck}\right) \exp(-i\omega\tau)\, \mathrm{d}\omega$$

$$= 2\pi i \frac{1}{2k} \{\exp(-ick\tau) - \exp(ick\tau)\} = -\frac{2\pi}{k} \sin(ck\tau) \quad \text{for} \quad \tau > 0,$$

where C is chosen as a closed path with a large radius $|\omega|$ to avoid these poles.

The integral factor with respect to φ is equal to $\int_0^{2\pi} d\varphi = 2\pi$, and the integral with θ can be calculated as

$$\int_{-1}^{+1} \exp(ikr\cos\theta)\,d(\cos\theta) = \frac{2\sin(kr)}{kr}, \quad \text{where} \quad r = |\boldsymbol{r}|.$$

Hence, we have

$$G(\boldsymbol{r}, \tau) = -\frac{1}{2\pi^2 r}\int_0^{\infty} \sin(kr)\sin(ck\tau)\,dk, \tag{12.11a}$$

where the integration should be carried out for positive k over the range $0 \le k < \infty$, and the integrand can be written as

$$\sin(kr)\sin(ck\tau)\,dk = \tfrac{1}{4}\{\exp ik(r - c\tau) - \exp ik(r + c\tau)\},$$

ignoring the complex exponentials for $-k$, because a negative k implies an advanced arrival of information at r, which is physically not acceptable. However, excluding the negative region $-\infty < k \le 0$ from the above calculation is mathematically incorrect, and hence we re-write (12.10a) for $-\infty < k < +\infty$ as

$$G(\boldsymbol{r}, \tau) = -\frac{1}{8\pi^2 r}\int_{-\infty}^{+\infty} \exp ik(r - c\tau)\,dk, \tag{12.11b}$$

after omitting physically incorrect exponentials. Noted that the integral in (12.10b) can be related to the delta-function as

$$\delta(r - c\tau) = \frac{1}{2\pi}\int_{-\infty}^{+\infty} \exp ik(r - c\tau)\,dk, \tag{12.12}$$

we obtain

$$G(\boldsymbol{r}, \tau) = -\frac{1}{4\pi r}\delta(r - c\tau), \tag{12.13a}$$

signifying a peak at a later time $\tau = r/c$. Returning to original space-time coordinates, the Green function can be expressed as

$$G(\boldsymbol{r} - \boldsymbol{r}', t - t') = -\frac{1}{4\pi|\boldsymbol{r} - \boldsymbol{r}'|}\delta\{\boldsymbol{r} - \boldsymbol{r}' - c(t - t')\}. \tag{12.13b}$$

Using (12.13b) in (12.10), the potential function can be calculated as

$$V(\boldsymbol{r}, t) = \frac{1}{4\pi\varepsilon_o}\int_{v(r')} \frac{\rho(\boldsymbol{r}', t')\,\delta(\boldsymbol{r} - \boldsymbol{r}')}{|\boldsymbol{r} - \boldsymbol{r}'|}d^3\boldsymbol{r}' \int_{-\infty}^{t - |\boldsymbol{r} - \boldsymbol{r}'|/c} \delta(t - t')\,dt',$$

in which

$$\int_{-\infty}^{t-|\boldsymbol{r}-\boldsymbol{r}'|/c} \rho\left(\boldsymbol{r}', t'\right)\delta(t-t')\,\mathrm{d}t' = \rho\left(\boldsymbol{r}', t - \frac{|\boldsymbol{r}-\boldsymbol{r}'|}{c}\right), \tag{12.14}$$

expressing the charge density at $\boldsymbol{r}'$ at a *retarded time* $t - |\boldsymbol{r} - \boldsymbol{r}'|/c$. Thus, a solution of the inhomogeneous equation (12.7b) at $(\boldsymbol{r}, t)$ is determined by the charge density (12.13) at a retarded time. Therefore, we can express the scalar potential as

$$V(\boldsymbol{r}, t) = \frac{1}{4\pi\varepsilon_o}\int_{v(\boldsymbol{r}')} \frac{\rho\left(\boldsymbol{r}', t - \dfrac{|\boldsymbol{r}-\boldsymbol{r}'|}{c}\right)}{|\boldsymbol{r}-\boldsymbol{r}'|}\delta(\boldsymbol{r}-\boldsymbol{r}')\,\mathrm{d}^3\boldsymbol{r}'. \tag{12.15a}$$

For a given current density $\boldsymbol{j}(\boldsymbol{r}', t')$, the inhomogeneous equation (12.8a) can also be solved for the vector potential $\boldsymbol{A}(\boldsymbol{r}, t)$ in a similar manner, resulting in a vector potential at a retarded time, i.e.,

$$\boldsymbol{A}(\boldsymbol{r}, t) = \frac{\mu_o}{4\pi}\int_{v(\boldsymbol{r}')} \frac{\boldsymbol{j}\left(\boldsymbol{r}', t - \dfrac{|\boldsymbol{r}-\boldsymbol{r}'|}{c}\right)}{|\boldsymbol{r}-\boldsymbol{r}'|}\delta(\boldsymbol{r}-\boldsymbol{r}')\,\mathrm{d}^3\boldsymbol{r}'. \tag{12.15b}$$

These potentials (12.15a) and (b) for a retarded time are called the *retarded potentials*.

12.5. Multipole Expansion

In practical applications, charges and currents are confined to a finite region of space and observed at a distant point away from their distribution. Under such a circumstance, it is significant that the electromagnetic field can approximately be described by a distance r from the center of distribution to the observing point at a retarded time $t - r/c$. In addition, the electromagnetic potentials $(\boldsymbol{A}, V)$ in (12.14a) and (b) are involved in the factor $1/|\boldsymbol{r} - \boldsymbol{r}'|$, which needs to expanded into a Taylor series, considering $r \gg r'$. Here, the retarded time is considered sufficiently accurate to describe the field at an approximate distance $|\boldsymbol{r} - \boldsymbol{r}'| \sim r$, and we write these source quantities as $\rho(\boldsymbol{r}', \tau_R)$ and $\boldsymbol{j}(\boldsymbol{r}', \tau_R)$, where $\tau_R = t - r/c$.

We apply the Taylor expansion to such a scalar function as $F(\boldsymbol{r}) = \dfrac{j_x\left(\boldsymbol{r}', \tau_R\right)}{|\boldsymbol{r}-\boldsymbol{r}'|}$, and write

$$F(\boldsymbol{r}-\boldsymbol{r}') = F(\boldsymbol{r}) - \boldsymbol{r}'\cdot\mathrm{grad}_{\boldsymbol{r}} F(\boldsymbol{r}) + \cdots,$$

in which the terms higher than second order are not particularly necessary because they are seldom used in practical applications. Components of the vector potential in (12.15a) can then be expanded in the form

$$A_x(\boldsymbol{r},t) = \frac{\mu_o}{4\pi} \int\limits_{v(\boldsymbol{r}')} \left(1 - \boldsymbol{r}' \cdot \mathrm{grad}_r + \cdots\right) \frac{j_x\left(\boldsymbol{r}', \tau_R\right)}{r} \mathrm{d}^3\boldsymbol{r}'. \tag{12.16}$$

The vector potential of components given by first term can be expressed as

$$\boldsymbol{A}_1(\boldsymbol{r},t) = \frac{\mu_o}{4\pi r} \int\limits_{v(\boldsymbol{r}')} \boldsymbol{j}\left(\boldsymbol{r}', \tau_R\right) \mathrm{d}^3\boldsymbol{r}' = \frac{\mu_o}{4\pi r} \int\limits_{r'} \mathrm{d}\boldsymbol{r}' \cdot \oint\limits_{S'} \boldsymbol{j}\left(\boldsymbol{r}', \tau_R\right) \mathrm{d}\boldsymbol{S}',$$

where $\mathrm{d}^3\boldsymbol{r}' = \mathrm{d}\boldsymbol{S}' \cdot \mathrm{d}\boldsymbol{r}'$, representing a volume element between thin layers of surfaces $\mathrm{d}\boldsymbol{S}'$. Using the Gauss theorem, the surface integral can be replaced by $-\int_{v(\boldsymbol{r}')} \mathrm{div}\,\boldsymbol{j}\left(\boldsymbol{r}', \tau_R\right) \mathrm{d}^3\boldsymbol{r}'$, where the inward direction of $\boldsymbol{j}$ to S' is considered positive, which is further modified by the charge-current continuity relation as div $\boldsymbol{j}(\boldsymbol{r}', \tau_R) = -\dfrac{\partial\rho\left(\boldsymbol{r}', \tau_R\right)}{\partial\tau_R}$. Thus, we obtain the expression

$$\boldsymbol{A}_1(\boldsymbol{r},t) = \frac{\mu_o}{4\pi r} \frac{\partial \boldsymbol{p}(\tau_R)}{\partial t}, \quad \text{where} \quad \boldsymbol{p}\left(\tau_R\right) = \int\limits_{v(r')} \rho\left(\boldsymbol{r}', \tau_R\right) \boldsymbol{r}' \mathrm{d}^3\boldsymbol{r}' \tag{12.17}$$

represents the electric dipole moment of the source.

The second term in expansion (12.16) is

$$A_{2x}(\boldsymbol{r},t) = -\frac{\mu_o}{4\pi} \int\limits_{v(\boldsymbol{r}')} \boldsymbol{r}' \cdot \mathrm{grad}_r \frac{j_x\left(\boldsymbol{r}', \tau_R\right)}{\boldsymbol{r}} \mathrm{d}^3\boldsymbol{r}'$$

$$= -\frac{\mu_o}{4\pi}\left[\frac{\partial}{\partial x}\left\{\frac{1}{r} \int\limits_{v(\boldsymbol{r}')} x' j_x\left(\boldsymbol{r}', \tau_R\right) \mathrm{d}^3\boldsymbol{r}'\right\} + \frac{\partial}{\partial y}\left\{\frac{1}{r} \int\limits_{v(\boldsymbol{r}')} y' j_x\left(\boldsymbol{r}', \tau_R\right) \mathrm{d}^3\boldsymbol{r}'\right\}\right.$$

$$\left. + \frac{\partial}{\partial z}\left\{\frac{1}{r} \int\limits_{v(\boldsymbol{r}')} z' j_x\left(\boldsymbol{r}', \tau_R\right) \mathrm{d}^3\boldsymbol{r}'\right\}\right].$$

To make the expression short, we write this with components of $\boldsymbol{r} = (x_i)$ and $\boldsymbol{r}' = (x'_i)$, where $i = x$, y, and z, and obtain

$$A_{2x}(\boldsymbol{r},t) = -\frac{\mu_o}{4\pi} \sum_j \frac{\partial}{\partial x_j}\left\{\frac{1}{r} \int\limits_{v(\boldsymbol{r}')} x'_j j_i\left(\boldsymbol{r}', \tau_R\right) \mathrm{d}^3\boldsymbol{r}'\right\}, \quad \text{where} \quad i = x.$$

Here, the integrand can be modified as

$$x'_i j_i = 1(x'_i j_i + j_j x'_i) + 1(x'_i j_i - j_j x'_i),$$

where the first symmetric part can be further re-written using the relations

$$\nabla'.(\boldsymbol{j} x'_i x'_j) = \boldsymbol{j}.\nabla'(x'_i x'_j) + (x'_i x'_j)\nabla'\boldsymbol{j} \quad \text{and} \quad \boldsymbol{j}.\nabla'(x'_i x'_j) = 1(x'_j j_i + j_j x'_i).$$

Using the Gauss theorem,

$$\int_{v(\boldsymbol{r}')} \nabla' . \left(\boldsymbol{j} x_i' x_j' \right) \mathrm{d}^3 \boldsymbol{r}' = \oint_S x_i' x_j' \boldsymbol{j} . \mathrm{d}\boldsymbol{S} = 0,$$

and the continuity relation,

$$\nabla' \boldsymbol{j}(\boldsymbol{r}', \tau_R) = -\frac{\partial \rho \left(\boldsymbol{r}', \tau_R \right)}{\partial t},$$

we arrive at the expression

$$A_{2i}(\boldsymbol{r}, t) = -\frac{\mu_o}{8\pi} \sum_j \frac{\partial}{\partial x_j} \frac{1}{r} \left[\frac{\partial}{\partial t} \int_{v(\boldsymbol{r}')} x_i' x_j' \rho \left(\boldsymbol{r}', \tau_R \right) \mathrm{d}^3 \boldsymbol{r}' + \int_{v(\boldsymbol{r}')} \left(x_j' j_i - x_i' j_j \right) \mathrm{d}^3 \boldsymbol{r}' \right]. \tag{12.18}$$

Defining the quadrupole tensor as

$$Q_{ij} \left(\tau_R \right) = \int_{r'} 3 x_i' x_j' \rho \left(\boldsymbol{r}', \tau_R \right) \mathrm{d}^3 \boldsymbol{r}', \tag{12.19}$$

the first term of (12.17), representing the quadrupolar contribution to the vector potential, can be expressed as

$$[\boldsymbol{A}_2(\boldsymbol{r}, t)]_{\text{quad}} = \sum_i \boldsymbol{e}_i A_{2i} \left(\boldsymbol{r}', t \right) = -\frac{\mu_o}{8\pi} \frac{\partial}{\partial t} \sum_{ij} \boldsymbol{e}_i \frac{\partial}{\partial x_j} \frac{Q_{ij} \left(\tau_R \right)}{3r}.$$

On the other hand, the second antisymmetric term in (12.17) is written as

$$[\boldsymbol{A}_2(\boldsymbol{r}, t)]_{\text{mag}} = -\frac{\mu_o}{8\pi} \sum_j \frac{\partial}{\partial x_j} \int_{v(\boldsymbol{r}')} \frac{x_j' \boldsymbol{j} - j_j \boldsymbol{r}'}{r} \mathrm{d}^3 \boldsymbol{r}' = \frac{\mu_o}{4\pi} \boldsymbol{\nabla} \times \frac{\boldsymbol{p}_m(\tau_R)}{r},$$

after a similar manipulation, where the vector $\boldsymbol{p}_m$ is defined as

$$\boldsymbol{p}_m(\tau_R) = \frac{1}{2} \int_{v(\boldsymbol{r}')} \boldsymbol{r}' \times \boldsymbol{j}(\tau_R) \mathrm{d}^3 \boldsymbol{r}', \tag{12.20}$$

which is the magnetic moment of the source current. Thus, the electric quadrupole and magnetic moment represents the charge-current source for the vector potential in the same order of expansion.

The scalar potential $V(\boldsymbol{r}, t)$ can be expanded in a similar way. Namely,

$$V(\boldsymbol{r}, t) = \frac{1}{4\pi\varepsilon_o} \int_{v(r')} \left(1 - \boldsymbol{r}' . \text{grad}_r + \cdots \right) \frac{\rho \left(\boldsymbol{r}', \tau_R \right)}{r} \mathrm{d}^3 \boldsymbol{r}',$$

where the first and second terms represent the charge-current source. The first term can be written as

$$V_o(\boldsymbol{r}, t) = \frac{1}{4\pi\varepsilon_o}\frac{q(\tau_R)}{r}, \quad \text{where} \quad q(\tau_R) = \int\limits_{v(r')} \rho(\boldsymbol{r}', \tau_R)\, d^3\boldsymbol{r}' \tag{12.21}$$

is the total charge in the source.

The second term is

$$V_1(\boldsymbol{r}, t) = -\frac{1}{4\pi\varepsilon_o}\text{grad}_{\boldsymbol{r}} \int\limits_{v(\boldsymbol{r}')} \frac{\boldsymbol{r}'\rho(\boldsymbol{r}', \tau_R)}{r} d^3\boldsymbol{r}' = -\frac{1}{4\pi\varepsilon_o}\nabla_r \frac{\boldsymbol{p}(\tau_R)}{r},$$

where

$$\boldsymbol{p}(\tau_R) = \int\limits_{v(\boldsymbol{r}')} \boldsymbol{r}'\rho(\boldsymbol{r}', \tau_R)\, d^3\boldsymbol{r}' \tag{12.22}$$

is the electric dipole moment of the source.

It is noted that the dipole moment, magnetic moment, and quadrupole tensor are all time-dependent, being involved in the time-dependent vector $\boldsymbol{r}'$ or its components, whereas the total charge $q(\tau_R)$ is constant of time. The dipole moment contributes to both scalar and vector potentials, playing a dominant role for electromagnetic radiation, as discussed in Chapter 13, whereas the total charge is only significant for the Coulomb potential.

13 Electromagnetic Radiation

13.1. Dipole Antenna

Hertz demonstrated that electromagnetic radiation could be emitted from a spark gap in an induction current, thereby giving evidence supporting the Maxwell theory. Then, Marconi (1895) invented wireless communication by means of electromagnetic waves at radio frequencies and opened the modern era of telecommunication technology. Using an *antenna*, electromagnetic waves can be transmitted through open space, and detected by another antenna at a distant location. Although not clearly established in Hertz's experiment, the spark gap played the role of an antenna for emitting a high-frequency radiation. Further, he demonstrated a persistent oscillation is required for continuous radiation from the antenna.

Figures 13.1(a) and (b) show schematically an antenna attached at the end of a transmission line, where oscillating voltage and current are characterized by the wavelength λ, which related to the frequency ν, as $\lambda = c/\nu$. The figure shows that a sizable standing wave can be set on the antenna if its length is equal to $\lambda/2 \times$ an odd integer, depending on the reflecting condition at the input and at the antenna. For short waves the antenna of length $\lambda/2$, called a *half-wave antenna*, is most practical, serving as an idealized model for dipole radiation. This is discussed in Section 13.2. Also, a loop antenna is in practical use, which is signified by an oscillating magnetic moment for magnetic radiation.

Orbiting electrons in atoms and molecules can also be analyzed for optical radiations, to which a model similar to an antenna can be applied, supporting the Maxwell theory.

13.2. Electric Dipole Radiation

An oscillating dipole moment defined in (12.17) acts as a source of radiation from the antenna in Figures 13.1(a) and (b), where the internal coordinate $\boldsymbol{r}'$ of charge $\rho(\boldsymbol{r}', \tau_R)$ is considered as varying in sinusoidal motion, corresponding to the voltage standing wave of $\lambda/2$. Assuming that the antenna is represented by an oscillating dipole $\boldsymbol{p} = \boldsymbol{p}_o \exp(-i\omega t)$, it can be observed at a distant point $\boldsymbol{r}$ by the

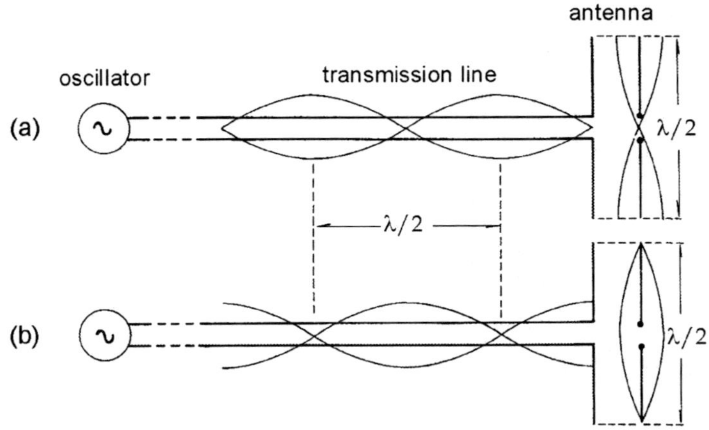

FIGURE 13.1. A transmission from an oscillator to a linear antenna. (a) Voltage maximum at the two ends. (b) Voltage is maximum at the center of antenna.

emitted radiation as

$$\boldsymbol{p}_{\mathrm{o}} \exp(-i\omega\tau_{\mathrm{R}}) = \boldsymbol{p}_{\mathrm{o}} \exp\left\{-i\omega\left(t - \frac{r}{c}\right)\right\} = \boldsymbol{p}_{\mathrm{o}} \exp i\,(kr - \omega t),$$

where $k = \frac{\omega}{c} = \frac{2\pi}{\lambda}$. Therefore, the scalar and vector potentials at a distant point $\boldsymbol{r}$ and time t can be expressed as

$$V_1(\boldsymbol{r}, t) = -\frac{1}{4\pi\varepsilon_{\mathrm{o}}} \boldsymbol{p}_{\mathrm{o}} . \boldsymbol{\nabla}_r \left(\frac{\exp i\boldsymbol{k}.\boldsymbol{r}}{\boldsymbol{r}}\right) \exp(-i\omega t)$$

and

$$\boldsymbol{A}_1(\boldsymbol{r}, t) = \frac{\mu_{\mathrm{o}}}{4\pi r} \frac{\partial}{\partial t} \boldsymbol{p}_{\mathrm{o}} \exp i\,(kr - \omega t) = -\frac{i\omega\mu_{\mathrm{o}}\boldsymbol{p}_{\mathrm{o}}}{4\pi} \frac{\exp ikr}{r} \exp(-i\omega t).$$

Assuming that $\boldsymbol{p}$ is parallel to the z-axis,

$$\begin{aligned} V_1(\boldsymbol{r}, t) &= -\frac{p_{\mathrm{o}}}{4\pi\varepsilon_{\mathrm{o}}} \frac{\partial}{\partial r} \left(\frac{\exp ikr}{r}\right) \frac{\partial r}{\partial z} \exp(-i\omega t) \\ &= -\frac{p_{\mathrm{o}} \cos\theta}{4\pi\varepsilon_{\mathrm{o}}} \frac{\partial}{\partial r} \left(\frac{\exp ikr}{r}\right) \exp(-i\omega t) \\ &= -\frac{p_{\mathrm{o}}}{4\pi\varepsilon_{\mathrm{o}} r} \left(ik - \frac{1}{r}\right) \cos\theta \exp i(kr - \omega t) \end{aligned} \qquad \text{(i)}$$

and

$$\boldsymbol{A}_1(\boldsymbol{r}, t) = -\frac{i\omega\mu_{\mathrm{o}}}{4\pi} p_{\mathrm{o}} \boldsymbol{u}_3 \frac{\exp i(kr - \omega t)}{r}, \qquad \text{(ii)}$$

where $\boldsymbol{u}_3$ is the unit vector along the z-axis, and θ is the polar angle.

From (ii), the magnetic flux density vector can be calculated by $\boldsymbol{B} = \text{curl}\,\boldsymbol{A}_1$, which is signified by the azimuthal component B_φ, i.e.,

$$B_r = B_\theta = 0, \qquad B_\varphi = \frac{1}{r}\frac{\partial (rA_{1\theta})}{\partial r} - \frac{1}{r}\frac{\partial A_{1r}}{\partial \theta},$$

where $A_{1r} = A_1 \cos\theta$ and $A_{1\theta} = -A_1 \sin\theta$ are the components of $\boldsymbol{A}_1$. Using (ii) for the magnitude A_1, we arrive at

$$B_\varphi = \frac{i\omega\mu_o}{4\pi}\left(ik - \frac{1}{r}\right)\frac{p_o \sin\theta}{r}\exp i(kr - \omega t), \tag{13.1}$$

which is in the azimuthal direction of $\boldsymbol{u}_\varphi$.

The electric field vector $\boldsymbol{E}$ can also be calculated by $\boldsymbol{E} = -\text{grad}\,V_1 - \dfrac{\partial \boldsymbol{A}_1}{\partial t}$, for which we write

$$\frac{\partial \boldsymbol{A}_1}{\partial t} = \boldsymbol{u}_r \frac{\partial A_r}{\partial t} + \boldsymbol{u}_\theta \frac{\partial A_\theta}{\partial t}$$

and

$$\text{grad}\,V_1 = \boldsymbol{u}_r \frac{\partial V_1}{\partial \boldsymbol{r}} + \boldsymbol{u}_\theta \frac{1}{r}\frac{\partial V_1}{\partial \theta} + \boldsymbol{u}_\varphi \frac{1}{r\sin\theta}\frac{\partial V_1}{\partial \varphi}.$$

Using (i) and (ii) for $\boldsymbol{A}_1$ and V_1, the $\boldsymbol{E}$ vector can be expressed as

$$E_r = -\frac{p_o \cos\theta}{2\pi\varepsilon_o r}\left(\frac{ik}{r} - \frac{1}{r^2}\right)\exp i\,(kr - \omega t),$$

$$E_\theta = \frac{p_o \sin\theta}{4\pi\varepsilon_o r}\left(-k^2 - \frac{ik}{r} + \frac{1}{r^2}\right)\exp i\,(kr - \omega t),$$

$$E_\varphi = 0, \tag{13.2}$$

indicating the vector $\boldsymbol{E}$ lies on the *meridian plane*, to which the vector $\boldsymbol{B}$ is perpendicular.

For a large r signified by $kr \gg 1$, the radial component E_r becomes negligibly small, so that the electromagnetic field at a distant point is dominated by E_θ and B_φ which are perpendicular to each other, as shown in Figure 13.2. Further, the plane of $\boldsymbol{E}$ and $\boldsymbol{B}$ (or $\boldsymbol{H}$ in vacuum space) signified by a phase $\phi = kr - \omega t$, is perpendicular to the radius r if $kr \gg 1$. Away from the source, such a field characterized by the phase ϕ propagates along the normal direction $\boldsymbol{n}$ to a plane $\phi = \text{const}$. Therefore, we define a vector $\boldsymbol{k} = k\boldsymbol{n}$ for propagation along the direction $\boldsymbol{n}$, which is normal to a plane determined by $\phi = \boldsymbol{k}.\boldsymbol{r} - \omega t = \text{const}$. Such an electromagnetic field is called a *plane wave*, where the vectors $\boldsymbol{E}$, $\boldsymbol{H}$, and $\boldsymbol{k}$ are mutually perpendicular, as sketched in Figure 13.2.

As discussed in Section 12.2 for the Poynting theorem, the electromagnetic field occupying a finite volume v is characterized by the energy $\int_v \left(\dfrac{\varepsilon_o \boldsymbol{E}^2 + \mu_o \boldsymbol{H}^2}{2}\right) dv$, emitting radiation $\int_S (\boldsymbol{E} \times \boldsymbol{H}).d\boldsymbol{S}$ from the surface per unit time. The Poynting

FIGURE 13.2. Radiation fields from a dipole $\boldsymbol{p}$ at a distant point P.

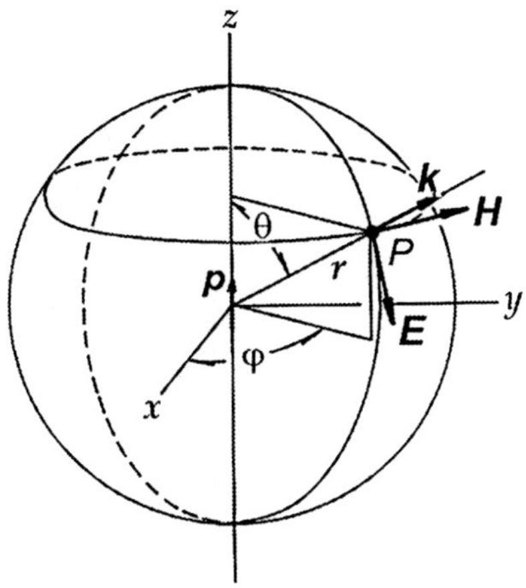

vector $\boldsymbol{P} = \boldsymbol{E} \times \boldsymbol{H}$ represents the radiation, expressing the power for outgoing energy per unit time. To calculate the radiation from a dipolar source, we consider the field at $kr \gg 1$, or $r \gg \lambda/\pi$, using the definition of wavelength λ, where

$$E_r = E_\varphi 0, \quad E_\theta = -\frac{p_o \sin\theta}{4\pi\varepsilon_o r} k^2 \exp i\phi,$$

and

$$B_r = B_\theta = 0, \quad B_\varphi = -\frac{\omega\mu_o k}{4\pi r} p_o \sin\theta \exp i\phi.$$

Since $\omega = ck$ and $c = \dfrac{1}{\sqrt{\varepsilon_o\mu_o}}$, we can derive that $\sqrt{\varepsilon_o}E_\theta = \sqrt{\mu_o}H_\varphi$, and therefore

$$\boldsymbol{u}_r \times \boldsymbol{E} = \sqrt{\frac{\mu_o}{\varepsilon_o}}\boldsymbol{H}. \tag{13.3}$$

Although field components are conveniently expressed by complex exponentials, the observable power should be calculated as a real quantity. We note that time averages of these complex quantities over the period are given with a factor $\frac{1}{2}$, so it is convenient to write that

$$\langle \boldsymbol{P} \rangle_t = \frac{1}{2}\mathrm{Re}(E \times H^*) = \frac{1}{2}\boldsymbol{n}\mathrm{Re}(E_\theta H_\varphi)$$

to express the radiation power per unit area. Using the above expressions for E_θ and H_φ, we obtain

$$\langle \boldsymbol{P} \rangle_t \approx \frac{cp_o^2 k^4}{32\pi^2\varepsilon_o r^2}\sin^2\theta\, \boldsymbol{u}_r, \tag{13.4}$$

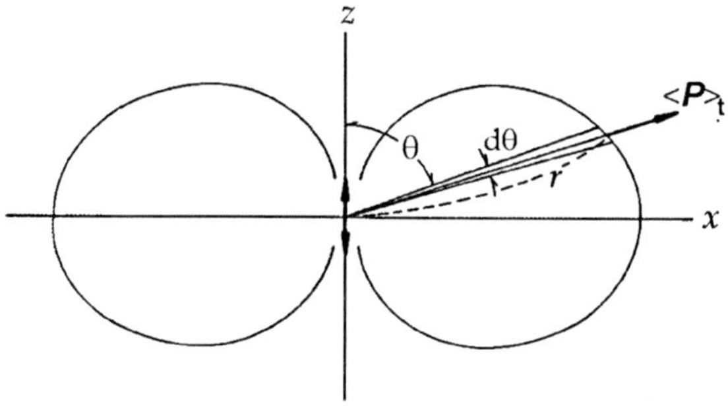

FIGURE 13.3. Angular pattern of the averaged Poynting vector $\langle \boldsymbol{P} \rangle_t$ from an oscillating electric dipole.

indicating that the intensity of radiation is maximum at $\theta = \frac{1}{2}\pi$, and minimum at $\theta = 0$, as shown by the lobe-like pattern for a constant $\langle \boldsymbol{P} \rangle_t$ in Figure 13.3. Also significant is that $\langle \boldsymbol{P} \rangle_t$ in (13.4) is proportional to k^4, implying that the larger k is, the higher is the *directivity* of radiation that can be obtained. The factor

$$Z_o = \sqrt{\frac{\mu_o}{\varepsilon_o}} = 377\Omega \tag{13.5}$$

in the relation (13.3) is a characteristic of the free space and is called the *wave impedance*.

In Figure 13.3, the radiation power is depicted as in three dimensions by the donut-shaped pattern around the z-axis, for which the total radiation power can be calculated with the formula $W = \int_0^{2\pi} d\varphi \int_{-1}^{+1} \langle \boldsymbol{P} \rangle_t \, d\cos\theta$. Numerically, the power is

$$W = \frac{cp_o^2 k^4}{12\pi\varepsilon_o} = \left(9.00 \times 10^{17}\right) p_o^2 k^4 \text{ watts.}$$

It is significant theoretically that $\langle \boldsymbol{P} \rangle_t$ is proportional to c, which is related to a relativistic interpretation of the *radiation pressure*, implying a *momentum* of the electromagnetic radiation.

13.3. The Hertz Vector

In the Maxwell theory the charge and current density represent a source for electromagnetic potentials V and $\boldsymbol{A}$, as related with ρ and $\boldsymbol{j}$ in equations (12.7b) and (12.7a), respectively. However, the differential equations are of the same type, whereas the source quantities ρ and $\boldsymbol{j}$ are related by the continuity relation (10.1b), for which the Lorentz condition (12.7d) relates V with $\boldsymbol{A}$ by the gauge transformation. Representing a source of field, it is logical to define a single variable $\boldsymbol{J}$ for these two quantities ρ and $\boldsymbol{j}$. Also, since two potentials V and $\boldsymbol{A}$ satisfy the same differential equation, Hertz introduced a single vector $\boldsymbol{Z}$ to relate with the source vector $\boldsymbol{J}$.

In (12.7a) and (b), the operator $\nabla^2 - \varepsilon_o\mu_o\partial^2/\partial t^2$ can be conveniently written as $\square^2$. With this symbol, (12.7a) and (b) can be written as

$$\square^2 \boldsymbol{A} = -\mu_o \boldsymbol{j} \quad \text{and} \quad \square^2 V = -\frac{\rho}{\varepsilon_o}, \tag{13.5a}$$

where

$$\operatorname{div} \boldsymbol{j} + \frac{\partial \rho}{\partial t} = 0 \tag{i}$$

and

$$\operatorname{div} \boldsymbol{A} + \varepsilon_o\mu_o \frac{\partial V}{\partial t} = 0. \tag{ii}$$

Hertz defined a vector $\boldsymbol{Z}$ as

$$\boldsymbol{A} = \mu_o \frac{\partial \boldsymbol{Z}}{\partial t} \quad \text{and} \quad V = -\frac{1}{\varepsilon_o} \operatorname{div} \boldsymbol{Z}, \tag{13.6}$$

which is consistent with the Lorentz condition (ii). On the other hand, the source vector $\boldsymbol{J}$ defined as

$$\boldsymbol{j} = \frac{\partial \boldsymbol{J}}{\partial t} \quad \text{and} \quad \rho = -\operatorname{div} J \tag{13.7}$$

is the same as the continuity relation (i). With these definitions of $\boldsymbol{Z}$ and $\boldsymbol{J}$, we can write the equation

$$\square^2 \boldsymbol{Z} = -\boldsymbol{J}, \tag{13.5b}$$

relating $\boldsymbol{Z}$, called the *Hertz vector*, directly to the source $\boldsymbol{J}$.

If the source vector $\boldsymbol{J}(\boldsymbol{r}', t)$ is given as

$$\boldsymbol{J}(\boldsymbol{r}', t') = \boldsymbol{p}(t')\delta^3(\boldsymbol{r} - \boldsymbol{r}'),$$

the solution of (13.5b) at a point $\boldsymbol{r}$ can be expressed, as discussed in Section 12.4, as

$$\boldsymbol{Z}(\boldsymbol{r}, t) = \frac{1}{4\pi} \int_{v(\boldsymbol{r}')} \frac{\boldsymbol{J}(\boldsymbol{r}', \tau_R)}{|\boldsymbol{r} - \boldsymbol{r}'|} \mathrm{d}^3\boldsymbol{r}', \tag{13.8}$$

where $\tau_R = t - \dfrac{|\boldsymbol{r} - \boldsymbol{r}'|}{c}$ is the retarded time. Expanding $\dfrac{1}{|\boldsymbol{r} - \boldsymbol{r}'|}$ with respect to $\left|\dfrac{\boldsymbol{r}'}{\boldsymbol{r}}\right| \ll 1$ into a Taylor series, the dominant first term is given by

$$\boldsymbol{Z}_1(\boldsymbol{r}, t) = \frac{1}{4\pi|\boldsymbol{r} - \boldsymbol{r}'|} \boldsymbol{p}\left(t - \frac{|\boldsymbol{r} - \boldsymbol{r}'|}{c}\right), \tag{13.9}$$

where

$$\boldsymbol{p}(\tau_R) = \frac{1}{2} \int_{v(\boldsymbol{r}')} \boldsymbol{r}' \times \boldsymbol{J}(\tau_R) \mathrm{d}^3\boldsymbol{r}'$$

is referred to as the *Hertz dipole*. The field vectors $\boldsymbol{B}$ and $\boldsymbol{E}$ can be computed from the vector $\boldsymbol{Z}$ as

$$\boldsymbol{B} = \mu_o \mathrm{curl}\left(\frac{\partial \boldsymbol{Z}_1}{\partial t}\right) \quad \text{and} \quad \boldsymbol{E} = \frac{1}{\varepsilon_o}\mathrm{grad}\,(\mathrm{div}\,\boldsymbol{Z}_1) - \mu_o^2 \frac{\partial^2 \boldsymbol{Z}_1}{\partial t^2}. \tag{13.10}$$

Example. Field of a Hertz Dipole Parallel to the z-Axis.

In this example we discuss the dipole radiation using the Hertz vector, although it is the same problem as discussed in Section 13.2. Here, we consider the case where the source vector $\boldsymbol{J}$ or the corresponding Hertz's dipole $\boldsymbol{p}(\tau_R)$ is in a fixed direction referred to as the z-axis. In this case the electromagnetic field of $\boldsymbol{p}(\tau_R)$ can be expressed by the corresponding Hertz vector $\mathbf{Z}$. In the following only $\mathbf{Z}_1$ in the first approximation is considered, from which the suffix 1 may be omitted for simplicity. Setting a polar coordinate system with respect to the z-axis, components of $\boldsymbol{Z}$ can be expressed as

$$Z_r = Z\cos\theta, \quad Z_\theta = -Z\sin\theta \quad \text{and} \quad Z_\varphi = 0.$$

To calculate $\boldsymbol{B}$ and $\boldsymbol{E}$, we use (13.10), where grad(div $\mathbf{Z}$) and curl $\mathbf{Z}$ need to be fully expressed by polar coordinates. Abbreviating derivatives of Z as $Z' = \partial Z/\partial r$ and $\dot{Z} = \partial Z/\partial t$,

$$\mathrm{div}\boldsymbol{Z} = \frac{1}{\boldsymbol{r}^2\sin\theta}\left\{\frac{\partial(r^2\sin\theta Z_r)}{\partial r} + \frac{1}{\boldsymbol{r}\sin\theta}\frac{\partial\,(\sin\theta Z_\theta)}{\partial\theta}\right\} = \frac{\partial Z_r}{\partial \boldsymbol{r}} = Z'\cos\theta,$$

and therefore

$$(\text{grad div } \mathbf{Z})_r = \frac{\partial}{\partial r}(Z'\cos\theta) = Z''\cos\theta,$$

$$(\text{grad div } \mathbf{Z})_\theta = \frac{1}{r}\frac{\partial}{\partial\theta}(Z'\cos\theta) = -\frac{Z'}{r}\sin\theta,$$

and

$$(\text{grad div}\mathbf{Z})_\varphi = 0.$$

On the other hand,

$$(\text{curl } \mathbf{Z})_r = \frac{1}{r\sin\theta}\left\{\frac{\partial\left(\sin\theta Z_\varphi\right)}{\partial\theta} - \frac{\partial Z_\theta}{\partial\varphi}\right\} = 0,$$

$$(\text{curl } \mathbf{Z})_\theta = \frac{1}{r}\left\{\frac{1}{\sin\theta}\frac{\partial Z_r}{\partial\varphi} - \frac{\partial(rZ_\varphi)}{\partial r}\right\} = 0,$$

and

$$(\text{curl}\mathbf{Z})_\varphi = \frac{1}{r}\left\{\frac{\partial(rZ_\theta)}{\partial r} - \frac{\partial Z_r}{\partial\theta}\right\} = -Z'\sin\theta.$$

Using these results in (13.10), we obtain

$$B_r = 0, \quad B_\theta = 0, \quad B_\varphi = \mu_o\dot{Z}'\sin\theta;$$

and

$$E_r = \left(\frac{Z''}{\varepsilon_o} - \mu_o^2 \ddot{Z}\right)\cos\theta,\ E_\theta = \left(-\frac{Z'}{\varepsilon_o r} + \mu_o^2 \ddot{Z}\right)\sin\theta, \quad E_\varphi = 0.$$

Also it is noted that $p' = -\frac{\dot{p}}{c}$, as derived from the definition$p = p\left(t - \frac{r}{c}\right)$. Accordingly, for the function $Z = \frac{p\,(t - r/c)}{4\pi r}$, we have

$$Z' = \frac{1}{4\pi}\left(-\frac{p}{r^2} - \frac{\dot{p}}{cr}\right) \quad \text{and} \quad Z'' = \frac{1}{4\pi}\left(\frac{2p}{r^3} + \frac{2\dot{p}}{cr^2} + \frac{\ddot{p}}{c^2 r}\right).$$

Thus, we finally arrive at the field components

$$B_r = B_\theta = 0,\ B_\varphi = \frac{\mu_o}{4\pi c r}\sin\theta\left(\frac{\dot{p}}{r} + \frac{\ddot{p}}{c}\right)$$

and

$$E_r = \frac{\cos\theta}{4\pi\varepsilon_o r^2}\left(\frac{2p}{r} + \frac{2\dot{p}}{c}\right), \quad E_\theta = \frac{\sin\theta}{4\pi\varepsilon_o r}\left(\frac{p}{r^2} + \frac{\dot{p}}{cr} + \frac{\ddot{p}}{c^2}\right), \quad E_\varphi = 0.$$

For a static field, $\dot{p} = 0$ and $\ddot{p} = 0$, and all magnetic components vanish, and the remaining electric components are given by

$$E_r = \frac{2p\cos\theta}{4\pi\varepsilon_o r^3} \quad \text{and} \quad E_\theta = \frac{p\sin\theta}{4\pi\varepsilon_o r^3}.$$

On the other hand, the time-dependent field at a large distance r, dominant components are

$$E_\theta = \frac{\ddot{p}\sin\theta}{4\pi\varepsilon_o c^2 r} \quad \text{and} \quad B_\varphi = \frac{\mu_o \ddot{p}\sin\theta}{4\pi c^2 r}.$$

The Poynting vector can be calculated as $\boldsymbol{P} = E_\theta H_\varphi \boldsymbol{u}_r$, where $\boldsymbol{u}_r$ is the unit vector in the radial direction, that is,

$$|\boldsymbol{P}| = \frac{\ddot{p}^2\sin^2\theta}{(4\pi)^2\,\varepsilon_o c^4 r^2} = \frac{\mu_o \ddot{p}^2 \sin^2\theta}{(4\pi)^2\,c^2 r^2},$$

indicating the rate at which the radiation energy is flowing out of a unit area on a spherical surface of radius $r \gg p$. Here, due to the factor $\ddot{p} \propto \omega^2$, we have $|\boldsymbol{P}| \propto \frac{\omega^4}{c^2} \propto \frac{c^2}{\lambda^4}$, indicating that the angular pattern of radiation $|\boldsymbol{P}|$ shown in Figure 13.2 exhibits a sharper pencil of rays for short wavelengths. The total loss rate by radiation through the spherical area $4\pi r^2$ is given by

$$-\frac{\mathrm{d}W}{\mathrm{d}t} = -c\left(\frac{\mu_o \ddot{p}^2 \sin^2\theta}{4\pi c^3}\right).$$

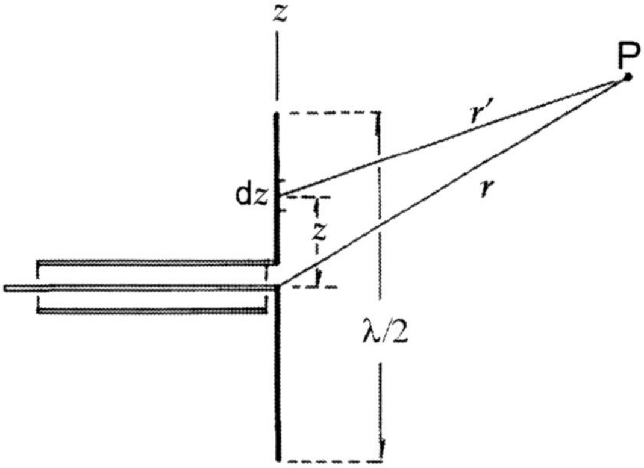

FIGURE 13.4. A half-wave antenna.

13.4. A Half-Wave Antenna

As indicated in the above antenna theory, the radiation intensity depends on the value of $\ddot{p}$, which should be maximized for a practical antenna system. Considering a linear antenna of length l as a part of a transmission line, as in Figure 13.1, maximum densities of charge and current ρ and $\boldsymbol{j}$ can be achieved, if l matches wavelengths of the standing wave on the antenna, i.e., $l = \frac{1}{2}\lambda \times$ odd integers, giving rise to a maximum amplitude for the Hertz dipole $p(\tau_R)$. Figures 13.1(a) and (b) show schematically maximum and minimum voltages of the standing wave corresponding to maximum charge densities at the ends and maximum current density at the center part of the linear antenna.

Figure 13.4 shows a linear antenna attached at the terminal of a transmission cable, where the antenna direction z is perpendicular to the cable whose center is $z = 0$. At a frequency ω of an oscillator, the charge and current densities on the antenna related by the continuity equation $\dfrac{\partial j\,(z,t)}{\partial z} + \dfrac{\partial \rho\,(z,t)}{\partial t} = 0$ can be expressed by

$$\rho(z,t) = \rho_o \sin\frac{\pi z}{l}\cos\omega t \quad \text{and} \quad j(z,t) = j_o \cos\frac{\pi z}{l}\sin\omega t.$$

By definition (13.7), the source vector, $j = \dfrac{\partial J}{\partial t}$ and $\rho = \dfrac{\partial J}{\partial z}$, can be written as

$$J(z,t) = \frac{j_o}{\omega}\cos\frac{\pi z}{l}\cos\omega t,$$

which is observed at a distant point P($\boldsymbol{r}$) as

$$Z_1(r,t) = \frac{1}{4\pi}\int_{-l/2}^{+l/2}\frac{J(z,\tau_R)}{r}A\mathrm{d}z,$$

where A is the cross-sectional area of the antenna line. The distance r' between the element $J\mathrm{d}z$ and P can be expressed as $r' = r - z\cos\theta + \dots$, if $r > z$, but at a very distant point $r \gg z$ we may assume $r' \approx r$. In this case, replacing $j_o A$ by I_o, $Z_1(r,t)$ can be written

$$Z_1(r,t) = \frac{I_o}{4\pi\omega r}\left(\int_{-l/2}^{+l/2} \cos\frac{\pi z}{l}\mathrm{d}z\right)\cos\omega\tau_R = \frac{I_o l}{2\pi^2\omega r}\cos\omega\tau_R,$$

corresponding to the Hertz dipole $p(\tau_R) = \dfrac{2I_o l}{\pi\omega}\cos\omega\tau_R$, which is given by $\dfrac{2cI_o}{\omega^2}\cos\omega\tau_R$ for a half-wave antenna $l = \frac{1}{2}\lambda$. The radiation from such an antenna can be calculated with $p(\tau_R)$, as discussed in the previous section.

13.5. A Loop Antenna

A circular loop antenna, as illustrated in Figure 13.5(a), is also used in practical applications. Considering such a circle of radius a in the x, y-plane, an oscillating current $I = I_o \sin\omega t$ on the loop will generate a radiating field. Under a condition of the wavelength $\lambda \gg a$ and $r \gg \lambda$, the field at a distant point r can be discussed with the Hertz vector.

Figure 13.4(b) shows the ring current in the x, y-plane, where the current density element is given as $Ia\mathrm{d}\varphi' = jAa\mathrm{d}\varphi'$ (A is the cross-sectional area of the wire). The source current density is then expressed as

$$j_x = -j\sin\varphi', \quad j_y = j\cos\varphi' \quad \text{and} \quad j_z = 0, \quad \text{where} \quad j = \frac{I_o}{A}\sin\omega t.$$

The source vector defined by (13.7) can be written as

$$J_x = J\sin\varphi', J_y = -J\cos\varphi' \quad \text{and} \quad J_z = 0, \quad \text{where} \quad J = \frac{I_o}{A\omega}\cos\omega t,$$

with which the Hertz vector can be expressed as

$$\boldsymbol{Z}(\boldsymbol{r},t) = \frac{1}{4\pi}\int_{v(\boldsymbol{r}')} \frac{\boldsymbol{J}(\boldsymbol{r}',\tau_R)}{|\boldsymbol{r}-\boldsymbol{r}'|}\mathrm{d}^3\boldsymbol{r}', \quad \text{where} \quad \mathrm{d}^3\boldsymbol{r}' = Aa\mathrm{d}\varphi'.$$

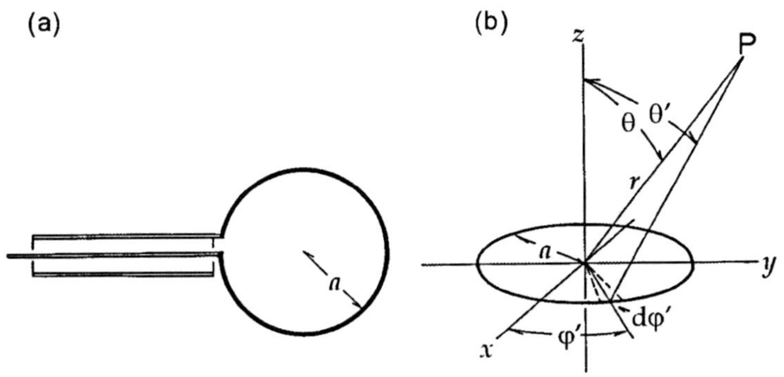

FIGURE 13.5. (a) A loop antenna. (b) Calculating the radiation from a loop antenna.

We calculate $\boldsymbol{Z}(\boldsymbol{r}, t)$ approximately as

$$|\boldsymbol{r} - \boldsymbol{r}'| = r - a\cos\Theta, \quad \frac{1}{|\boldsymbol{r} - \boldsymbol{r}'|} = \frac{1}{r}\left(1 + \frac{a}{r}\cos\Theta\right),$$

where Θ is the angle between $\boldsymbol{r}$ and $\boldsymbol{r}' = \boldsymbol{a}$, and given by

$$\cos\Theta = \sin\theta\cos(\varphi - \varphi'). \tag{i}$$

In this case, as the retarded time is given by

$$\tau_{\mathrm{R}} = \left(t - \frac{r}{c}\right) + \frac{a}{r}\cos\Theta,$$

and hence

$$\begin{aligned} J &= \frac{I_o}{A\omega}\cos\left\{\omega\left(t - \frac{r}{c}\right) + \frac{a\omega}{c}\cos\Theta\right\} \\ &= \frac{I_o}{A\omega}\left\{\cos\left(t - \frac{r}{c}\right) - \frac{a\omega}{c}\cos\Theta\sin\omega\left(t - \frac{r}{c}\right)\right\}, \end{aligned}$$

provided that $a\omega/c \ll 1$, or $a \ll \lambda$. Therefore,

$$\begin{aligned} Z_x = &-\frac{I_o a}{4\pi r}\int_0^{2\pi} \mathrm{d}\varphi'\left(1 + \frac{a}{r}\cos\Theta\right) \\ &\times\left\{\cos\omega\left(t - \frac{r}{c}\right) - \frac{a\omega}{c}\cos\Theta\sin\omega\left(t - \frac{r}{c}\right)\right\}\sin\varphi'. \end{aligned}$$

Using (i) to replace $\cos\Theta$ by $\sin\theta\cos(\varphi - \varphi')$ and noting that integrals for $0 \le \varphi' \le 2\pi$ will vanish, except for $\int_0^{2\pi}\cos^2\varphi' d\varphi' = \int_0^{2\pi}\sin^2\varphi'\mathrm{d}\varphi' = \pi$, we obtain

$$Z_x = -\frac{I_o\pi a^2\sin\theta}{4\pi c r}\sin\omega\left(t - \frac{r}{c}\right)\sin\varphi,$$

and similarly

$$Z_y = \frac{I_o\pi a^2\sin\theta}{4\pi c r}\sin\omega\left(t - \frac{r}{c}\right)\cos\varphi, \quad \text{and} \quad Z_z = 0.$$

Hence,

$$Z_\varphi = \frac{I_o}{4\pi c}\frac{\pi a^2\sin\theta}{r}\sin\omega\left(t - \frac{r}{c}\right),$$

and by (13.6)

$$A_\varphi = \frac{\mu_o I_o\pi a^2\omega}{4\pi c^2 r}\sin\theta\cos\omega\left(t - \frac{r}{c}\right).$$

The field vectors can be obtained from $\boldsymbol{E} = -\dfrac{\partial\boldsymbol{A}}{\partial t}$ and $\boldsymbol{B} = \mathrm{curl}\boldsymbol{A}$, that is,

$$E_\varphi = \frac{\mu_o I_o\pi a^2\omega^2}{4\pi c^2 r}\sin\theta\sin\omega\left(t - \frac{r}{c}\right)$$

and

$$H_\theta = -\frac{1}{\mu_o r}\frac{\partial (rA_\varphi)}{\partial r} \approx -\frac{I_o \pi a^2 \omega^2}{4\pi c^3 r}\sin\theta \sin\omega\left(t - \frac{r}{c}\right),$$

ignoring a term proportional to r^{-2}, which is valid for $r \gg \lambda$. In such a distant point, we have a simple relation

$$-\frac{E_\varphi}{H_\theta} = \mu_o c = \sqrt{\frac{\mu_o}{\varepsilon_o}},$$

which is characteristic of a plane wave that is emitted by a magnetic dipole moment, as can be defined by

$$p_m(\boldsymbol{r}', t) = I_o \pi a^2 \sin\omega t.$$

In this case, we notice that $\boldsymbol{E}$ and $\boldsymbol{H}$ vectors are perpendicular, though they are *out of phase*, exhibiting phase difference π. The time average of the Poynting vector can be expressed as

$$\langle \boldsymbol{P}_r \rangle_t = -\frac{\mu_o c}{(4\pi)^2}\left(\frac{p_m}{c}\right)^2\left(\frac{\omega}{c}\right)^4 \frac{\sin^2\theta}{r^2},$$

and the total radiation power is

$$\frac{dW}{dt} = -\int_0^{2\pi} d\varphi \int_{-1}^{+1} \langle \boldsymbol{P} \rangle_t r^2 d(\cos\theta) = -\frac{\mu_o c}{4\pi}\left(\frac{2}{3c^4} p_m^2 \omega^4\right).$$

13.6. Plane Waves in Free Space

In free space away from the source, the Maxwell equations are specified locally by $\rho = 0$ and $\boldsymbol{j} = 0$. We have shown in Section 12.1 that the field vectors $\boldsymbol{E}$ and $\boldsymbol{H}$ in free space are derived from the wave equation, i.e.,

$$\left(\nabla^2 - \frac{1}{c^2}\frac{\partial^2}{\partial t^2}\right)(\boldsymbol{E}, \boldsymbol{H}) = 0, \tag{13.11}$$

where $c = 1/\sqrt{\varepsilon_o \mu_o}$ is equal to the speed of light. It has been confirmed that light speed is identical to that of electromagnetic propagation at lower frequencies. Hence, (13.11) is believed to govern all electromagnetic waves, including optical light. Solutions of (13.11) express propagating waves from a remote source signified by oscillating charges at a specific frequency ω, which are expressed by a sinusoidal function of x and t.

As related to the dipolar source $\boldsymbol{J}(\mathbf{r}', t')$ that is observed as $\boldsymbol{J}\left(\boldsymbol{r}, t - \frac{|\boldsymbol{r} - \boldsymbol{r}'|}{c}\right)$, phases of the field vectors $\boldsymbol{E}$ and $\boldsymbol{H}$ are expressed as $\omega\left(t - \frac{r}{c}\right) + \phi$, where $r = |\boldsymbol{r} - \boldsymbol{r}'|$, and ϕ depends on the nature of the distant source. Complicated radiation fields near the source may be considered as trivial at distant points for a periodic

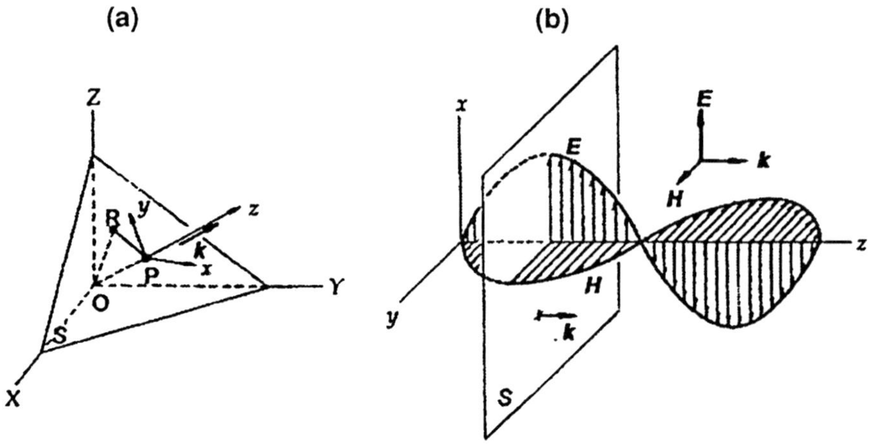

FIGURE 13.6. Plane waves. (a) A wavefront of a plane wave. (b) An example of plane waves ($\boldsymbol{E}$, $\boldsymbol{H}$) that are in phase.

repetition of $\boldsymbol{E}$ and $\boldsymbol{H}$, where we can only be concerned about the periodic variation in space-time $\boldsymbol{r}$ and t or in the phase $\omega t - \boldsymbol{k.r}$, where $|\boldsymbol{k}| = \omega/c$, with respect to a reference near the observing point. Yet, the phase difference between $\boldsymbol{E}$ and $\boldsymbol{H}$, that is 0 or πs is significant for a source: It is either an electric or a magnetic dipole, respectively. The solution of (13.11) can be expressed by sin or cos functions with arbitrary phases, however it is more convenient to treat them as complex functions of a phase, considering that only the function's real part is physically meaningful. With the ratio $\omega/c = k$, called the *wave number*, we can write

$$\boldsymbol{E} = \boldsymbol{E}_o \exp i(kr - \omega t) \quad \text{and} \quad \boldsymbol{H} = \boldsymbol{H}_o \exp i(kr - \omega t + \varphi), \tag{13.12}$$

where $\boldsymbol{E}_o$ and $\boldsymbol{H}_o$ are amplitude vectors, and $\boldsymbol{E}_o \perp \boldsymbol{H}_o$, as verified in Sections 13.3 and 13.4, at a large distance r. Figure 13.2 illustrates such vectors $\boldsymbol{E}$, $\boldsymbol{H}$ and $\boldsymbol{k}$ at a distant point (r, θ, φ).

Further, $\boldsymbol{E}$ and $\boldsymbol{H}$ are constant at all points on a plane of $kr - \omega t =$ const. Hence, k can be considered as a vector $\boldsymbol{k}$ parallel to the normal $\boldsymbol{n}$ of such a plane, i.e., $\boldsymbol{k} = k\boldsymbol{n}$, which is called the wave vector. Define an arbitrary point on the plane by the vector $\boldsymbol{r}$, the algebraic equation of the plane is expressed by $\phi = \boldsymbol{k.r} - \omega t +$ const, on which lie field vectors $\boldsymbol{E}$ and $\boldsymbol{H}$. Taking two such parallel planes where $\Delta\phi = 2\pi$ the perpendicular distance is given by $\lambda = 2\pi/k$, which is referred to as the *wavelength*. For such a *plane wave* we have relations

$$\omega = ck, \quad k = 2\pi/\lambda, \quad \lambda\omega = 2\pi c. \tag{13.13}$$

Figure 13.6(a) shows a plane with respect to the laboratory system X, Y, and Z. We indicated the direction of propagation along the z direction, and the plane is determined by $\boldsymbol{E}$ and $\boldsymbol{H}$, whose directions are designated by the x and y axes, as shown in Figure 13.6(b). In the figure, plane waves in phase are illustrated.

Differential operations of complex vectors can be written for $\boldsymbol{E}$, for instance, as

$$\operatorname{curl}\{\boldsymbol{E}_o \exp i(\boldsymbol{k}.\boldsymbol{r} - \omega t)\} = i\boldsymbol{k} \times \boldsymbol{E}_o \exp i(\boldsymbol{k}.\boldsymbol{r} - \omega t),$$
$$\operatorname{div}\{\boldsymbol{E}_o \exp i(\boldsymbol{k}.\boldsymbol{r} - \omega t)\} = i\boldsymbol{k}.\boldsymbol{E}_o \exp i(\boldsymbol{k}.\boldsymbol{r} - \omega t)$$

and

$$\frac{\partial}{\partial t}\left[\boldsymbol{E}_o \exp i(\boldsymbol{k}.\boldsymbol{r} - \omega t)\right] = -i\omega \exp i(k.r - \omega t).$$

Using these differentiations in the Maxwell equation (i) for $\boldsymbol{j} = 0$ in Section 12.1, we can write

$$ik \times \boldsymbol{H}_o \exp i(\boldsymbol{k}.\boldsymbol{r} - \omega t + \varphi) = -i\omega\varepsilon_o \boldsymbol{E}_o \exp\ i(\boldsymbol{k}.\boldsymbol{r} - \omega t),$$

therefore

$$\boldsymbol{k} \times \boldsymbol{H}_o \exp i\varphi = -\omega\varepsilon_o \boldsymbol{E}_o. \tag{13.14a}$$

Similarly, from (ii) in Section 12.1, we obtain

$$\boldsymbol{k} \times \boldsymbol{E}_o = \omega\mu_o \boldsymbol{H}_o \exp i\varphi. \tag{13.14b}$$

It is clear from these relations that three vectors $\pm\, \boldsymbol{k}$, $\boldsymbol{E}$, and $\boldsymbol{H}$ are orthogonal. From (iii) and (iv), we can derive that

$$\boldsymbol{k}.\boldsymbol{E}_o = 0 \quad \text{and} \quad \boldsymbol{k}.\boldsymbol{H}_o = 0, \tag{13.15}$$

further confirming the orthogonality.

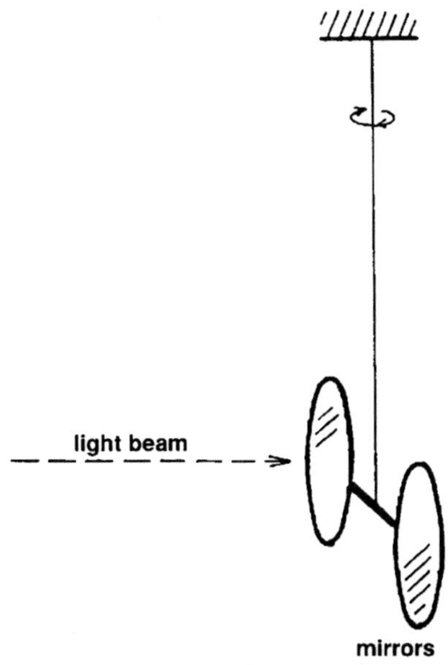

FIGURE 13.7. A demonstration of "radiation pressure."

The plane wave is characterized by its phase $\phi = \boldsymbol{k}.\boldsymbol{r} - \omega t = \text{const}$, which is nothing but the algebraic equation of a plane $k_x x + k_y y + k_z z = -\phi(t) + \text{const}$, where the components of the wavevector $\boldsymbol{k}$ are proportional to the direction cosines of the normal $\boldsymbol{n}$. As discussed already,

$$Z_o = \frac{|E_o|}{|H_o \exp i\varphi|} = \sqrt{\frac{\mu_o}{\varepsilon_o}} = 377.6\ \Omega,$$

which is the *wave impedance* in free space.

The energy density of plane electromagnetic waves is evaluated as

$$u(\boldsymbol{r}) = \frac{\varepsilon_o}{2}\boldsymbol{E}_o.\boldsymbol{E}_o^* + \frac{\mu_o}{2}\boldsymbol{H}_o.\boldsymbol{H}_o^* = \varepsilon_o \boldsymbol{E}_o^2,$$

with which the energy in a given volume is expressed as $U = \int_v u(\boldsymbol{r})\,\mathrm{d}^3\boldsymbol{r}$. According to the Poynting theorem, $-\frac{\mathrm{d}U}{\mathrm{d}t} = \int_S (\boldsymbol{E} \times \boldsymbol{H}).\mathrm{d}\boldsymbol{S}$, indicating that electromagnetic field energy in the volume v is decreased as a function of time. Applying the theorem to plane waves propagating at a constant wavevector $\boldsymbol{k}$, the Poynting vector can be calculated as

$$\boldsymbol{E} \times \boldsymbol{H}^* = \frac{\boldsymbol{E}_o^2}{\omega\mu_o}\boldsymbol{k} = \frac{c^2 u(\boldsymbol{r})}{\omega}k\boldsymbol{n} = cu(\boldsymbol{r})\boldsymbol{n}.$$

On the other hand, $\frac{\mathrm{d}U}{\mathrm{d}t}$ can be interpreted mechanically as the power Fc, where F is effectively a force exerting on the surface S, we can write the relation $Fc = c\bar{u}S_n$. Therefore we can regard the quantity $P = F/S_n = \bar{u}$ as the *radiation pressure*. Writing $F = \frac{\mathrm{d}p}{\mathrm{d}t}$ to define the momentum of the field confined to v, we obtain the energy-momentum relation as $U = cp$, that is consistent with the relativistic equation for a zero-mass particle. Substantiated by a simple experiment illustrated in Figure 13.7, the mechanical properties of a plane wave are expressed quantum-theoretically by photons characterized by energy $U = \mathrm{n}\hbar\omega$ and momentum $p = \mathrm{n}\hbar\omega/c$, where n is the number of photons.

14
The Special Theory of Relativity

14.1. Newton's Laws of Mechanics

In Newton's mechanics, a particle of mass m in motion under a force $\boldsymbol{F}$ is described by the equation

$$m\frac{\mathrm{d}^2\boldsymbol{r}}{\mathrm{d}t^2} = \boldsymbol{F}. \tag{i}$$

The mass m and the force $\boldsymbol{F}$ represent, respectively, the mechanical property of a particle and the cause of motion, quantities whose values are independent of the *observer*, whereas the acceleration $\mathrm{d}^2\boldsymbol{r}/\mathrm{d}t^2$ describes the motion of a particle with respect to the observer. On the other hand, the equation of motion is unchanged by a velocity transformation, i.e., $\mathrm{d}\boldsymbol{r}/\mathrm{d}t = \mathrm{d}\boldsymbol{r}'/\mathrm{d}t + \boldsymbol{v}$, between two observers, where v is a constant relative speed, implying that $\boldsymbol{r}(t)$ and $\boldsymbol{r}'(t)$ in different references are identical. In this context, Newton's law of motion is *invariant* for independent observers moving with a constant relative speed, whereas the time t remains as a *universal* variable for these observers. Therefore, a particle at *absolute rest* cannot be identified in such a framework of space-time of Newton's mechanics.

Considering two reference systems $\mathrm{S}(x, y, z)$ and $\mathrm{S}'(x', y', z')$ moving at a constant relative velocity v along the x and x' directions assumed to be parallel, equation (i) is unchanged under a transformation

$$x' = x - vt, \quad y' = y, \quad z' = z \quad \text{and} \quad t' = t, \tag{14.1a}$$

which is referred to as the *Galilean transformation*, and for which we have the velocity relation

$$V' = V - v. \tag{14.1b}$$

In contrast, the laws of electromagnetism are *not* invariant under the Galilean transformation because the Maxwell equations involve both static and moving charges. These equations are signified by the speed of light c that is *constant* for all observers, which contradicts (14.1b). On the other hand, Maxwell's equations *are* verified as invariant under a coordinate transformation called the *Lorentz transformation*, as will be discussed later. Nevertheless, the Galilean transformation

(14.1a) is consistent with the Lorentz transformation in the limit of $c \to \infty$. In his theory of relativity, Einstein proposed that *all* physics laws—including the laws of electromagnetism—must be *invariant* under the Lorentz transformation. Newton's law is regarded as approximately invariant, as can be verified in the limit of $c = \infty$.

Faraday and Maxwell considered that the electromagnetic field represents a physical object that occupies a volume in space encompassing an empty space and materials as well. Electromagnetically, the vacuum can be regarded as a medium characterized by constants ε_o and μ_o, in which the speed of light is given by $c = 1\sqrt{\varepsilon_o\mu_o}$. It was once thought that the whole universe was filled with a hypothetical material called *ether* through which electromagnetic waves were assumed to propagate. In this postulate absolute rest would be attributable to ether; this notion was, however, proved incorrect by the Michelson–Morley experiment (1887), which provided convincing evidence against the ether theory.

14.2. The Michelson–Morley Experiment

Michelson and Morley carried out an optical experiment to determine if "ether" really existed or not; they obtained a negative result for the ether hypothesis. Figure 14.1 shows their experimental layout, which consisted of a large turntable on which their optical apparatus was installed.

If light travels at the speed c with respect to ether that was considered as absolutely at rest, light should travel at a speed $c + v$ for a moving observer at a speed v, according to (14.1b). These authors considered that v could be the speed of rotating

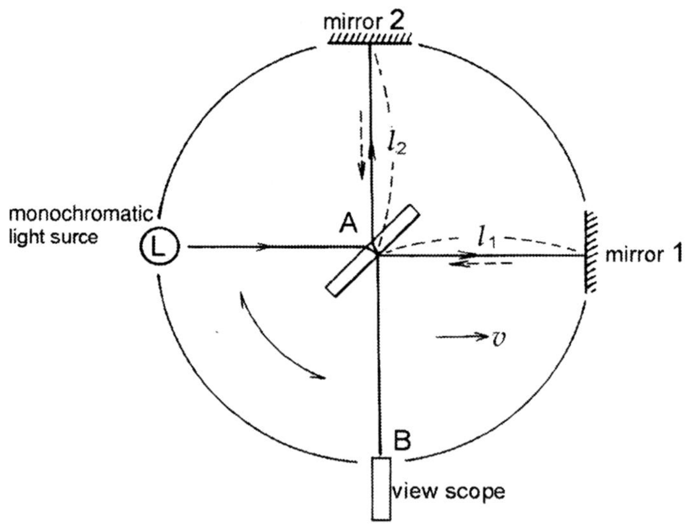

FIGURE 14.1. The Michelson—Morley experiment.

Earth, on which their laboratory is steadily located. The turntable was utilized to set the light beam in the direction of Earth's rotation, where monochromatic light from a steady source on the table was split by a semi-transparent mirror A toward mirrors 1 and 2, located in parallel and perpendicular directions to Earth's rotation, respectively. Reflected light beams from 1 and 2 were viewed by the scope B in the figure. Although not shown in the figure, another glass plate similar to A was installed in the experiment to adjust the path length.

Adjusting the light path from A to the mirror 1 in parallel with the direction of the rotating Earth, the light propagated toward the mirror at the speed $c + v$, whereas the light reflected from the mirror 1 propagated backward at the speed $c - v$. Denoting the distance between A and the mirror 1 by l_1, the time t_1 for the return trip A $\rightarrow$ mirror 1 $\rightarrow$ A is given by

$$t_1 = \frac{l_1}{c+v} + \frac{l_1}{c-v} = \frac{2cl_1}{c^2 - v^2}$$

On the other hand, on the path from A to the mirror 2, the light propagated at the speed $\sqrt{c^2 - v^2}$. Therefore, letting the distance be l_2, the time t_2 for the return trip A $\rightarrow$ mirror 2 $\rightarrow$ A can be expressed as

$$t_2 = \frac{2l_2}{\sqrt{c^2 - v^2}}.$$

Accordingly, returning back to A from mirrors 1 and 2 by reflection, the two light beams have a time difference

$$\Delta t = t_2 - t_1 = \frac{2}{c}\left(\frac{l_2}{\sqrt{1 - \frac{v^2}{c^2}}} - \frac{l_1}{1 - \frac{v^2}{c^2}}\right).$$

Next, the whole apparatus was rotated by 90° so that the light from A to mirror 2 was parallel to v; the other beam from A to mirror 1 was perpendicular to v. When the two beams returned back to A, the time difference $\Delta t'$ can be written as

$$\Delta t' = \frac{2}{c}\left(\frac{-l_1}{\sqrt{1 - \frac{v^2}{c^2}}} + \frac{l_1}{1 - \frac{v^2}{c^2}}\right).$$

If $\Delta t \neq \Delta t'$ in these experiments, an interference pattern should be observed at B, due to a small but detectable phase difference, which can be expressed as

$$\omega\left(\Delta t - \Delta t'\right) \approx \frac{\omega v^2 (l_1 + l_2)}{c^3},$$

assuming $v \ll c$. In the apparatus, typically $l_1 = l_2 = 1.2$ m, $v = 3 \times 10^4$ m/sec and $2\pi/\omega = 2 \times 10^{-15}$ sec. If the ether theory were correct, Michelson and Morley expected to observe a shift of 0.04 fringe at a time interval of $\Delta t' - \Delta t = 8 \times 10^{-17}$ s with a visible light, which, in their experiment, was a detectable order of magnitude. Yet, no fringe shift was detected within optical accuracy, thus providing evidence that refuted the ether theory. Accordingly, the speed of light must be considered to be exactly the same in all inertial systems.

On the basis of the Michelson–Morley experiment, Lorentz and Fitzgerald proposed that the length of a moving object measured by a stationary observer appears contracted by a factor $\sqrt{1 - v^2/c^2}$, whereas its length perpendicular to the moving direction remains unchanged. Although proposed before Einstein's discovery, their hypothesis was proven to be a direct consequence of the relativity theory. Such an effect on length of a moving object is known as the *Lorentz–Fitzgerald contraction*.

14.3. The Lorentz Transformation

In Newton's mechanics, time is considered an absolute measure for any event in progress, and hence a variable independent from an observer. In the relativity theory, on the other hand, it is a variable between observers in relative motion.

A "clock" set in a system S indicates a time t for an event in progress. However, for the same event observed in a different system S′ moving at a relative speed v, the time t' is not necessarily the same as t measured in S. The reason is that information transmitted by means of the *light signal* takes *finite* time to reach the observer. Therefore, it is significant to realize that no event can be observed "simultaneously" by different observers.

In this context, there is no *absolute* time; clocks can be set to one time when their individual systems coincide, but they show different times if they are in motion relative to each other. With an individual time set in each system, we use *four* coordinates, x, y, z, and t, in the reference system S to describe a physical event progressing in space-time. On the other hand, as confirmed in the Michelson–Morley experiment, a light signal propagates at a constant speed c in any system S moving at a constant relative speed v with respect to another system S′.

Observed in two systems S and S′ in relative motion at a speed v, we consider an event described as $(\Delta x, \Delta t)$ and $(\Delta x', \Delta t')$, respectively. For a light signal propagates in all directions in space at the speed c, the frontal spheres are given by

$$(\Delta x)^2 + (\Delta y)^2 + (\Delta z)^2 = c^2(\Delta t)^2 \quad \text{and} \quad (\Delta x')^2 + (\Delta y')^2 + (\Delta z')^2 = c^2(\Delta t')^2 \tag{14.1}$$

in S and S′, respectively. However, assuming parallel motion of S and S′ in their respective x and x' directions, we may consider $\Delta y = \Delta z = 0$ and $\Delta y' = \Delta z' = 0$, for which (14.1) can be simplified as

$$\Delta x = c\Delta t \quad \text{and} \quad \Delta x' = c\Delta t'. \tag{14.2}$$

On the other hand, for such S and S′ the Galilean formula (14.1a) should be modified by introducing factors γ and γ', as

$$\Delta x' = \gamma(\Delta x - v\Delta t) \quad \text{and} \quad \Delta x = \gamma'(\Delta x' + v\Delta t'); \tag{14.3}$$

however, we should have $\gamma = \gamma'$ in order for the Newton equation (i) to be invariant. Using (14.2) in these equations, we can obtain the relation between the time intervals Δt and $\Delta t'$, i.e.,

$$c\Delta t' = \gamma(c\Delta t - v\Delta t) \quad \text{and} \quad c\Delta t = \gamma'(c\Delta t' + v\Delta t').$$

Therefore,

$$\gamma^2 = \frac{1}{1 - \frac{v^2}{c^2}}.$$

Writing $\beta = \frac{v}{c}$, the transformation formula can be elaborated from (14.2) and (14.3), resulting in

$$\Delta x' = \frac{\Delta x - v\Delta t}{\sqrt{1-\beta^2}}, \quad \Delta y' = \Delta y, \quad \Delta z' = \Delta z \quad \text{and} \quad \Delta t' = \frac{\Delta t - \frac{v\Delta x}{c^2}}{\sqrt{1-\beta^2}}, \tag{14.4a}$$

and the inverse relations

$$\Delta x = \frac{\Delta x' + v\Delta t'}{\sqrt{1-\beta^2}}, \quad \Delta y = \Delta y', \quad \Delta z = \Delta z' \quad \text{and} \quad \Delta t = \frac{\Delta t' + \frac{v\Delta x'}{c^2}}{\sqrt{1-\beta^2}}, \tag{14.4b}$$

which are collectively known as the *Lorentz transformation*.

We note that (14.4a) can be reduced to

$$\Delta x' = \Delta x - v\Delta t, \quad \Delta y' = \Delta y, \quad \Delta z' = \Delta z \quad \text{and} \quad \Delta t' = \Delta t,$$

if $\beta \ll c$ or $v \ll c$, which are identical to (14.1a). Therefore, Newton's law can be regarded as a valid approximation for $v \ll c$, a condition that is valid for most terrestrial phenomena. For example, the speed of the Earth orbiting around the Sun is about 30 km/s, and hence $\gamma = 1.000000005 \approx 1$, so for this Newton's equation is sufficiently accurate, whereas for such fast-moving particles as *muons* in cosmic rays, $\beta = 0.98$, for which the relativistic effects are very significant. It is clear from the definition of γ that the speed v cannot exceed the speed of light c, which is considered the fastest speed in the universe. In the special theory of relativity, Einstein proposed that all physical laws must be invariant under the Lorentz transformation. Here, by "special" he meant that invariant laws are held in all systems of observation moving at *constant* relative speed, which is generally called the *inertial system*.

Emitted from a point x, y, z at a time t in an inertial system S, a light signal propagates spherically in all directions in space at speed c. Including t as an additional coordinate with x, y, z, a set of four coordinates x, y, z, t can specify such a physical event in *four-dimensional space*, and we write $x_1 = x$, $x_2 = y$, $x_3 = z$ and $x_4 = ict$, for convenience, where $i = \sqrt{-1}$. An inertial system in four dimensions can therefore be characterized by the invariance of the *metric* defined as

$$|\Delta s|^2 = \sum_i \Delta x_i^2 = \Delta x^2 + \Delta y^2 + \Delta z^2 - c^2 \Delta t^2,$$

where Δs is a four-dimensional distance vector with components Δx, Δy, Δz, and $ic\Delta t$ in the system S. In fact, for an *accelerating* system Einstein elaborated the *general theory of relativity*; the metric invariance is violated in an accelerating system, thus representing a distorted four-dimensional space. Supported by extraterrestrial evidence, his general theory plays an important role in astrophysics.

For a particular event stationary in S′, i.e., $\Delta x' = \Delta y' = \Delta z' = 0$, the time interval $\Delta t'$ is Lorentz invariant because $|\Delta s|^2$ is invariant. Hence, this $\Delta t' = \tau$ is referred to as a *proper time interval* in S′, indicating the timescale of a clock attached to S′. On the other hand, for an event at a fixed time in S′, i.e., $\Delta t' = 0$, the corresponding $\Delta x'$ represents (if $\Delta y' = \Delta z' = 0$) a *proper length* l_o of an object attached to S′. Therefore, in a system S moving at a relative speed v, such events can be observed as

$$\Delta t = \gamma\tau \quad \text{and} \quad l_o = \gamma \Delta x$$

from (14.3), assuming $v < c$. Hence, writing these as

$$\Delta t = \frac{\tau}{\sqrt{1-\beta^2}} > \tau \quad \text{and} \quad \Delta x = l_o\sqrt{1-\beta^2} < l_o \tag{14.5}$$

which are known as the *Lorentz dilatation* and the *Lorentz contraction*, respectively. Such relativistic relations are substantial in many high-speed phenomena, where v is close to c. For example, a *muon* particle has life of the order of 2×10^{-6} s, which can be considered as the proper time τ in the system S′ moving with cosmic ray. Therefore, as observed in the system S on the ground, the distance for a muon to travel between the moments of creation and disintegration can be estimated by $c\Delta t = c\gamma\tau \sim 6 \times 10^2$ m if its speed is considered close to c. The second relation in (14.5) is exactly Lorentz–Fitzgerald's hypothesis proposed for the Michelson–Morley experiment.

14.4. Velocity and Acceleration in Four-Dimensional Space

In this section we discuss transformations of velocity and acceleration components between inertial systems. We assume that S and S′ are in parallel, and v is in the direction of the x and x' axes. In this case, the coordinate transformation is written as

$$\Delta x_1 = \gamma\left(\Delta x_1' + v\Delta t'\right), \quad \Delta x_2 = \Delta x_2', \quad \Delta x_3 = \Delta x_3',$$
$$\Delta t = \gamma\left(\frac{v}{c^2}\Delta x_1' + \Delta t'\right).$$

Three-dimensional components u_1, u_2, u_3 of the four-dimensional velocity can be defined as

$$u_1 = \lim_{\Delta t \to 0} \frac{\Delta x_1}{\Delta t} = \lim_{\Delta t \to 0} \frac{\dfrac{\Delta x_1'}{\Delta t'} + v}{1 + \dfrac{v}{c^2}\dfrac{\Delta x_1'}{\Delta t'}} = \frac{u_1' + v}{\gamma\left(1 + \dfrac{v}{c^2}u_1'\right)},$$

$$u_2 = \frac{u_2'}{\gamma\left(1 + \dfrac{v}{c^2}u_1'\right)}, \quad \text{and} \quad u_3 = \frac{u_3'}{\gamma\left(1 + \dfrac{v}{c^2}u_1'\right)},$$

and the relation between Δt and $\Delta t'$ can be expressed as

$$\Delta t = \Delta t'\left(1 + \frac{v}{c^2}u_1'\right).$$

The temporal advance of a stationary event in an inertial system can be described by the proper time τ, which is, however, measured as $\gamma\tau$ by an observer in a system moving at a relative speed v. Therefore, for a given system of reference, we can logically define a speed, $\mathrm{d}x_1/\mathrm{d}\tau$ for an object, referring to the proper time τ. From the general relation $\Delta t = \gamma\tau$, we have $\mathrm{d}t/\mathrm{d}\tau = \gamma$. So, writing a position vector in the four-dimensional space as a row matrix

$$\langle \boldsymbol{r}| = (x_1, x_2, x_3, x_4) = (x, y, z, ict),$$

the four-dimensional velocity can be expressed as

$$\langle \mathsf{v}| = (\mathsf{v}_1, \mathsf{v}_2, \mathsf{v}_3, \mathsf{v}_4),$$

where

$$\mathsf{v}_1 = \frac{dx_1}{\mathrm{d}\tau} = \frac{\mathrm{d}x_1}{\mathrm{d}t}\frac{\mathrm{d}t}{\mathrm{d}\tau} = \gamma u_1,$$

$$\mathsf{v}_2 = \frac{\mathrm{d}x_2}{\mathrm{d}\tau} = \gamma u_2, \quad \mathsf{v}_3 = \frac{\mathrm{d}x_3}{\mathrm{d}\tau} = \gamma u_3,$$

and

$$\mathsf{v}_4 = \frac{\mathrm{d}x_4}{\mathrm{d}\tau} = \frac{\mathrm{d}(ict)}{\mathrm{d}\tau} = ic\gamma.$$

in the system S′. Here, a so-called *bra* vector $\langle \boldsymbol{r}|$ is defined to facilitate matrix multiplication with the corresponding *ket* vector as a column matrix, i.e.

$$\langle \boldsymbol{r}| = \begin{pmatrix} r_1 \\ r_2 \\ r_3 \\ r_4 \end{pmatrix}.$$

Because the proper time is Lorentz invariant, the velocity vector $\langle \boldsymbol{v}|$ is transformed as $\langle \boldsymbol{r}|$, that is,

$$\mathsf{v}_1 = \gamma\left(\mathsf{v}_1' - \frac{iv}{c}\mathsf{v}_4'\right), \quad \mathsf{v}_2 = \mathsf{v}_2', \quad \mathsf{v}_3 = \mathsf{v}_3' \quad \text{and} \quad \mathsf{v}_4 = \gamma\left(\frac{iv}{c}\mathsf{v}_1' + \mathsf{v}_4'\right).$$

Similarly, the four-dimensional acceleration can be defined as

$$\langle \boldsymbol{a}| = \frac{\mathrm{d}\,\langle \mathbf{v}|}{\mathrm{d}\tau},$$

which transforms similarly to the velocity $\langle v|$. It is, however, neither simple nor useful for practical applications for us to elaborate the derivation here. Nevertheless, we can derive the following relation

$$\langle \mathsf{v} \mid \mathsf{v}\rangle = \sum_{\mathrm{i}=1}^{4} \mathsf{v}_{\mathrm{i}}^{2} = \gamma^{2}v^{2} - \gamma^{2}c^{2} = -\,c^{2},$$

and hence

$$\frac{\mathrm{d}}{\mathrm{d}\tau}\langle \mathbf{v} \mid \mathbf{v}\rangle = 2\,\langle \mathbf{v} \mid \boldsymbol{a}\rangle = 0, \tag{14.6}$$

which is a useful relation.

14.5. Relativistic Equation of Motion

Newton's law is fundamental in all inertial systems and sufficiently accurate, provided that a system's relative speed v is sufficiently small compared with the speed of light c. Under this condition , the equation of motion is regarded as invariant under the Lorentz transformation.

Newton's equation of motion for a particle of mass m_{o} in the three-dimensional space is

$$\boldsymbol{F} = m_{\mathrm{o}}\frac{\mathrm{d}\boldsymbol{u}}{\mathrm{d}t},$$

and the corresponding equation in four-dimensional space-time can be written as

$$\langle \mathsf{F}| = m_{\mathrm{o}}\frac{\mathrm{d}\,\langle \mathbf{v}|}{\mathrm{d}\tau},$$

where the four-dimensional vector $\langle \mathsf{F}| = (\mathsf{F}_1, \mathsf{F}_2, \mathsf{F}_3, \mathsf{F}_4)$ are defined by

$$(\boldsymbol{\mathsf{F}}_1, \boldsymbol{\mathsf{F}}_2, \boldsymbol{\mathsf{F}}_3) = m_{\mathrm{o}}\gamma\frac{\mathrm{d}\left(\mathbf{v}_1, \mathbf{v}_2, \mathbf{v}_3\right)}{\mathrm{d}t} = \gamma\boldsymbol{F},$$

implying that the effective mass $m = m_{\mathrm{o}}\gamma$ is not a constant and is written as

$$m = \frac{m_{\mathrm{o}}}{\sqrt{1-\beta^{2}}}. \tag{14.7}$$

For the fourth component F_4 we use the relation (14.7), which can be written as

$$\mathsf{v}_1\mathsf{F}_1 + \mathsf{v}_2\mathsf{F}_2 + \mathsf{v}_3\mathsf{F}_3 + \mathsf{v}_4\mathsf{F}_4 = 0,$$

and hence

$$\gamma^{2}\boldsymbol{u}\cdot\boldsymbol{F} + ic\gamma\mathsf{F}_4 = 0,$$

from which we obtain

$$F_4 = \frac{i\gamma}{c}\boldsymbol{u}.\boldsymbol{F}.$$

Here, writing that $\boldsymbol{u}.\boldsymbol{F} = \frac{\mathrm{d}K}{\mathrm{d}t}$, the quantity K is the kinetic energy, for which we have the relation

$$\frac{\mathrm{d}K}{\mathrm{d}t} = \frac{c}{i\gamma}F_4 = \frac{c}{i\gamma}m_o\gamma\frac{\mathrm{d}v_4}{\mathrm{d}t} = \frac{\mathrm{d}\left(m_o c^2\gamma\right)}{\mathrm{d}t}$$

because of $v_4 = ic\gamma$ integrating this equation we obtain

$$K = \frac{m_o c^2}{\sqrt{1-\beta^2}} + K_o = mc^2 + K_o, \tag{14.8}$$

where K_o is constant. Expanding $1/\sqrt{1-\beta^2}$ for a small β, i.e., $v \ll c$, (14.8) can be written as

$$K - K_o = \Delta K = mc^2 = m_o c^2 + \frac{1}{2}m_o v^2 + \cdots \tag{14.9a}$$

For a particle of $v = 0$, $m = m_o$, which is therefore called the *rest mass*. In this case we have the relation

$$\Delta K = m_o c^2, \tag{14.9b}$$

signifying the equivalence of mass and energy, which is substantiated by a *mass defect* of an atomic nucleus. The nucleus of an atom of atomic number Z and mass number A is composed of Z protons and $(A - Z)$ neutrons, but the nuclear mass m is empirically given by

$$m = Zm_\mathrm{p} + (A - Z)m_\mathrm{n} - \Delta,$$

where Δ is interpreted as related to the energy released when these protons and neutrons are bound to form nuclear matter. Hence, the *mass defect* Δ is a useful measure for the binding energy Δc^2 of a nucleus. According to (14.9a), besides the mass energy $m_o c^2$, ΔK represents the classical kinetic energy $\frac{1}{2}m_o v^2$ in the limit of $v \to c$.

The relativistic momentum in four-dimensional space can be defined as

$$\langle \boldsymbol{p}| = m_o \langle \mathbf{V}| = (p_1, p_2, p_3, p_4),$$

where

$$p_1 = m_o\gamma u_1,\ p_2 = m_o\gamma u_2,\ p_3 = m_o\gamma u_3,\ p_4 = im_o c\gamma = \frac{i\Delta K}{c}.$$

We can see that

$$\langle \boldsymbol{p} \mid \boldsymbol{p}\rangle = \sum_{\mathrm{i}=1}^{4} \boldsymbol{p}_\mathrm{i}^2 = \boldsymbol{p}_1^2 + \boldsymbol{p}_2^2 + \boldsymbol{p}_3^2 - \frac{\Delta K^2}{c^2} = m_o \langle \mathbf{V} \mid \mathbf{V}\rangle = m_o^2 c^2,$$

which is constant and Lorentz invariant. Also, the result indicates that

$$\Delta K^2 = c^2 \boldsymbol{p}.\boldsymbol{p} + m_o^2 c^4 \quad \text{where} \quad \boldsymbol{p} = m\boldsymbol{u} = \frac{\Delta K}{c^2}\boldsymbol{u}. \tag{14.10}$$

We considered a particle of a finite rest mass m_o in the above argument. On the other hand, to travel at the speed c, a particle should have zero mass $m_o = 0$, for which $\Delta K = cp$ from (14.10). It is noted that an electromagnetic radiation is characterized by a pressure described by the relation $U = cp$, as discussed in the previous chapter. Therefore, such a pressure p should be originated from the intrinsic nature of the electromagnetic field, unrelated to the gravitational character, which may therefore be expressed effectively by $m_o = 0$. To be consistent with such a relativistic argument, the electromagnetic radiation can be considered as consisting of a number of *massless* particles whose momentum is $p = \Delta K/c$.

14.6. The Electromagnetic Field in Four-Dimensional Space

Maxwell's equations constitute the fundamental law for electromagnetism in all inertial systems, which should therefore be invariant under the Lorentz transformation. Physically, the charge and current density are conserved as a source, obeying the equation of continuity. It is fundamental that such a conservation law should govern Maxwell's equations under the Lorentz transformation. Writing the continuity equation in two systems S and S′ as

$$\frac{\partial \rho}{\partial t} + \operatorname{div}(\rho v) = 0 \qquad \text{and} \qquad \frac{\partial \rho'}{\partial t'} + \operatorname{div}\left(\rho' v'\right) = 0,$$

the variables $(\rho, \rho\boldsymbol{v})$ and $(\rho', \rho'\boldsymbol{v}')$ can be transformed in four dimensions similar to a four velocity vector $(\boldsymbol{v}, ic)$. We can therefore define the four-dimensional vector

$$\langle \boldsymbol{s} | = (\rho v_1, \rho v_2, \rho v_3, ic\rho),$$

thereby expressing the continuity relation as

$$\sum_i \frac{\partial s_i}{\partial x_i} = 0 \quad \text{and} \quad \sum_i \frac{\partial s'_i}{\partial x'_i} = 0. \tag{14.11}$$

These s_i and s'_i are linearly related, so that

$$s'_i = \alpha_{ik} s_k,$$

for which the product $\langle \boldsymbol{s} | \boldsymbol{s} \rangle$ should be invariant. Namely, we have

$$\sum_i s_i s_i = \sum_i s'_i s'_i,$$

hence

$$\rho^2 (v_1^2 + v_2^2 + v_3^2 - c^2) = \text{constant}.$$

Expressing this constant as ρ_o^2 and writing $\rho = \gamma\rho_o$, we obtain

$$\langle s| = \rho_o \langle \boldsymbol{v}| , \tag{14.12}$$

resulting in the linear relation between the four charge-current density vector and the four velocities, as expected.

The electric field vector $\boldsymbol{E}$ in three-dimensional space exerts a force $\boldsymbol{F} = q\boldsymbol{E}$ on a charge q. The force in four-dimensional space can be written as $\langle \boldsymbol{F}| = (\gamma q E_1, \gamma q E_2, \gamma q E_3, \frac{iq\gamma}{c} v.\boldsymbol{E})$, for which the four-dimensional velocity is $\langle \boldsymbol{v}| = (\gamma v_1, \gamma v_2, \gamma v_3, ic\gamma)$. In this context, the electric field in four-dimensional space can be defined as a 4×4 tensor that relates $\langle \boldsymbol{F}|$ with $\langle \boldsymbol{v}|$. Therefore, expressing such a tensor by $\mathbf{E}$, we have the expression

$$\langle \boldsymbol{F}| = \frac{q}{c} \langle \boldsymbol{v}| \, \mathbf{E}. \tag{14.13}$$

On the other hand, writing

$$\langle \boldsymbol{F}| = \frac{q}{c}(ic\gamma)(-iE_1), \frac{q}{c}(ic\gamma)(-iE_2), \frac{q}{c}(ic\gamma)(-iE_3), \frac{q}{c} i\gamma \boldsymbol{u}.\boldsymbol{E},$$

the tensor $\mathbf{E}$ can be given explicitly as

$$\mathbf{E} = \begin{pmatrix} 0 & 0 & 0 & -iE_1 \\ 0 & 0 & 0 & -iE_2 \\ 0 & 0 & 0 & -iE_3 \\ iE_1 & iE_2 & iE_3 & 0 \end{pmatrix}, \tag{14.14}$$

which is asymmetric with respect to the indexes, i.e., $\mathsf{E}_{ij} = -E_{ji}$.

The Lorentz transformation between four-dimensional vectors $\langle \boldsymbol{r}| = (x, y, z, ict)$ and

$$\langle \boldsymbol{r}'| = \begin{pmatrix} x' \\ y' \\ z' \\ ict' \end{pmatrix}$$

can be expressed as

$$\mathbf{T} = \begin{pmatrix} \gamma & 0 & 0 & -iv\gamma/c \\ 0 & 1 & 0 & 0 \\ 0 & 0 & 1 & 0 \\ iv\gamma/c & 0 & 0 & \gamma \end{pmatrix}, \tag{14.15}$$

for which we have the relation $\mathbf{T}^{-1}\mathbf{T} = \mathbf{e}$, where $\mathbf{e}$ the unit matrix in four dimensions, thereby transforming force $\langle \boldsymbol{F}|$ as

$$\langle \boldsymbol{F}'| = \quad \langle \boldsymbol{F}|\mathbf{T}^{-1} \quad \text{and} \quad |\boldsymbol{F}'\rangle = \mathbf{T}|\boldsymbol{F}\rangle .$$

Using (14.13), the latter can be expressed as

$$|\boldsymbol{F}'\rangle = \frac{q}{c}\mathbf{TE}|\boldsymbol{v}\rangle = \frac{q}{c}\mathbf{TET}^{-1}|\boldsymbol{v}'\rangle ,$$

and therefore the electric field tensor (14.12) in the S′ system is transformed to S′ as

$$\mathsf{E}' = \mathsf{TET}^{-1}. \tag{14.16}$$

Calculating the detail with (14.13),

$$\mathsf{E}' = \begin{pmatrix} \gamma & 0 & 0 & iv\gamma/c \\ 0 & 1 & 0 & 0 \\ 0 & 0 & 1 & 0 \\ -iv\gamma/c & 0 & 0 & \gamma \end{pmatrix} \times \begin{pmatrix} 0 & 0 & 0 & -iE_1 \\ 0 & 0 & 0 & -iE_2 \\ 0 & 0 & 0 & -iE_3 \\ iE_1 & iE_2 & iE_3 & 0 \end{pmatrix}$$

$$\times \begin{pmatrix} \gamma & 0 & 0 & -iv\gamma/c \\ 0 & 1 & 0 & 0 \\ 0 & 0 & 1 & 0 \\ iv\gamma/c & 0 & 0 & \gamma \end{pmatrix}.$$

$$= \begin{pmatrix} 0 & -v\gamma E_2/c & -v\gamma E_3 & -iE_1 \\ v\gamma E_2/c & 0 & 0 & -i\gamma E_2 \\ v\gamma E_3/c & 0 & 0 & -i\gamma E_3 \\ iE_1 & i\gamma E_2 & i\gamma E_3 & 0 \end{pmatrix}.$$

Changing notation as

$$E_1' = E_1, \quad E_2' = \gamma E_2, \quad E_3' = \gamma E_3; \quad B_2' = -\frac{v\gamma E_3}{c^2}, \quad B_3' = \frac{v\gamma E_2}{c^2},$$

we can write that

$$\mathsf{E}' = \begin{pmatrix} 0 & cB_3' & -cB_2' & -iE_1' \\ -cB_3' & 0 & 0 & -iE_2' \\ cB_2' & 0 & 0 & -iE_3' \\ iE_1' & iE_2' & iE_3' & 0 \end{pmatrix}. \tag{14.17}$$

Accordingly,

$$|\boldsymbol{F}\rangle = \frac{q}{c}\mathsf{E}'\,|\boldsymbol{v}\rangle = q\gamma \begin{pmatrix} u_2'B_3' - u_3'B_2' + E_1' \\ -u_1'B_3' + E_2' \\ u_1'B_2' + E_3' \\ i\boldsymbol{u}' \cdot \boldsymbol{E}'/c \end{pmatrix},$$

which can be interpreted in three-dimensional space in S′ system as

$$\boldsymbol{F}' = \frac{F_1'\boldsymbol{i} + F_2'\boldsymbol{j} + F_3'\boldsymbol{k}}{\gamma} = q\left(\boldsymbol{E}' + \boldsymbol{u}' \times \boldsymbol{B}'\right),$$

where

$$\boldsymbol{E}' = E_1'\boldsymbol{i} + E_2'\boldsymbol{j} + E_3'\boldsymbol{k} \quad \text{and} \quad \boldsymbol{B}' = B_2'\boldsymbol{j} + B_3'\boldsymbol{k}$$

are conventional expressions in three dimensions, and where the vector $\boldsymbol{B}'$ should be the magnetic field observed in S′ that is transformed from the electric field in S.

Thus, the tensor $\mathbf{E}'$ is not purely electric, indicating that an electric field in a moving system S′ also consists of a magnetic feature, as observed in S. Needless to say, this is equivalent to considering that a moving charge at a velocity $\boldsymbol{v}$ in a magnetic field $\boldsymbol{B}$ will be deflected by the dynamic electric field given by $\boldsymbol{v} \times \boldsymbol{B}$.

In four-dimensional space-time, the electromagnetic field $(\boldsymbol{E}, \boldsymbol{B})$ is therefore considered as a 4×4 tensor in a form generalized from (14.17) in S′, that is

$$\mathbf{G} = \begin{pmatrix} 0 & cB_3 & -cB_2 & -iE_1 \\ -cB_3 & 0 & cB_1 & -iE_2 \\ cB_2 & -cB_1 & 0 & -iE_3 \\ iE_1 & iE_2 & iE_3 & 0 \end{pmatrix}, \qquad (14.18)$$

which is anti-symmetrical, i.e., $G_{ij} = -G_{ji}$, being composed of

$$(G_{23}, G_{31}, G_{12}) = c\boldsymbol{B} \quad \text{and} \quad (G_{11}, G_{22}, G_{33}) = \boldsymbol{E}. \qquad (14.19)$$

Calculating the transformed $\mathbf{G'} = \mathbf{T}^{-1}\mathbf{GT}$ from the elements G'_{ij} we obtain

$$E'_1 = E_1, \quad E'_2 = \gamma\,(E_2 - vB_3), \quad E'_3 = \gamma\,(E_3 + vB_2)$$
$$B'_1 = B_1, \quad B'_2 = \gamma\left(B_2 + \frac{v}{c^2}E_3\right), \quad B'_3 = \gamma\left(B_3 - \frac{v}{c^2}E_2\right).$$

Solving these for components in the S system,

$$E_1 = E'_1, \quad E_2 = \gamma\left(E'_2 + v_3 B'_3\right), \quad E_3 = \gamma\left(E'_3 - vB'_2\right)$$
$$B_1 = B'_1, \quad B_2 = \gamma\left(B'_2 - \frac{v}{c^2}E'_3\right), \quad B_3 = \gamma\left(B'_3 + \frac{v}{c^2}E'_2\right),$$

which can also obtained by exchanging the primed and unprimed components, e.g., E_1 and E'_1, B_2 and B'_2, etc., and replacing v by $-\,v$ in the former set. Considering $\boldsymbol{v} = (v, 0, 0)$, from the latter set we can show that

$$\boldsymbol{E} = \gamma\boldsymbol{E'} + (1-\gamma)\frac{v\left(v.\boldsymbol{E'}\right)}{v^2} + \gamma\left(v \times \boldsymbol{B'}\right)$$
$$\boldsymbol{B} = \gamma\boldsymbol{B'} + (1-\gamma)\frac{v\left(v.\boldsymbol{B'}\right)}{v^2} + \gamma(v \times \boldsymbol{E'}),$$

which are a valid form by the Lorentz transformation between S and S′.

The Maxwell equations are composed by first derivatives of field vectors with respect to space-time coordinates. Therefore, such derivatives should be transformed as a four-vector matrix in the four-dimensional space-time. We notice that derivatives of the field tensor like $\partial G_{ij}/\partial x_k$ should be combined to construct a four-vector tensor. Calculating $\sum_{i=1}^{4}\frac{\partial G_{ij}}{\partial x_i}$ for $j = 1, 2, 3, 4$ with (14.17), we obtain

$$\sum_{i=1}^{4}\frac{\partial G_{i1}}{\partial x_i} = \frac{\partial G_{11}}{\partial x_1} + \frac{\partial G_{21}}{\partial x_2} + \frac{\partial G_{31}}{\partial x_3} + \frac{\partial G_{41}}{\partial x_4}$$
$$= -c\frac{\partial B_3}{\partial x_2} + c\frac{\partial B_2}{\partial x_3} + i\frac{\partial E_1}{\partial x_4} = -c\mu_o\,(\text{curl}\,\boldsymbol{H})_1 + \frac{i}{ic\varepsilon_o}\frac{\partial D_1}{\partial t},$$

which can be expressed as

$$= \mu_o c \left\{ -(\text{curl}\,\boldsymbol{H})_1 + \frac{\partial D_1}{\partial t} \right\}.$$

Similarly,

$$\sum_{i=1}^{4} \frac{\partial G_{i2}}{\partial x_i} = \mu_o c \left\{ -(\text{curl}\,\boldsymbol{H})_2 + \frac{\partial D_2}{\partial t} \right\},$$

$$\sum_{i=1}^{4} \frac{\partial G_{i3}}{\partial x_i} = \mu_o c \left\{ -(\text{curl}\,\boldsymbol{H})_3 + \frac{\partial D_3}{\partial t} \right\}$$

and

$$\sum_{i=1}^{4} \frac{\partial G_{i4}}{\partial x_i} = i\,(\text{div}\,\boldsymbol{E}).$$

Hence, these derivatives constitute a four-vector that can be written as

$$\left(\sum_{i=1}^{4} \frac{\partial G_{i1}}{\partial x_i}, \sum_{i=1}^{4} \frac{\partial G_{i2}}{\partial x_i}, \sum_{i=1}^{4} \frac{\partial G_{i3}}{\partial x_i}, \sum_{i=1}^{4} \frac{\partial G_{i4}}{\partial x_i} \right) \propto \rho_o\,|\boldsymbol{v}\rangle, \tag{14.20}$$

corresponding to the Maxwell equations

$$\text{curl}\,\boldsymbol{H} - \frac{\partial \boldsymbol{D}}{\partial t} = \boldsymbol{j} \quad \text{and} \quad \text{div}\,\boldsymbol{D} = \rho. \tag{i}$$

The field tensor **G** per se is clearly insufficient for all of the Maxwell equations in space-time. The other Maxwell equations,

$$\text{curl}\,\boldsymbol{E} + \frac{\partial \boldsymbol{B}}{\partial t} = 0 \quad \text{and} \quad \text{div}\,\boldsymbol{B} = 0, \tag{ii}$$

can formally be obtained from (i) by substituting $(\boldsymbol{H}, \boldsymbol{D})$ by $(\boldsymbol{E}, -\boldsymbol{B})$, replacing the electrical source by a zero magnetic source, i.e., $(\boldsymbol{j}, \rho) \to (0, 0)$. In this context, instead of **G** defined by (14.18) we consider another field tensor $\mathbf{G}^*$ that is obtained by replacing $(\boldsymbol{B}, \boldsymbol{E})$ by $(\boldsymbol{E}, -\boldsymbol{B})$, that is

$$\mathbf{G}^* = \begin{pmatrix} 0 & -iE_3 & iE_2 & cB_1 \\ iE_3 & 0 & -iE_1 & cB_2 \\ -iE_2 & iE_1 & 0 & cB_3 \\ -cB_1 & -cB_2 & -cB_3 & 0 \end{pmatrix}, \tag{14.21}$$

which is known as the *dual* of the tensor **G**.

By similar calculation on the dual $\mathbf{G}^*$, we can show that

$$\sum_{i=1}^{4} \frac{\partial G^*_{ik}}{\partial x_i} = -i\,(\text{curl}\,\boldsymbol{E})_k - i\frac{\partial B_k}{\partial t}, \quad \text{where k} = 1, 2, 3$$

and

$$\sum_{i=1}^{4} \frac{\partial G_{i4}^{*}}{\partial x_i} = -c \operatorname{div} \boldsymbol{B}.$$

Therefore, we have a four-vector

$$\left(\sum_{i=1}^{4} \frac{\partial G_{i1}^{*}}{\partial x_i}, \sum_{i=1}^{4} \frac{\partial G_{i2}^{*}}{\partial x_i}, \sum_{i=1}^{4} \frac{\partial G_{i3}^{*}}{\partial x_i}, \sum_{i=1}^{4} \frac{\partial G_{i4}^{*}}{\partial x_i} \right) = 0, \qquad (14.22)$$

which is consistent with the Maxwell equations (ii).

Although the mathematical detail for the tensor **G** and its dual $\mathbf{G}^*$ and their transformation was not discussed, the Lorentz invariance of the Maxwell theory was verified in the above argument. Those readers interested in the formal analysis should refer to books on the relativity theory, such as *The Theory of Relativity* by C. Møller.

15 Waves and Boundary Problems

15.1. Skin Depths

Although a static conductor is characterized by $\boldsymbol{E} = 0$, the electric field $\boldsymbol{E}$ can penetrate into a conducting medium if it is varying as a function of time. In a conductor, charge carriers are mobilized by an applied electric field so that currents flow through a conductor under time-dependent conditions. Distributed currents, called *eddy currents*, flow at high densities in the close vicinity of surfaces to certain depth, called the *spin-depth*. In free space, electromagnetic waves are reflected by dielectric boundaries as well as conducting surfaces where surface currents play a significant role.

A magnetic field can also penetrate into a conductor. In a uniform wire of conductivity σ of circular cross-section of radius R, the magnetic field originates from a current through the wire. For a magnetic field outside a current, we have considered a thin wire with the Ampère law, but the magnetic field also exists *inside* the wire corresponding to the distributed current. Such a conduction current is related to an applied field $\boldsymbol{E}$ as described by the Ohm relation $\boldsymbol{j} = \sigma \boldsymbol{E}$, and it is also related to the magnetic field $\boldsymbol{H}$ by the Maxwell equations, that is

$$\operatorname{curl} \boldsymbol{H} = \boldsymbol{j} + \varepsilon \frac{\partial \boldsymbol{E}}{\partial t} \quad \text{and} \quad \operatorname{curl} \boldsymbol{E} = -\mu \frac{\partial \boldsymbol{H}}{\partial t}, \tag{15.1}$$

where constants ε and μ of the conductor can be assumed as close to ε_0 and μ_0 in vacuum space.

Consider a sinusoidal current $I = I_0 \exp(i\omega t)$ along the z-axis of a long cylindrical conductor. Using cylindrical coordinates (r, θ, z), the circular component of the magnetic field H_θ inside and outside the conductor is, by symmetry, a function of radius r, varying at the frequency ω; and the total current density is contributed by conduction and displacement currents and hence given by $\sigma E_z + i\omega\varepsilon E_z$, where E_z is related to H_θ. Under normal circumstances however, the second term is negligibly small as compared with the first. A metallic conductivity σ is typically of the order of $10^8 \Omega^{-1}/\text{m}$, which is much greater than $\omega\varepsilon$ even at a frequency ω in

MHz. In this case, combining two equations in (15.1), we write

$$\operatorname{curl}\boldsymbol{j} = -i\omega\sigma\mu_o\boldsymbol{H}, \tag{i}$$

and

$$\operatorname{curl}(\operatorname{curl}\boldsymbol{j}) = \alpha^2\boldsymbol{j}, \text{ where } \alpha^2 = -\omega\sigma\mu_o. \tag{ii}$$

Considering that the $\boldsymbol{H}$ vector is dominated by H_θ, the corresponding current density is signified only by $j_z(r)$, where r is the radial coordinate, as determined by (i), i.e.,

$$H_\theta = \frac{1}{\alpha^2 r}\frac{\mathrm{d}}{\mathrm{d}r}(rj_z), \tag{iii}$$

and from (ii) we have

$$-\frac{\mathrm{d}}{\mathrm{d}r}\left[\frac{1}{r}\frac{\mathrm{d}}{\mathrm{d}r}(rj_z)\right] = \alpha^2 j_z.$$

This equation can be expressed as

$$\frac{\mathrm{d}^2 j_z}{\mathrm{d}r^2} + \frac{1}{r}\frac{\mathrm{d}j_z}{\mathrm{d}r} + \left(\alpha^2 - \frac{1}{r^2}\right) j_z = 0,$$

which is a Bessel's equation, whose solution can be written as

$$j_z(r) = j_o J_1(\alpha r). \tag{iv}$$

Using (iv) in (iii), we have

$$H_\theta = \frac{1}{\alpha^2}\left(\frac{j_z}{r} + \frac{\mathrm{d}j_z}{\mathrm{d}r}\right) = \frac{j_o}{\alpha^2}\left(\frac{J_1(\alpha r)}{r} + \frac{\mathrm{d}J_1(\alpha r)}{\mathrm{d}r}\right) = \frac{j_o}{\alpha}J_o(\alpha r), \tag{v}$$

where the last expression was obtained by a recurrence formula for Bessel functions. This expression of H_θ in (v) must be consistent with the Ampère law at $r = R$, and hence $j_o = \dfrac{\alpha I}{2\pi R J_o(\alpha R)}$. Therefore, from (iv) and (v) the current density and magnetic field inside the wire are expressed by

$$j_z(r) = \frac{\alpha I}{2\pi R}\frac{J_1(\alpha r)}{J_o(\alpha R)} \quad \text{and} \quad H_\theta = \frac{I}{2\pi R}\frac{J_o(\alpha r)}{J_o(\alpha R)} \quad \text{for } r < R,$$

respectively.

From (ii), the constant α is a complex parameter, and written as

$$\alpha = \frac{1-i}{\delta}, \quad \text{where} \quad \delta = \sqrt{\frac{2}{\omega\sigma\mu_o}} \tag{vi}$$

For $r \gg \delta$ the current density $j_z(r,t)$ can be expressed asymptotically as

$$j_z(r,t) = \frac{I}{2\pi R\delta}\exp\left(\frac{r-R}{\delta}\right)\exp\left[i\left(\omega t + \frac{\pi}{4} + \frac{R-r}{\delta}\right)\right],$$

and the magnetic field is

$$H_\theta = \frac{I}{2\pi R}\exp\left(\frac{r-R}{\delta}\right)\exp\left[i\left(\omega t + \frac{R-r}{\delta}\right)\right].$$

In this approximation, the parameter $\delta(0 < \delta < R)$ signifies the effective depth to which the current density must penetrate into the conductor, which is, however, limited to a thin area close to the surface of radius R. It is noted that at $r = 0$, $j_z = 0$, but $H_z \neq 0$, as given by the Bessel functions $J_1(\alpha r)$ and $J_o(\alpha r)$, taking zero and non-zero values at $r = 0$, respectively.

15.2. Plane Electromagnetic Waves in a Conducting Medium

In this section, we consider a plane electromagnetic wave propagating through a uniform medium characterized by a conductivity σ. Mobile electrons dominate conducting properties, although the wave's dielectric and magnetic properties determined, respectively, by ε and μ depend on the host medium not including its electrons, which are insignificant in most conductors. In normal conductors the Maxwell equations are written as

$$\text{curl}\,\boldsymbol{H} = \sigma\boldsymbol{E} + \varepsilon_o\frac{\partial \boldsymbol{E}}{\partial t} \quad \text{and} \quad \text{curl}\,\boldsymbol{E} = -\mu_o\frac{\partial \boldsymbol{H}}{\partial t}.$$

In this case, it is convenient to use the vector potential $\boldsymbol{A}$ instead of $\boldsymbol{E}$ and $\boldsymbol{H}$, namely

$$\boldsymbol{E} = -\frac{\partial \boldsymbol{A}}{\partial t} \quad \text{and} \quad \boldsymbol{H} = \frac{1}{\mu_o}\,\text{curl}\,\boldsymbol{A},$$

and we obtain the equation

$$\nabla^2\boldsymbol{A} - \varepsilon_o\mu_o\frac{\partial^2\boldsymbol{A}}{\partial t^2} - \mu_o\sigma\frac{\partial \boldsymbol{A}}{\partial t} = 0. \tag{15.2}$$

For a plane wave, writing $\boldsymbol{A}(\boldsymbol{r},t) = \boldsymbol{A}_o\exp[i(\boldsymbol{k}\,.\boldsymbol{r} - \omega t)]$ in (15.2), we obtain the relation

$$-\boldsymbol{k}^2 + \omega^2\varepsilon_o\mu_o + i\omega\mu_o\,\sigma = 0.$$

Hence, the wave vector $\boldsymbol{k}$ is a complex variable, that is, $k = k' + ik''$, where the real and imaginary parts k' and k'' are expressed by the relations

$$k'^2 - k''^2 = \omega^2\varepsilon_o\mu_o \quad \text{and} \quad 2k'k'' = \sigma\mu_o\omega.$$

Solving these equations for k' and k'', we obtain

$$k' = \omega\sqrt{\frac{\varepsilon_o\mu_o}{2}}\left\{1+\sqrt{1+\left(\frac{\sigma}{\omega\varepsilon_o}\right)^2}\right\}^{1/2} \quad \text{and}$$

$$k'' = \sqrt{\frac{\varepsilon_o\mu_o}{2}}\left\{\sqrt{1+\left(\frac{\sigma}{\omega\varepsilon_o}\right)^2}-1\right\}^{1/2}.$$

For a normal conductor, the condition $\sigma \gg \omega\varepsilon_o$ an be applied at a typical frequency ω, so that these k' and k'' can be approximately given by

$$k' \approx k'' \approx \omega\sqrt{\frac{\varepsilon_o\mu_o}{2}\frac{\sigma}{\omega\varepsilon_o}} = \sqrt{\frac{\omega\sigma\mu_o}{2}}. \tag{15.3}$$

In this case, the vector potential can be written as

$$\boldsymbol{A}(\boldsymbol{r},t) = \boldsymbol{A}_o\exp(-\boldsymbol{k}''.\boldsymbol{r})\exp[i(\boldsymbol{k}'.\boldsymbol{r}-\omega t)], \tag{15.4}$$

indicating that the amplitude decreases as propagating along the direction of $\boldsymbol{k}$. We can define a characteristic distance for the amplitude decay by

$$\delta = \frac{1}{k''} = \sqrt{\frac{2}{\omega\sigma\mu_o}},$$

which is identical to the skin depth defined in Section 15.1.

Writing $\boldsymbol{k} = (k'+ik'')\boldsymbol{u} = k'(1+i)\boldsymbol{u}$, where the unit vector $\boldsymbol{u}$ signifies a *linearly polarized* wave, the field vectors can be calculated for the vector potential of (15.3), that is,

$$\boldsymbol{H}(\boldsymbol{r},t) = \frac{\text{curl}\,\boldsymbol{A}}{\mu_o} = \frac{ik'(1+i)}{\mu_o}\boldsymbol{u}\times\boldsymbol{A}(\boldsymbol{r},t) \approx i\sqrt{\frac{\omega\sigma}{\mu_o}}\exp\left\{\frac{\pi i}{4}[\boldsymbol{u}\times\boldsymbol{A}(\boldsymbol{r},t)]\right\}$$
$$= i\sqrt{\frac{\omega\sigma}{\mu_o}}(\boldsymbol{u}\times\boldsymbol{A}_o)\exp\left(-\boldsymbol{k}''.\boldsymbol{r}\right)\exp\left[i\left(\boldsymbol{k}'.\boldsymbol{r}-\omega t+\frac{\pi}{4}\right)\right]$$

and

$$\boldsymbol{E}(\boldsymbol{r},t) = i\omega\boldsymbol{A}(\boldsymbol{r}.t) = i\omega\boldsymbol{A}_o\exp(-\boldsymbol{k}''.\boldsymbol{r})\exp[i(\boldsymbol{k}'.\boldsymbol{r}-\omega t)].$$

It is noted that these amplitudes are related as $\sqrt{\frac{\varepsilon_o}{\mu_o}}\frac{|\boldsymbol{E}|}{|\boldsymbol{H}|} = \sqrt{\frac{\omega\varepsilon_o}{\sigma}} \ll 1$, and hence $|\boldsymbol{E}| \ll |\boldsymbol{H}|$ in normal conductors. In fact, by considering $\sigma = \infty$, an idealized conductor may be characterized by $\boldsymbol{E} = 0$, as in a static case, even at high frequencies. In this context we can set up boundary conditions for the electromagnetic field to satisfy $\boldsymbol{E} = 0$ on conducting surfaces, which is consistent to assuming the skin depth δ as zero.

15.3. Boundary Conditions for Propagating Waves

The plane wave is a useful model for electromagnetic radiation propagating away from the source. In practice, a plane wave propagates through media of conducting and non-conducting materials for which the media boundaries play a significant role in transmitting energies to carry information by the wave. In the presence of boundaries waves can no longer be "free" to propagate, as in isolated space, and the field should necessarily be modified, as expressed by a linear combination of appropriate waves to satisfy boundary conditions.

Because of a high conductivity σ, a normal conductor surface can be characterized by a thin layer of the skin depth δ, where eddy currents flow to keep the field from penetrating deeper, whereas the boundary between two uniform dielectric materials 1 and 2 is characterized by different dielectric constants ε_1 and ε_2. Such boundaries can be expressed by a mathematical surface, but physically signified by surface currents and charges, although on dielectric boundaries no boundary charges are normally considered.

The basic equation curl $\boldsymbol{H} = \boldsymbol{j}$ should determine the surface condition for the $\boldsymbol{H}$ field that arises from a surface current. Figure 5.1 shows a flat conducting surface, where we consider a rectangular path for integration perpendicular to the current $\boldsymbol{j}$. Here, the lengths of 1-2 and 3-4 are denoted as l, and those of 2-3 and 4-1 are 2δ, where the surface current is assumed to flow through the bottom half of this rectangular path. Denoting the rectangular area by $\mathbf{A}$, the current can be expressed as $\boldsymbol{j}(l\delta) = \boldsymbol{K}l$, where $\boldsymbol{K} = \boldsymbol{j}\delta$ describes the surface density. For the equation curl $\boldsymbol{H} = \boldsymbol{j}$, the left side can be calculated as (curl $\boldsymbol{H}$) . $\mathbf{A} = (\boldsymbol{n} \times \boldsymbol{H})l$ because the $\boldsymbol{H}$ vector is all zero, except for the path 1-2 above the conductor. Hence, as the boundary condition we obtain the relation

$$\boldsymbol{n} \times \boldsymbol{H} = \boldsymbol{K}. \tag{15.4a}$$

Accordingly, the tangential component of $\boldsymbol{H}$ is discontinuous across the surface.

Applying curl $\boldsymbol{E} = 0$ to a similar path, we can obtain another condition

$$\boldsymbol{n} \times \boldsymbol{E} = 0 \tag{15.4b}$$

for the $\boldsymbol{E}$ vector. This formula (15.4b) indicates that the tangential component of $\boldsymbol{E}$ should be equal to zero at any outside point of conducting surface.

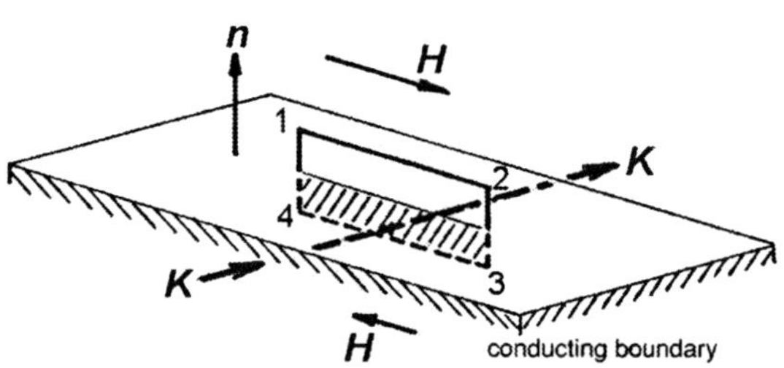

FIGURE 15.1. A boundary condition for tangential components of $\boldsymbol{H}$ vectors on a conducting surface. $\boldsymbol{n} \times (\boldsymbol{H}_1 - \boldsymbol{H}_2) = \boldsymbol{K}$, where $\boldsymbol{K} = \boldsymbol{j}\delta$ is the surface current density, $\boldsymbol{n}$ the normal vector, and δ the skin depth.

Next, we consider a pill-box shaped short cylindrical volume across a conducting boundary in parallel with its flat top and bottom, similar to Figure 3.5, where the basic equations div $\boldsymbol{B} = 0$ and div $\boldsymbol{D} = \rho$ are analyzed in integral form using the Gauss theorem. Assuming that the side area is negligible as compared with the top and bottom, we obtain that the normal component of $\boldsymbol{B}$ is continuous, and hence

$$(\boldsymbol{H}_\mathrm{n})_1 = (\boldsymbol{H}_\mathrm{n})_2, \tag{15.4c}$$

where we considered that $\mu_1 = \mu_2$. Inside a conducting medium 2, $(\boldsymbol{H}_\mathrm{n})_2$ is equal to zero, except in the skin depth, and hence, $\boldsymbol{H}$ should be *tangential* to the surface.

In contrast, because of distributed charge ρ, the $\boldsymbol{E}$ vector cannot be continuous, as we already discussed for the static case; namely from $(\boldsymbol{D}_\mathrm{n})_1 = \sigma$, we have $(\boldsymbol{E}_\mathrm{n})_1 = \sigma/\varepsilon_\mathrm{o}$.

Summarizing these conditions, for propagating waves the conducting surface demands the boundary conditions that

$$\boldsymbol{E}_\mathrm{tangential} = 0, \quad \boldsymbol{E}_\mathrm{normal} = \sigma/\varepsilon_\mathrm{o}; \quad \boldsymbol{H}_\mathrm{tangential} = \boldsymbol{K} \quad \text{and} \quad \boldsymbol{H}_\mathrm{normal} = 0. \tag{15.5}$$

It is clear that such conditions cannot be met by a single propagating wave specified by the wavevector $\boldsymbol{k}$ for which a superposition with another wave at different $\boldsymbol{k}'$ is required, resulting in a reflected wave from the surface.

On a dielectric boundary, in contrast, two different dielectric constants ε_1 and ε_2 are explicit, for which we can write conditions

$$\varepsilon_1(\boldsymbol{E}_\mathrm{n})_1 = \varepsilon_2(\boldsymbol{E}_\mathrm{n})_2 \quad \text{and} \quad (\boldsymbol{H}_\mathrm{n})_1 = (\boldsymbol{H}_\mathrm{n})_2, \tag{15.6a}$$

assuming that $\mu_1 = \mu_2$. On the other hand, as long as the absence of boundary current and charge signifies the dielectric properties, tangential components $\boldsymbol{E}_\mathrm{t}$ and $\boldsymbol{H}_\mathrm{t}$ should be continuous. Namely,

$$(\boldsymbol{E}_\mathrm{t})_1 = (\boldsymbol{E}_\mathrm{t})_2 \quad \text{and} \quad (\boldsymbol{H}_\mathrm{t})_1 = (\boldsymbol{H}_\mathrm{t})_2. \tag{15.6b}$$

Therefore, the boundary conditions are involved only in discontinuous $\boldsymbol{E}_\mathrm{n}$, as expressed by (15.6a), and the $\boldsymbol{H}$ vector has nothing to do with a dielectric boundary. However, the continuity of $\boldsymbol{E}_\mathrm{t}$ in (15.6b) cannot be disregarded, because at a dielectric boundary the $\boldsymbol{E}$ vector is modified as specified by discontinuous $\boldsymbol{E}_\mathrm{n}$ together with continuous $\boldsymbol{E}_\mathrm{t}$.

15.4. Reflection from a Conducting Boundary

Although expressible by optical analogy, the reflection and refraction of waves need to be interpreted in terms of electromagnetic propagation at boundaries. For a free wave we deal with the field vectors $\boldsymbol{E}$ and $\boldsymbol{H}$ in orthogonal relation with the wave vector $\boldsymbol{k}$ as given by (13.4) and (13.5). In optics, on the other hand, these vector relations represent *polarization of light* to indicate a direction of the electric field. For a plane electromagnetic wave such a polarization arises from the source current, and the natural optical light is normally "unpolarized," originating from atomic or molecular transitions in r and om phases.

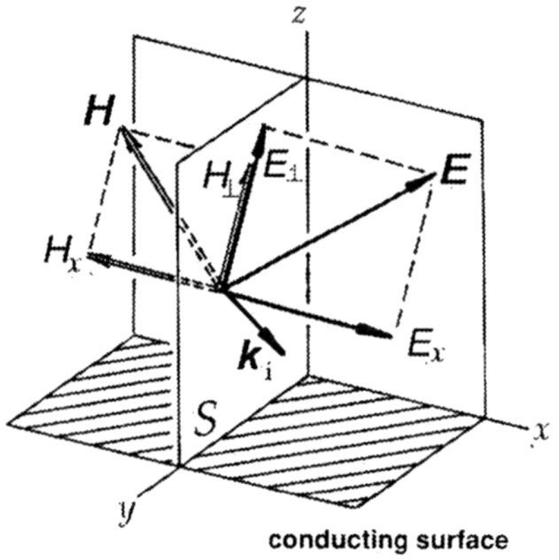

FIGURE 15.2. Plane free waves incident on a conducting surface. The y, z-plane is the incident plane S on which the wave vector k_i, $E_\perp$, and $H_\perp$ lie, and the other components E_x and H_x are perpendicular to S.

Figure 15.2 shows a plane wave incident upon a conducting surface, on which we set the x,y-coordinate plane, with the z-axis in the normal direction. It is always possible to consider that the wave vector $\boldsymbol{k}$ lies on a plane perpendicular to the x,y-plane, and hence we can consider the vector $\boldsymbol{k}$ on the y,z-plane for convenience, which is called the *incident plane S*. The $\boldsymbol{E}$ and $\boldsymbol{H}$ vectors associated with the incident wave can be in arbitrary directions with respect to S, and hence their components (E_x, $E_\perp$) and (H_x, $H_\perp$) are considered for the boundary condition, where E_x, H_x are perpendicular to S, and $E_\perp$, $H_\perp$ are in the plane of S, as illustrated in the figure.

The plane wave is signified by parallel lines along $\boldsymbol{k}$, with the corresponding phases $\phi = \boldsymbol{k} \,.\, \boldsymbol{r} - \omega t + 2\pi \mathrm{m}$, where m can be 0 or $\pm$integers, indicating solutions of the wave equation are identical at all phases ϕ_m that are separated by $\Delta r = 2\pi\mathrm{m}/k$ along and between these lines. As illustrated by Figure 15.3, we consider

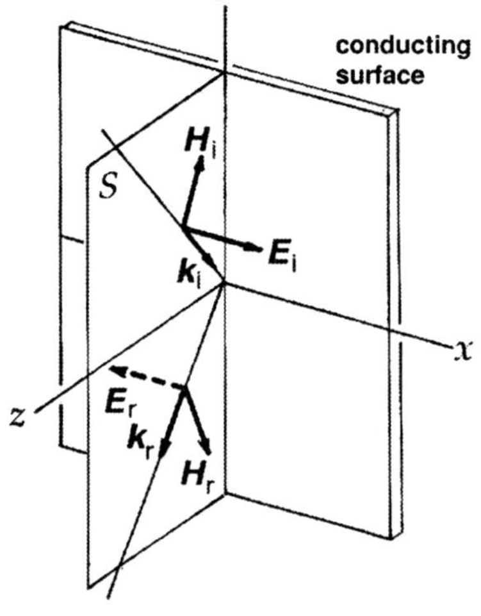

FIGURE 15.3. Incident and reflected plane waves on a conducting surface. ($\boldsymbol{k}_i$, $\boldsymbol{H}_i$) and ($\boldsymbol{k}_r$, $\boldsymbol{H}_r$) are in the incident plane S, whereas $\boldsymbol{E}_i$ and $\boldsymbol{E}_r$ are perpendicular to S.

an incident plane wave ($\boldsymbol{E}_\mathrm{i}$, $\boldsymbol{H}_\mathrm{i}$, $\boldsymbol{k}_\mathrm{i}$) and a *reflected* wave ($\boldsymbol{E}_\mathrm{r}$, $\boldsymbol{H}_\mathrm{r}$, $\boldsymbol{k}_\mathrm{r}$), which can be linearly combined at the same phase to satisfy the boundary conditions (15.5) on the conducting surface. Using notation of Figure 15.2, if we take $\boldsymbol{E}_{\mathrm{i}x} = -\boldsymbol{E}_{\mathrm{r}x}$, the corresponding $\boldsymbol{H}_{\mathrm{i}\perp}$ and $\boldsymbol{H}_{\mathrm{r}\perp}$ can both be in the incident plane, making the vector sum $\boldsymbol{H}_{\mathrm{i}\perp} + \boldsymbol{H}_{\mathrm{r}\perp} = \boldsymbol{K}$, as specified by the boundary condition (15.5). In this case, the wavevector $\boldsymbol{k}_\mathrm{r}$ of the reflected wave is on the incident plane as symmetrical to $\boldsymbol{k}_\mathrm{i}$ with respect to the z-axis that is normal to the conducting surface. Angles θ_i, θ_r that vectors $\boldsymbol{k}_\mathrm{i}$, $\boldsymbol{k}_\mathrm{r}$ make with the normal to the conducting surface $\boldsymbol{n}$ are called the *angle of incidence* and *angle of reflection*, respectively, for which we should consider that

$$\theta_\mathrm{i} = \theta_\mathrm{r} \tag{15.7}$$

to satisfy the boundary conditions. Equation (15.7) represents the law of reflection.

The oscillating surface current density $\boldsymbol{K}$ is responsible for the reflected wave. If the incident $\boldsymbol{H}_\mathrm{i}$ wave is polarized along the x-axis, $\boldsymbol{H}_\mathrm{r}$ must be chosen as $\boldsymbol{H}_\mathrm{i} + \boldsymbol{H}_\mathrm{r} = 0$, implying that $\boldsymbol{K} = 0$, hence generating no reflection. Accordingly, to obtain reflection, the incident wave should have a component along the x-axis. In this context, if $\boldsymbol{E}_\mathrm{i}$ is arbitrarily polarized, only its x component is responsible for reflection. Consequently, a natural light reflected from a conducting surface is linearly polarized.

In the above treatment the incident and reflected waves are mixed in phase and connected with the penetrated wave at the conducting boundary. As we discussed in Section 15.2, the conductivity σ makes the penetration limited to the skin depth δ. During penetration, the electromagnetic energy is converted to heat in the conducting layer.

15.5. Dielectric Boundaries

Dielectric boundaries are signified by different dielectric constants characterized by no surface charges and currents under normal circumstances. In this case, the general boundary conditions (15.6a) and (15.6b) can be rephrased by continuity of $\boldsymbol{E}_\mathrm{tangential}$ and $\boldsymbol{D}_\mathrm{normal}$. Magnetic properties are insignificant for most dielectrics, which are specified by the same magnetic permeability μ_o as in a vacuum.

More important, at normal boundaries of *isotropic* dielectric media there is no particular mechanism for rotating the polarization vector. Therefore, the vector $\boldsymbol{E}$, and hence $\boldsymbol{D}$, does not change direction at boundaries; reflected and refracted waves both propagate with no change in polarization in the plane of incidence. As is clear from the illustration in Figure 15.2, where ($\boldsymbol{E}_x$, $\boldsymbol{H}_x$) and ($\boldsymbol{E}_\perp$, $\boldsymbol{H}_\perp$) $\sin\theta$ represent tangential and normal components of ($\boldsymbol{E}$, $\boldsymbol{H}$), respectively. Therefore, as indicated by such boundary conditions, the wave vectors $\boldsymbol{k}_\mathrm{i}$, $\boldsymbol{k}_\mathrm{r}$, and $\boldsymbol{k}_\mathrm{p}$ for incident, reflected, and penetrated waves are all in the incident plane, as shown in Figure 15.4, where θ_r and θ_p are angles of reflection and refraction (or penetration). Thus, we have a

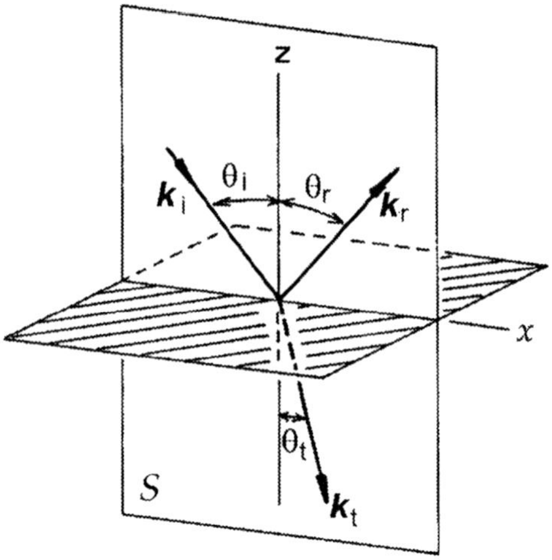

FIGURE 15.4. The Snell law of reflection and refraction.

relation

$$\frac{\sin\theta_i}{\sin\theta_p} = \frac{k_i}{k_p} = \frac{k_1}{k_2} = \sqrt{\frac{\varepsilon_1}{\varepsilon_2}}.$$

For optical light, a dielectric material is specified by the *refractive index* that is defined by $n = \sqrt{\varepsilon/\varepsilon_o}$, and the equation

$$\frac{\sin\theta_1}{\sin\theta_2} = \frac{n_1}{n_2} \tag{15.8}$$

is called the *Snell law* in *geometrical optics*, indicating that $\theta_1 > \theta_2$ if $n_1 > n_2$ or $\varepsilon_1 > \varepsilon_2$.

If, on the other hand, $n_1 < n_2$, from the Snell law we have $\sin\theta_2 = \dfrac{n_2}{n_1}\sin\theta_1$, where the angle of refraction cannot be obtained unless $|\sin\theta_2| \leq 1$. Therefore, in this case if the incident angle θ_1 is larger than a specific angle θ_B determined by $\dfrac{n_2}{n_1}\sin\theta_B \leq 1$, there is no refraction, as shown in Figure 15.5. Such an angle defined by $\theta_B = \sin^{-1}\dfrac{n_1}{n_2}$ is known as the *Brewster angle*, and the wave is totally reflected if an incidence angle exceeds θ_B. At optical frequencies waves propagate normally as a collimated beam carrying a well-defined energy; these waves are split into reflected and refracted beams at the boundary. If $\theta_1 > \theta_B$, in particular, all the beam energy is carried away in the medium 1 by *total reflection*. In *laser optics* light can be transmitted through an insulating cable of dielectric material where $\varepsilon > \varepsilon_o$ with virtually no loss of energy due to the total reflection condition on the dielectric wall. Such a cable, known as *fiberoptic*, is widely used for optical communication in modern technology.

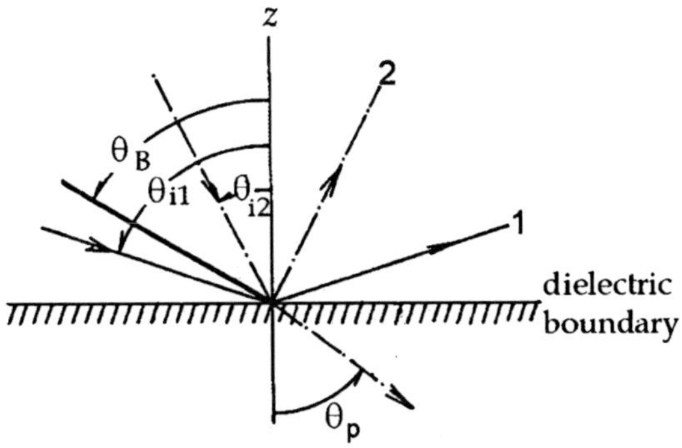

FIGURE 15.5. Total reflection and the Brewster angle θ_B.

15.6. The Fresnel Formula

In Section 15.4 we considered the phase of a wave variable for boundary conditions in order to determine directions for propagation. In this section we derive the relations among amplitudes of incident, reflected waves, and refracted waves at a dielectric boundary, which are known collectively as the *Fresnel formula* in optics.

Figure 15.6 shows the field vectors of waves in the vicinity of a dielectric boundary, where the incident wave is polarized in such a way that the ***E*** vector is perpendicular to the plane of incidence S. In this case, if we take the z-axis parallel to the normal to the conducting surface, the x-axis is in parallel to the ***E*** vector. A wave for such a specific polarization shall be called the "mode I," to distinguish it from another mode, "mode II," discussed shortly. Since the boundary conditions for these ***E*** and ***H*** vectors are determined by continuity of the tangential

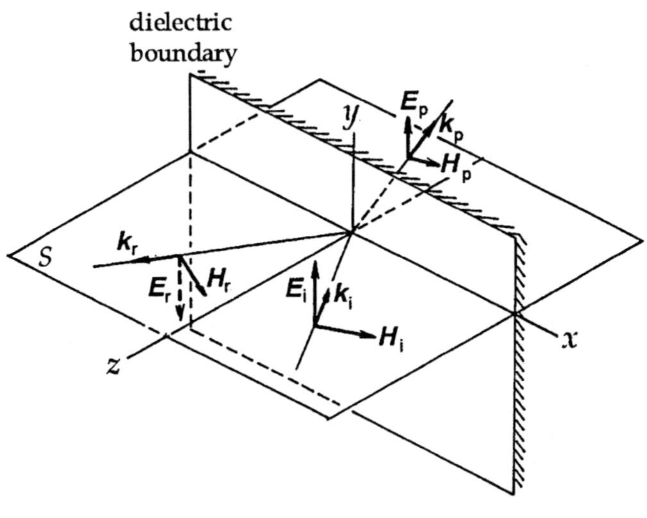

FIGURE 15.6. Directions of $(\boldsymbol{k}, \boldsymbol{E}, \boldsymbol{H})$ for incident, reflective, and refractive plane waves at a dielectric boundary.

components, we can write

$$E_i - E_r = E_p \quad \text{along the } x \text{ direction} \tag{i}$$

and

$$H_i \cos\theta_i - H_r \cos\theta_r = H_p \cos\theta_p \quad \text{along the } z \text{ direction.} \tag{ii}$$

These plane electromagnetic waves are characterized separately by

$$\frac{E_i}{H_i} = \frac{\omega\mu_o}{k_i}, \quad \frac{E_r}{H_r} = \frac{\omega\mu_o}{k_r} \quad \text{and} \quad \frac{E_p}{H_p} = \frac{\omega\mu_o}{k_p},$$

and the $\boldsymbol{H}$ fields in (ii) can be expressed as

$$k_i(E_i - E_r) = k_p E_p. \tag{iii}$$

Using the law of reflection, i.e., $\theta_i = \theta_r$ in (i), (ii), and (iii), we obtain

$$\left(\frac{E_r}{E_i}\right)_I = \frac{n_1\cos\theta_i - n_2\cos\theta_p}{n_1\cos\theta_i + n_2\cos\theta_p} \quad \text{and} \quad \left(\frac{E_p}{E_i}\right)_I = \frac{2n_1\cos\theta_i}{n_1\cos\theta_i + n_2\cos\theta_p} \tag{15.9a}$$

for mode I, where n_1 and n_2 given by $\sqrt{\varepsilon_1/\varepsilon_o}$ and $\sqrt{\varepsilon_2/\varepsilon_o}$ are refractive indexes of the media 1 and 2. Further applying the Snell law, i.e., $\dfrac{n_1}{n_2} = \dfrac{\sin\theta_i}{\sin\theta_p}$, these expressions can be written as

$$\left(\frac{E_r}{E_i}\right)_I = \frac{\sin(\theta_p - \theta_i)}{\sin(\theta_p + \theta_i)} \quad \text{and} \quad \left(\frac{E_p}{E_i}\right)_I = \frac{2\sin\theta_p\sin\theta_i}{\sin(\theta_p + \theta_i)}. \tag{15.9b}$$

For mode II we consider that the boundary conditions can be specified as

$$H_i - H_r = H_p \quad \text{along the } x\text{-axis}$$

and

$$E_i\cos\theta_i - E_r\cos\theta_r = E_p\cos\theta_p \quad \text{in the } y,z\text{-plane.}$$

After a similar manipulation to mode I, we can derive the following relations for mode II, that is,

$$\left(\frac{E_r}{E_i}\right)_{II} = \frac{n_2\cos\theta_i - n_1\cos\theta_p}{n_2\cos\theta_i + n_1\cos\theta_p} \quad \text{and} \quad \left(\frac{E_p}{E_i}\right)_{II} = \frac{2n_1\cos\theta_i}{n_2\cos\theta_i + n_1\cos\theta_p}, \tag{15.10a}$$

which are very similar to (15.9a), but noted that n_1 and n_2 are exchanged in (15.9a). Using the Snell law, the equation (15.10a) is led to a form different from (15.9b). Namely,

$$\left(\frac{E_r}{E_i}\right)_{II} = -\frac{\tan(\theta_i - \theta_p)}{\tan(\theta_i + \theta_p)} \quad \text{and} \quad \left(\frac{E_p}{E_i}\right)_{II} = \frac{\sin(\theta_i + \theta_p) - \sin(\theta_i - \theta_p)}{\sin(\theta_i + \theta_p)\sin(\theta_i - \theta_p)}. \tag{15.10b}$$

Equations (15.9) and (15.10) are known as the *Fresnel formula* in optics.

It is interesting to note that in mode II the amplitude of reflection E_r vanishes, if $\theta_i + \theta_p \to \pi/2$, and hence such a specific incident angle $\theta_i = 1\pi - \theta_p$ is called the *polarizing angle* θ_P. When $\theta_i = \theta_p$, from the Snell law we can write

$$\tan\theta_p = \frac{n_1}{n_2}, \tag{15.11}$$

where the polarizing angle is determined by the ratio of refractive indexes, but not exactly the same as the Brewster angle $\theta_B = \sin^{-1}\left(\frac{n_1}{n_2}\right)$.

For the perpendicular incidence,we have $\theta_i = \theta_r = \theta_p$, and the Fresnel formula can be written specifically as

$$\left(\frac{E_r}{E_i}\right)_{I,II} = \pm\frac{n_1 - n_2}{n_1 + n_2} \quad \text{and} \quad \left(\frac{E_p}{E_i}\right)_{I,II} = \frac{2n_1}{n_1 + n_2}, \tag{15.12}$$

indicating that if $n_1 < n_2$ the reflected wave is out of phase with respect to the incident phase.

16
Guided Waves

16.1. Propagation Between Parallel Conducting Plates

Wave equations have a simple plane-wave solution that can be obtained in free space and used for a field bounded by rectangular boundaries. In such cases, as a consequence of the linear differential equations involved, solutions can be expressed by a Fourier series of sinusoidal functions of the phase variable $\phi = \boldsymbol{k}.\boldsymbol{r} - \omega t$. Normally, a light beam propagates as a free wave but is coupled with a reflected wave at a boundary point, as discussed in the previous section. Nevertheless, such a reflection in optics is a significant concept for general electromagnetic waves in open and bound space. For electromagnetic applications the radiation needs to be guided in a desired direction, where it is necessary to specify directions of the propagating and reflecting power of the wave from a target object. In this chapter we discuss the basic principle for such guided waves in modern communication and guidance.

For the purpose of illustration we consider two parallel conducting plates of a large area, where plane electromagnetic waves can pass through the space, reflecting back and forth between the plates, as shown in Figure 16.1. Plane waves are transmitted zigzag in the x direction, a situation for which we consider two specific cases, namely, those where the $\boldsymbol{E}$ and $\boldsymbol{H}$ vectors are polarized in perpendicular to the z direction, respectively, as in Figure 16.1(a) and (b). The $\boldsymbol{k}$ vector changes the direction in the xz-plane of incidence at every reflection in a manner similar to a single reflection. Such polarized waves are referred to as being in *transverse electric* (TE) and *transverse magnetic* (TM) modes, for $\boldsymbol{E}$ and $\boldsymbol{H}$, respectively; both are transversal to the x-axis.

Denoting the two conductors as plates 1 and 2, reflections of a propagating wave can be described by the wavevector $\boldsymbol{k}$ that changes the direction zigzag in the plane S between 1 and 2. Such reflections take place at any phase $\phi = \boldsymbol{k}.\boldsymbol{r} - \omega t + 2\pi \mathrm{m}$, where m is an integer, and therefore we can apply the same boundary condition at every phase separated by 2π. At each point of reflection on the plates 1 and 2, we can write incident and reflected waves, referring to the $\boldsymbol{E}$ vector, as

$$\boldsymbol{E}_{\mathrm{i}}(1) = \boldsymbol{E}_{\mathrm{io}} \exp[i(\boldsymbol{k}_{\mathrm{i1}}.\boldsymbol{r} - \omega t)], \quad \boldsymbol{E}_{\mathrm{r}}(1) = \boldsymbol{E}_{\mathrm{ro}} \exp[i(\boldsymbol{k}_{\mathrm{r1}}.\boldsymbol{r} - \omega t)],$$

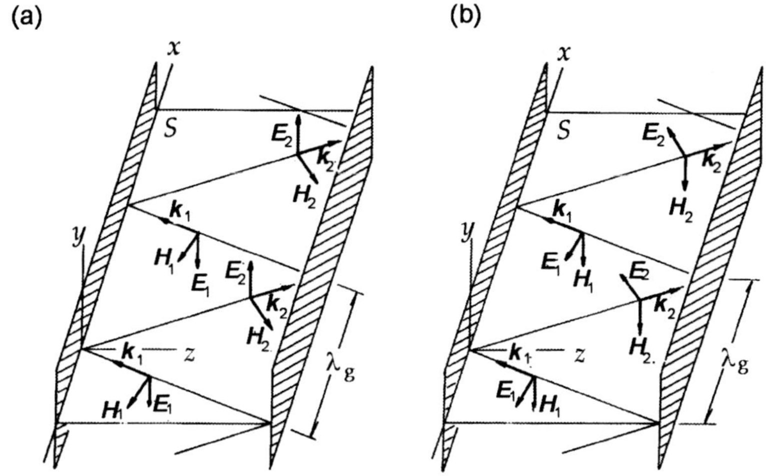

FIGURE 16.1. Multiple reflections between parallel conducting planes. (a) TE mode and (b) TM mode. λ_g is the apparent wavelength.

and

$$\boldsymbol{E}_i(2) = \boldsymbol{E}_{io} \exp[i(\boldsymbol{k}_{i2}.\boldsymbol{r} - \omega t)], \quad \boldsymbol{E}_r(2) = \boldsymbol{E}_{ro} \exp[i(\boldsymbol{k}_{r2}.\boldsymbol{r} - \omega t)].$$

Assuming no energy loss on the plates, these $\boldsymbol{E}$ and $\boldsymbol{k}$ are equal in magnitude at all reflections, and hence can be written as $|\boldsymbol{E}_{io}| = |\boldsymbol{E}_{\rho o}| = E_o$, and $|\boldsymbol{k}_{i1}| = |\boldsymbol{k}_{r1}| = |\boldsymbol{k}_{i2}| = |\boldsymbol{k}_{r2}| = k$. In the plane of incidence S, i.e., the xz-plane, $\boldsymbol{k}_i$ and $\boldsymbol{k}_r$ vectors make the same angles of incidence and reflection with the z-axis, i.e., $\theta_i = \theta_r = \theta$.

Figure 16.1(a) shows the transverse electric (TE) mode, where the boundary conditions for $\boldsymbol{E}_{\text{tangential}} = 0$ on the plates 1 and 2 give rise to

$$\boldsymbol{E}_i(1) = -\boldsymbol{E}_r(1) \quad \text{and} \quad \boldsymbol{E}_i(2) = -\boldsymbol{E}_r(2)$$

at all reflection points $(x, 0)$ and (x, a), respectively, at which we can write

$$\boldsymbol{E}_i(1) + \boldsymbol{E}_r(1) = \boldsymbol{E}_o[\exp(i\boldsymbol{k}_i.\boldsymbol{r}) + \exp(i\boldsymbol{k}_r.\boldsymbol{r})]_{z=0} \exp[i(kx\sin\theta - \omega t)] = 0$$

and

$$\begin{aligned}\boldsymbol{E}_i(2) + \boldsymbol{E}_r(2) &= \boldsymbol{E}_o[\exp(i\boldsymbol{k}_i.\boldsymbol{r}) + \exp(i\boldsymbol{k}_r.\boldsymbol{r})_{z=a} \exp[i(kx\sin\theta - \omega t)] \\ &= \boldsymbol{E}_o \times 2\sin[(k\cos\theta)\alpha] \exp[i(kx\sin\theta - \omega t)] = 0 \quad (16.1\text{a})\end{aligned}$$

at any x where $z = a$ represents the gap between the plates.

The latter condition requires that k satisfy

$$\sin[(k\cos\theta)a] = 0$$

at an arbitrary x, signifying $(k \cos\theta)a = \mathrm{m}\pi$, where m is any integer. Using the wavelength defined by $\lambda = 2\pi/k$, we obtain

$$\lambda = \frac{2a}{\mathrm{m}} \cos\theta = \lambda_c \cos\theta, \tag{16.1b}$$

which should be shorter than $\lambda_c = 2a/\mathrm{m}$ for such a wave to propagate through between the plates. If $\lambda > \lambda_c$, the angle θ in (16.1b) cannot be obtained; such a wave cannot go through. The parameter λ_c is called the *cut-off wavelength*, signifying that the gap a between the plates limits the propagation by $\lambda < \lambda_c$. The index m specifies various modes of TE waves, among which the mode $\mathrm{m} = 1$ is signified by the longest cut-off wavelength $\lambda_c = 2a$.

The corresponding $\boldsymbol{H}$ vectors in the TE mode are all in the plane S, as shown in Figure 16.1(a). Using similar notation as we did for the $\boldsymbol{E}$ vectors, at every point of reflection the $\boldsymbol{H}$ vectors are expressed as

$$\boldsymbol{H}_i(1) = [-\mathbf{i}H_o \cos\theta - \mathbf{k}H_o \sin\theta]\exp(-i\omega t),$$
$$\boldsymbol{H}_r(1) = [-\mathbf{i}H_o \cos\theta + \mathbf{k}H_o \sin\theta]\exp(-i\omega t),$$

and

$$\boldsymbol{H}_i(2) = [-\mathbf{i}H_o \cos\theta + \mathbf{k}H_o \sin\theta]\exp[i(kx\sin\theta + ka\cos\theta - \omega t)],$$
$$\boldsymbol{H}_r(2) = [-\mathbf{i}H_o \cos\theta - \mathbf{k}H_o \sin\theta]\exp[i(kx\sin\theta - ka\cos\theta - \omega t)].$$

Accordingly, at a point $(x, 0)$

$$\boldsymbol{H}_i(2) + \boldsymbol{H}_r(2) = -\mathbf{i}[2H_o \cos\theta\cos(ka\cos\theta)]\exp[i(kx\sin\theta - \omega t)] = 0 \tag{16.2}$$

because of (16.1b). It is noted from (16.1a) and (16.2) that both tangential components of $\boldsymbol{E}$ and $\boldsymbol{H}$ should vanish periodically at x_m, so that $kx_m\sin\theta - kx_{m-1}\sin\theta = 2\pi$. In this context, $x_m - x_{m-1} = \lambda_g$ is an *apparent wavelength* for propagation along the x direction, namely,

$$\lambda_g = \frac{\lambda}{\sin\theta}. \tag{16.3}$$

For a practical waveguide, λ_g is also known as the *guide wavelength*, serving as a significant measure for a standing wave. By definition, λ_c and λ_g are related, i.e.,

$$\frac{1}{\lambda^2} = \frac{1}{\lambda_c^2} + \frac{1}{\lambda_g^2}. \tag{16.4}$$

Figure 16.1(b) shows a transverse magnetic (TM) mode for which a calculation similar to TE mode can be carried out. In this case,

$$\boldsymbol{H}(1) = \boldsymbol{H}_i(1) + \boldsymbol{H}_r(1) = 0 \quad \text{and} \quad \boldsymbol{H}(2) = \boldsymbol{H}_i(2) + \boldsymbol{H}_r(2) = 0$$

on the plates 1 and 2. Hence, the $\boldsymbol{H}$ and $\boldsymbol{E}$ vectors at $(x, 0, a)$ near a reflection point $(x, 0, a)$ can be written as

$$\boldsymbol{H}(x) = \mathbf{j}\frac{E_o}{Z_o}\cos(ka\cos\theta)\exp[i(kx\sin\theta - \omega t)]$$

and

$$\boldsymbol{E}(x) = -\mathbf{i}[2iE_o \cos\theta \sin(ka\cos\theta)] \exp[i(kx\sin\theta - \omega t)]$$
$$-\mathbf{k}[2E_o \sin\theta \cos(ka\cos\theta)] \exp[i(kx\sin\theta - \omega t)].$$

Therefore, from the boundary condition for the $\boldsymbol{H}(x)$ we obtain $\cos(ka\cos\theta) = 0$, leading to the cut-off wavelength $\lambda_c = 4a/m$ for TM mode. On the other hand, the phase factor is the same as in TE mode, so that the apparent wavelength is given by $\lambda_g = \lambda/\sin\theta$.

16.2. Uniform Waveguides

Referring to the coordinate axes in Figure 16.1, the electromagnetic fields between parallel conducting plates are independent of the y-axis in both TE and TM modes. We can therefore place a conducting plate at any place in parallel with the xz-plane, thereby leaving the field-lines unchanged. In fact, such polarized waves can propagate through a rectangular *tube* of a cross-sectional area $a \times b$ as long as the wavelength is shorter than the cut-off wavelength determined by $2a$. Such a tubing of uniform cross-sectional area is in use for modern high-frequency circuitry and is known as a *waveguide*. For practical applications, rectangular and circular tubing is common, although theoretically the shape of the cross-section may be arbitrary, provided that it is uniformly made.

16.2.1. Transversal Modes of Propagation (TE and TM Modes)

We set a rectangular coordinate system where the x-axis represents the axis of a waveguide, and the transversal area is described by y and z coordinates. The propagating fields can then be expressed as

$$\boldsymbol{E} = \boldsymbol{E}_t + E_x\mathbf{i} \quad \text{and} \quad \boldsymbol{H} = \boldsymbol{H}_t + H_x\mathbf{i}, \tag{16.5}$$

where (E_x, H_x) and $(\boldsymbol{E}_t, \boldsymbol{H}_t)$ are called *longitudinal* and *transversal* components. We can show that these longitudinal components cannot be simultaneously zero and the propagating wave is characterized by either one of these unvarnished E_x, H_x, i.e., $H_x = 0$ or $E_x = 0$. These modes are referred to, respectively, as the TE and TM modes. In a coaxial cable, in contrast, there exists such a propagating wave characterized by both vanishing longitudinal components called the *transversal electric-magnetic* (*TEM*) *mode*, being analogous to a plane wave in free space.

For such guided waves, boundaries at both ends have to be specified by their respective connecting devices: e.g., for an isolated long waveguide the boundaries are considered open at $x = \pm\infty$, whereas in infinite parallel plates, the boundaries are closed (or short-circuited) at $z = \pm\infty$, but open at $x = \pm\infty$. We can show that such a TEM mode cannot occur in a waveguide with a cross-section of a single boundary, whereas this transmission is possible in a waveguide with two closed boundaries.

Equations (16.5) indicate that the field vectors can be factorized into products of separate functions of (y, z, t) and (x, t). Hence,

$$\boldsymbol{E}_\mathrm{t} = (E_y, E_z) \quad \text{and} \quad \boldsymbol{H}_\mathrm{t} = (H_y, H_z),$$

and all the field components are expressed to have a factor $\exp[i(k_x x - i\omega t)]$.

The basic equations div $\boldsymbol{E} = 0$ and div $\boldsymbol{H} = 0$ can be written as

$$\frac{\partial E_y}{\partial y} + \frac{\partial E_z}{\partial z} + ik_x E_x = 0 \tag{ia}$$

and

$$\frac{\partial H_y}{\partial y} + \frac{\partial H_z}{\partial z} + ik_x H_x = 0. \tag{ib}$$

On the other hand, equations curl $\boldsymbol{E} = i\omega\mu_\mathrm{o}\boldsymbol{H}$ and curl $\boldsymbol{H} = -i\omega\varepsilon_\mathrm{o}\boldsymbol{E}$ are expressed, respectively, as

$$\frac{\partial E_z}{\partial y} - \frac{\partial E_y}{\partial z} = i\omega\mu_\mathrm{o} H_x, \tag{iia}$$

$$ik_x E_z - \frac{\partial E_x}{\partial z} = i\omega\mu_\mathrm{o} H_\mathrm{y}, \tag{iib}$$

and

$$\frac{\partial E_x}{\partial y} - ik_x E_y = i\omega\mu_\mathrm{o} H_z; \tag{iic}$$

and

$$\frac{\partial H_z}{\partial y} - \frac{\partial H_y}{\partial z} = -i\omega\varepsilon_\mathrm{o} E_x, \tag{iiia}$$

$$ik_x H_z - \frac{\partial H_x}{\partial z} = -i\omega\varepsilon_\mathrm{o} E_y, \tag{iiib}$$

and

$$\frac{\partial H_x}{\partial y} - ik_x H_y = -i\omega\varepsilon_\mathrm{o} E_z. \tag{iiic}$$

Combining (iib) and (iiic), we can derive the relation between E_y and H_z, while E_z and H_y can be related using (iic) and (iiib). Namely,

$$E_y = -\frac{i}{k_c{}^2}\left(\omega\mu_\mathrm{o}\frac{\partial H_x}{\partial z} - k_x\frac{\partial E_x}{\partial y}\right), \tag{iva}$$

$$E_z = -\frac{i}{k_c{}^2}\left(\omega\mu_\mathrm{o}\frac{\partial H_x}{\partial y} + k_x\frac{\partial E_x}{\partial z}\right) \tag{ivb}$$

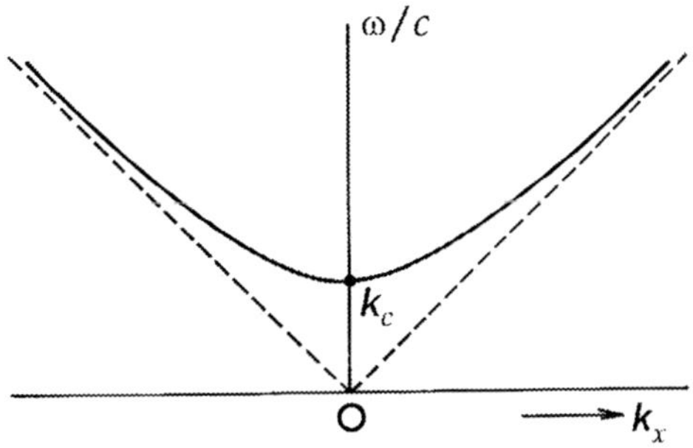

FIGURE 16.2. A plot of ω/c vs. k_x, where k_c is the cut-off wavevector. All frequencies $\omega < ck_c$ are cut-off (forbidden), while $\omega > ck_c$ are allowed for propagation.

and

$$H_y = \frac{i}{k_c^2}\left(\omega\varepsilon_o \frac{\partial E_x}{\partial z} + k_x \frac{\partial H_x}{\partial y}\right), \tag{va}$$

$$H_z = -\frac{i}{k_c^2}\left(\omega\varepsilon_o \frac{\partial E_x}{\partial y} - k_x \frac{\partial H_x}{\partial z}\right). \tag{vb}$$

Here the parameter k_c is defined by the relation

$$k_c^2 = \omega^2\varepsilon_o\mu_o - k_x^2, \tag{16.6}$$

where $k_c = 2\pi/\lambda_c$ is related to the cut-off wavelength λ_c, and $k_x = 2\pi/\lambda_g$ represents a propagating wave signified by the guide wavelength λ_g. The first term on the right is the squared wavevector $k = \omega/c$ in free space, and hence (16.6) is identical to (16.4). Figure 16.2 shows the curve of ω/c against k_x, where frequencies for $\omega < ck_c$ are not permitted for propagation.

It is noted from the relations (iv) and (v) that the transversal components depend on non-vanishing longitudinal components, for which one of E_x and H_x is at least essential for TE and TM modes, behaving as if scalar functions. For a TE mode, the transversal fields $\boldsymbol{E}_t$ and $\boldsymbol{H}_t$ can thus be derived from the longitudinal H_x, whereas for a TM mode transversal components are determined by E_x.

For a TE mode, H_x is given as a solution of the wave equation

$$\left(\boldsymbol{\Delta}_t - k_x^2 + k^2\right) H_x = \left(\boldsymbol{\Delta}_t + k_c^2\right) H_x = 0, \tag{16.7}$$

where $\boldsymbol{\Delta}_t \equiv \dfrac{\partial^2}{\partial y^2} + \dfrac{\partial^2}{\partial z^2}$ is the two-dimensional Laplacian operator in the yz-plane. Solving (16.7) (known as the Helmholtz equation) for H_x, and using $E_x = 0$ in relations (iv) and (v), we obtain that

$$E_y = \frac{i\omega\mu_o}{k_c^2}\frac{\partial H_x}{\partial z}, \qquad E_z = \frac{i\omega\mu_o}{k_c^2}\frac{\partial H_x}{\partial y}$$

and

$$H_y = \frac{ik_x}{k_c^2}\frac{\partial H_x}{\partial y}, \qquad H_z = \frac{ik_x}{k_c^2}\frac{\partial H_x}{\partial z}.$$

These results can be written with vector notation as

$$\boldsymbol{H}_{\mathrm{t}} = \frac{ik_x}{k_c{}^2}\operatorname{grad}_{\mathrm{t}} H_x \quad \text{and} \quad \boldsymbol{E}_{\mathrm{t}} = -\frac{ikZ_{\mathrm{o}}}{k_c{}^2}\left[\mathbf{i} \times \operatorname{grad}_{\mathrm{t}} H_x\right],$$

where $Z_{\mathrm{o}} = \sqrt{\mu_{\mathrm{o}}/\varepsilon_{\mathrm{o}}}$ is the wave impedance, as previously defined in (13.6), and the component H_x is regarded as a scalar function. These transversal fields are related by

$$\boldsymbol{E}_{\mathrm{t}} = -\frac{kZ_{\mathrm{o}}}{k_x}\left[\mathbf{i} \times \boldsymbol{H}_{\mathrm{t}}\right] \quad \text{and} \quad \boldsymbol{H}_{\mathrm{t}} = \frac{k_x}{kZ_{\mathrm{o}}}\left[\mathbf{i} \times \boldsymbol{E}_{\mathrm{t}}\right] \tag{16.8}$$

in the TE mode, where the $\boldsymbol{H}$ vector is not transversal, as given by $\boldsymbol{H} = \boldsymbol{H}_{\mathrm{t}} + H_x\mathbf{i}$.

For the TM mode where $H_x = 0$, we can obtain the following expressions after a similar calculation to the TE mode:

$$\left(\boldsymbol{\Delta}_{\mathrm{t}} - k_c{}^2\right) E_x = 0, \tag{16.9}$$

$$\boldsymbol{E}_{\mathrm{t}} = \frac{ik_x}{k_c{}^2}\operatorname{grad}_{\mathrm{t}} E_x, \quad \boldsymbol{H}_{\mathrm{t}} = \frac{ik}{Z_{\mathrm{o}}k_c{}^2}\left[\mathbf{i} \times \operatorname{grad}_{\mathrm{t}} E_x\right],$$

and

$$\boldsymbol{E}_{\mathrm{t}} = -\frac{k_x Z_{\mathrm{o}}}{k}\left[\mathbf{i} \times \boldsymbol{H}_{\mathrm{t}}\right] \tag{16.10}$$

16.2.2. Transversal Electric-Magnetic Modes (TEM)

In free space the $\boldsymbol{E}$ and $\boldsymbol{H}$ vectors of a propagating wave are both transversal and perpendicular to the wavevector $\boldsymbol{k}$. For a guided wave such a TEM mode can also be transversal, depending on the boundaries. Since both axial components E_x and H_x are zero in TEM mode, the equations (ia), (ib), (iia), and (iiia) can be written as

$$\frac{\partial E_y}{\partial y} + \frac{\partial E_x}{\partial x} = 0, \qquad \frac{\partial H_y}{\partial y} + \frac{\partial H_z}{\partial z} = 0$$

and

$$\frac{\partial E_z}{\partial y} - \frac{\partial E_y}{\partial z} = 0, \qquad \frac{\partial H_z}{\partial y} - \frac{\partial H_y}{\partial z} = 0.$$

Hence, these transversal components can be expressed by such a scalar function Φ that satisfies the Helmholtz equation

$$\boldsymbol{\Delta}_{\mathrm{t}}\Phi = 0. \tag{16.11}$$

Therefore, if such a scalar Φ can be obtained to signify the transversal vectors, the TEM mode is characterized by $\boldsymbol{k}_{\mathrm{t}} = 0$, and hence $k_x = k$ or $\lambda_{\mathrm{g}} = \lambda$ from (16.6), indicating the same kind of propagation as in free space along the x direction.

However, such a function Φ is singular on the x-axis in a waveguide that is tube shaped, which makes the propagation impossible. On the other hand, if it consists of two separate boundaries—one inside the other as in a coaxial cable—the function Φ can take unique values on the inner and outer surfaces, and hence the Laplace equation (16.11) can possess *regular* solutions. A coaxial cable is an example of waveguides consisting of two conductors where the TEM mode is utilized for signal transmission.

16.3. Examples of Waveguides

Example 1. A Rectangular Waveguide (TE Mode).

Waveguides of a rectangular cross-section $a \times b$ are commonly used for microwave transmission in TE mode. Illustrated in Figure 16.3(a), a normally indicates the narrow edge that is shorter than b for practical use, where the $\boldsymbol{E}$ vector is polarized in parallel to a. In this case, H_x is the basic function from which all other components are derived, and it is expressed as

$$H_x(x, y, z, t) = H_o \exp[i(k_x x - \omega t)] \exp(ik_y y) \exp(ik_z z),$$

where

$$k_x{}^2 + k_y{}^2 + k_z{}^2 = \frac{\omega^2}{c^2} = k^2.$$

Such waves as determined by $\pm k_x$, $\pm k_y$, $\pm k_z$ are elementary plane waves that can be linearly superposed to satisfy the boundary condition for $\boldsymbol{H}_{\text{tangential}} = 0$ on the conducting surface. Therefore, taking the origin of coordinates at the corner of the waveguide, we can consider a combination

$$H_x = H_o \sin(k_y y) \sin(k_z z) \exp[i(k_x x - \omega t)],$$

which is equal to zero on all conducting walls, i.e., $y = 0$ and b, and $z = 0$ and a. That is,

$$k_y b = \text{n}\pi \quad \text{and} \quad k_z a = \text{m}\pi,$$

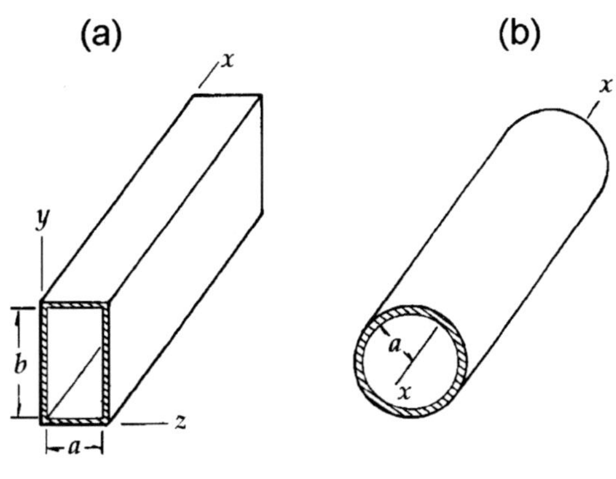

FIGURE 16.3. (a) A rectangular waveguide. (b) A circular waveguide.

where m and n are 0 or $\pm$ integers, and

$$k_c^{\,2} = k_y^{\,2} + k_z^{\,2} = \left(\frac{\mathrm{m}\pi}{a}\right)^2 + \left(\frac{\mathrm{n}\pi}{b}\right)^2.$$

Using this result, the transversal vectors calculated with (16.11) are given by

$$\boldsymbol{H}_\mathrm{t} = -\frac{ik_xk_y}{k_c^{\,2}} H_\mathrm{o}\exp[i(k_xx - \omega t)][\cos(k_yy)\sin(k_zz)\mathbf{j} + \sin(k_yy)\cos(k_zz)\mathbf{k}]$$

and

$$\boldsymbol{E}_\mathrm{t} = -\frac{i\omega\mu_\mathrm{o}k_y}{k_c^{\,2}} H_\mathrm{o}\exp[i(k_xx - \omega t)][\sin(k_yy)\cos(k_zz)\mathbf{j} - \cos(k_yy)\sin(k_zz)\mathbf{k}].$$

Specified by indexes m and n, a particular mode is denoted by $\mathrm{TE_{mn}}$. In practice, those specified by low indexes are in use; for example, in $\mathrm{TE_{10}}$ mode for m = 1 and n = 0, we have $\lambda_c = 2a$, as already discussed in Section 16.1. Waves of a wavelength shorter than the cut-off λ_c can be transmitted through such an "oversized" waveguide, but its energy may be dispersed into many modes of allowed transmission. In this context it is a normal practice that a waveguide is operated for a wavelength in the range $\lambda_c\ (\mathrm{TE_{10}}) > \lambda > \lambda_c\ (\mathrm{TE_{20}})$, i.e., $2a > \lambda > a$, to avoid such an energy loss.

Field-lines of $\boldsymbol{E}$ and $\boldsymbol{H}$ in $\mathrm{TE_{10}}$ mode are sketched at a given x and t in Figure 16.4, showing that the pattern consists of well-defined magnetic loops and concentrated electric flux periodically at every $1\lambda_\mathrm{g}$ in the xy- and xz-planes, respectively.

The TM mode in a rectangular waveguide can be discussed similar to the TE. In this case we can start with the longitudinal component E_x that is expressed in exactly the same way as H_x for a TE mode. Namely,

$$E_x = E_\mathrm{o}\sin\left(k_yy\right)\sin\left(k_zz\right)\exp\left[i\left(k_xx - \omega t\right)\right],$$

which satisfies $E_x = 0$ at the boundaries at y: $z = 0$, a, and b, resulting in the relations

$$k_yb = \mathrm{n}\pi \quad \text{and} \quad k_za = \mathrm{m}\pi,$$

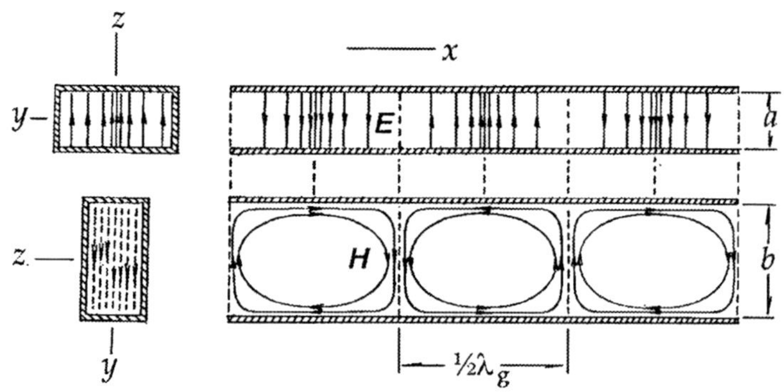

FIGURE 16.4. The $\mathrm{TE_{10}}$ mode in a rectangular waveguide: (a) $\boldsymbol{E}$ lines; (b) $\boldsymbol{H}$ lines.

leading to the formula

$$k_c{}^2 = k_y{}^2 + k_z{}^2 = \left(\frac{\text{m}\pi}{a}\right)^2 + \left(\frac{\text{n}\pi}{b}\right)^2,$$

which is the same as TE mode. However, in TM mode the transversals $\boldsymbol{E}_\text{t}$ and $\boldsymbol{H}_\text{t}$ are expressed in a form different from TE, and their derivation is left for interested readers as an exercise problem.

Example 2. A Circular Waveguide.

In a circular waveguide sketched in Figure 16.3(b), propagating waves can also be in TE and TM modes. In this case, because the y and z directions are identical by symmetry, a linearly polarized wave along the z direction, for example, can be easily mixed with a y-polarized wave, if there is a perturbing obstacle in the guide. However, because $a \neq b$ in a rectangular waveguide, the TE_{10} and TE_{01} modes are clearly distinct from each other. In this context, a circular waveguide is adequate for practical use; nevertheless, it is an interesting theoretical problem.

For the TM mode we solve the Helmholtz equation for E_x, which can be written with cylindrical coordinates (ρ, θ, x), that is,

$$\frac{1}{\rho}\frac{\partial}{\partial \rho}\left(\rho \frac{\partial E_x}{\partial \rho}\right) + \frac{1}{\rho^2}\frac{\partial^2 E_x}{\partial \theta^2} + \frac{\partial^2 E_x}{\partial x^2} - \frac{1}{c^2}\frac{\partial^2 E_x}{\partial t^2} = 0.$$

Writing that $E_x(\rho, \theta, x, t) = E_\text{o} R(\rho)\Theta(\theta)\exp[i(k_x x - \omega t)]$, this equation can be separated into ordinary differential equations for individual variables:

$$\frac{\text{d}^2\Theta}{\text{d}\theta^2} = -\text{m}^2\Theta$$

and

$$\frac{1}{\rho}\frac{\text{d}}{\text{d}\rho}\left(\rho\frac{\text{d}R}{\text{d}\rho}\right) + \frac{\text{m}^2}{\rho^2}R + \left(k^2 - k_x{}^2\right)R = 0,$$

where $\text{m} = 0, \pm$ integers. Letting $k^2 - k_x{}^2 = k_c{}^2$, the equation for $R(\rho)$ can be written as the standard form of the Bessel equation with a characteristic parameter k_c, i.e.,

$$\frac{\text{d}}{\text{d}\rho}\left(\rho\frac{\text{d}J_\text{m}(k_\text{c}\rho)}{\text{d}\rho}\right) + \left(k_\text{c}{}^2\rho - \frac{\text{m}^2}{\rho}\right)J_\text{m}(k_\text{c}\rho) = 0,$$

where $R(\rho) = J_\text{m}(k_c\rho)$ is Bessel's function of the mth order. Accordingly, the propagating mode is generated from E_x as given by

$$E_x(\rho, \theta, x, t) = E_\text{o} J_\text{m}(k_\text{c}\rho)\exp(\pm i\text{m}\theta)\exp[i(k_x x - \omega t)].$$

Applying the boundary condition $\boldsymbol{E}_\text{tangential} = 0$ at $\rho = a$ on the inner surface of

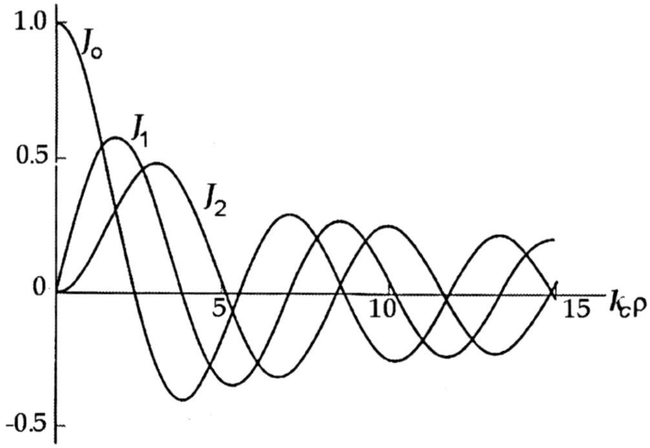

FIGURE 16.5. Bessel's functions $J_o(k_c\rho)$, $J_1(k_c\rho)$, and $J_2(k_c\rho)$. These are periodic, and $J_o(0) \neq 0$, but $J_1(0) = 0$ and $J_2(0) = 0$.

the tube, we have $J_m(k_c a) = 0$, from which k_c can be numerically determined as roots. Figure 16.5 shows curves of the Bessel functions that are quasi-oscillatory, having many values of $k_c a$ for J_m to vanish. Table 16.1 shows such roots for representative m = 0, 1, and 2. Indexing these sequential roots of $J_m = 0$ as r_{mn}, i.e., $J_m(r_{mn}) = 0$, we can specify propagating modes as TM_{mn} that are signified by characteristic $k_c = r_{mn}/a$ or by the corresponding cut-off wavelength $\lambda_c = 2\pi a/r_{mn}$.

For transversal components, we calculate $\text{grad}_t E_x \equiv \boldsymbol{u}_\rho \dfrac{\partial E_x}{\partial \rho} + \boldsymbol{u}_\theta \dfrac{1}{\rho}\dfrac{\partial E_x}{\partial \theta}$, and obtain

$$\left(E_\rho\right)_{mn} = \frac{ik_x E_o}{k_c{}^2} \frac{dJ_m(k_c\rho)}{d\rho} \exp(\pm im\theta) \exp[i(k_x x - \omega t)],$$

$$(E_\theta)_{mn} = \mp \frac{imk_x}{k_c{}^2} \frac{J_m(k_c\rho)}{\rho} \exp(\mp im\theta) \exp[i(k_x x - \omega t)]$$

TABLE 16.1. Roots of $J_m(x_n) = 0$ and $J'_m(x_n) = 0$.

	Roots of $J_m(x_n) = 0$			
m	0	1	2	3
n = 1	3.832	1.841	3.054	4.021
2	7.016	5.331	6.076	8.015
3	10.173	8.536	9.969	11.346
	Roots of $J'_m(x_n) = 0$			
m	0	1	2	3
n = 1	2.405	3.832	5.136	6.380
2	5.520	7.016	8.417	9.761
3	8.653	10.173	11.620	13.015

and

$$\left(H_{\rho}\right)_{\mathrm{mn}} = \pm\frac{i\mathrm{m}\omega\varepsilon_{\mathrm{o}}E_{\mathrm{o}}}{k_c{}^2}\frac{J_{\mathrm{m}}(k_c\rho)}{\rho}\exp\left(\mp i\mathrm{m}\theta\right)\exp\left[i(k_x x-\omega t)\right],$$

$$(H_{\theta})_{\mathrm{mn}} = \frac{i\omega\varepsilon_{\mathrm{o}}E_{\mathrm{o}}}{k_c{}^2}\frac{\mathrm{d}J_{\mathrm{m}}(k_c\rho)}{\mathrm{d}\rho}\exp\left(\pm i\mathrm{m}\theta\right)\exp\left[i(k_x x-\omega t)\right],$$

where k_{c} and k_{x} depend notably on the indexes m and n.

For the TE mode, we can carry out calculation similar to the TM mode. Here, the resulting expressions are listed.

$$H_x(\rho,\theta,x,t) = H_{\mathrm{o}}J_{\mathrm{m}}(k_c\rho)\exp\left(\pm i\mathrm{m}\theta\right)\exp\left[i(k_x x-\omega t)\right],$$

$$\left(H_{\rho}\right)_{\mathrm{mn}} = \frac{ik_x H_{\mathrm{o}}}{k_c{}^2}\frac{\mathrm{d}J_{\mathrm{m}}(k_c\rho)}{\mathrm{d}\rho}\exp\left(\pm i\mathrm{m}\theta\right)\exp\left[i(k_x x-\omega t)\right],$$

$$(H_{\theta})_{\mathrm{mn}} = \frac{\pm i\mathrm{m}k_x H_{\mathrm{o}}}{k_c{}^2}\frac{J_{\mathrm{m}}(k_c\rho)}{\rho}\exp\left(\pm i\mathrm{m}\theta\right)\exp\left[i(k_x x-\omega t)\right],$$

$$\left(E_{\rho}\right)_{\mathrm{mn}} = -\frac{i\omega\mu_{\mathrm{o}}H_{\mathrm{o}}}{k_c{}^2}\frac{J_{\mathrm{m}}(k_c\rho)}{\rho}\exp\left(\pm i\mathrm{m}\theta\right)\exp\left[i(k_x x-\omega t)\right],$$

and

$$(E_{\theta})_{\mathrm{mn}} = \frac{i\omega\mu_{\mathrm{o}}H_{\mathrm{o}}}{k_c{}^2}\frac{\mathrm{d}J_{\mathrm{m}}(k_c\rho)}{\mathrm{d}\rho}\exp\left(\pm i\mathrm{m}\theta\right)\exp\left[i(k_x x-\omega t)\right].$$

In this case the boundary condition $(E_{\theta})_{\rho=\mathrm{a}} = 0$ demands that $\dfrac{\mathrm{d}J_{\mathrm{m}}(k_c\rho)}{\mathrm{d}\rho} = 0$ at $\rho = a$, and k_c should be determined as roots. Representative values of such $k_c a$ as given by $J_{\mathrm{m}}{}'(k_c a) = 0$ are also listed in Table 16.1. It is notable that $\mathrm{TE}_{0\mathrm{n}}$ and $\mathrm{TM}_{1\mathrm{n}}$ modes share their cut-off k_c in common, hence they are *degenerate*, since $\dfrac{\mathrm{d}J_{\mathrm{o}}(u)}{\mathrm{d}u} = J_1(u)$ for an arbitrary variable u. Also, the mode m $= 0$ (TE_{01}, for example) is centrally symmetrical, and m $= 1$ (TE_{11}, in particular) is a polarized mode across the circular section, as illustrated in Figure 16.6. Such a TE_{11} mode in a circular waveguide can be regarded as equivalent to a rectangular TE_{01} if the boundary is circularly deformed.

Example 3. TEM Mode in a Coaxial Cable.

In Section 16.2 we have discussed the TEM mode for a propagating wave through a uniform tubing of double conductors for which the transverse components should be characterized by a scalar function that satisfies the Helmholtz equation. In this example, such a scalar function is determined for a coaxial cable sketched in Figure 16.7.

In general, the field vectors $\boldsymbol{E}$ and $\boldsymbol{H}$ can be represented by vector potential $\boldsymbol{A}$ and vector potential V, obeying inhomogeneous wave equations as related, respectively, to the source $\boldsymbol{j}$ and ρ, under the Lorentz condition (12.6d). In the present case, characterized by no source of radiation, both potentials can be obtained from

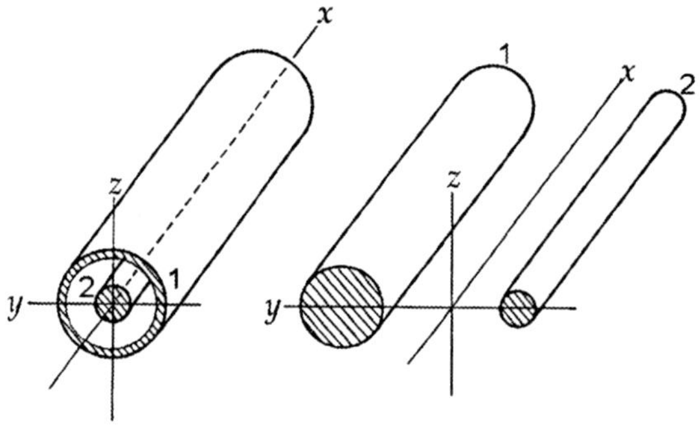

FIGURE 16.6. Transmission lines with two conductors. (a) A coaxial cable. (b) Parallel cylinders.

homogeneous wave equations, as restricted by

$$\mathrm{div}\boldsymbol{A} + \frac{1}{c^2}\frac{\partial V}{\partial t} = 0. \tag{12.6d}$$

In a TEM mode the $\boldsymbol{E}$ and $\boldsymbol{H}$ vectors are both transversal and in the yz-plane, while signified by a longitudinal phase $k_x x - \omega t$. Accordingly, we may write $V = V_o \exp[i(k_x x - \omega t)]$ and $A_x = A_o \exp[i(k_x x - \omega t)]$, leaving A_y and A_z as functions of y and z. Then (12.6d) can be expressed as

$$\frac{\partial A_y}{\partial y} + \frac{\partial A_z}{\partial z} + i\left(k_x A_o - \frac{\omega V_o}{c^2}\right)\exp\left[i(k_x x - \omega t)\right] = 0.$$

The third term vanishes when $k_x = \dfrac{\omega}{c} = k$, provided that $A_o = \dfrac{V_o}{c}$, and we obtain

$$\frac{\partial A_y}{\partial y} + \frac{\partial A_z}{\partial z} = 0 \quad \text{and} \quad A_x = \frac{V_o}{c}\exp\left[i(kx - \omega t)\right].$$

Therefore, from the basic relation $\boldsymbol{B} = \mathrm{curl}\,\boldsymbol{A}$,

$$B_y = -\frac{\partial A_x}{\partial z} = -\frac{1}{c}\frac{\partial V_o}{\partial z} \quad \text{and} \quad B_z = -\frac{\partial A_x}{\partial y} = -\frac{1}{c}\frac{\partial V_o}{\partial y},$$

hence the scalar function $V_o/\mu_o c$ can be considered responsible for the transverse vector $\boldsymbol{H}_t$. Further, as A_y and A_z are time-independent, we can write $\boldsymbol{E}_t = -\mathrm{grad}_t V_o$ in the yz-plane. Thus, for a TEM mode, the scalar potential $V = V_o \exp[i(kx - \omega t)]$

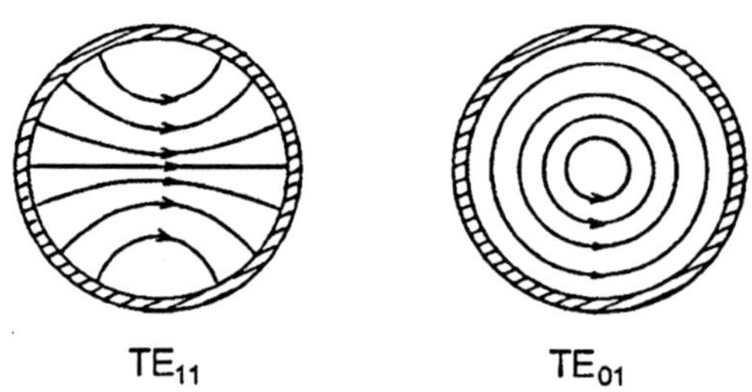

FIGURE 16.7. TE_{11} and TE_{01} modes in a cylindrical waveguide.

is essential for propagation at $\omega = ck$, although the transversal fields are different from the free plane waves.

To determine the transversal field, we need to solve the Helmholtz equation

$$\Delta_t V_o = 0.$$

For a coaxial cylindrical cable, it is expressed as

$$\frac{1}{\rho}\frac{\partial}{\partial \rho}\left(\rho \frac{\partial V_o}{\partial \rho}\right) + \frac{1}{\rho^2}\frac{\partial^2 V_o}{\partial \theta^2} = 0.$$

Practically, the most important mode has a circular symmetry, being independent of θ. In this case we only deal with the equation

$$\frac{d}{d\rho}\left(\rho \frac{dV_o}{d\rho}\right) = 0,$$

whose solution is expressed as

$$V_o(\rho) = c_1 \ln \rho + c_2,$$

where c_1 and c_2 are constants of integration. Because of the two separate conductors (radii $a < b$), the potential is signified by the potential difference $\Delta V = V_o(b) - V_o(a)$ as $c_1 = \Delta V/\ln(b/a)$, and

$$E_\rho = -\frac{kc_1}{\omega\varepsilon_o\mu_o}\frac{\exp[i(kx - \omega t)]}{\rho}, \quad E_\theta = 0$$

and

$$H_\rho = 0, \quad H_\theta = -\frac{c_1}{\mu_o}\frac{\exp[i(kx - \omega t)]}{\rho}.$$

A coaxial cable is useful for applications in a wide range of frequencies, as there is no cut-off in the TEM mode. However, because of double conductors, the attenuation due to conductive loss of energy is relatively heavier than it is for waveguides of a single conductor.

Part 4
Coherent Waves and Radiation Quanta

Generated by oscillatory charges, the electromagnetic wave propagates at the speed of light, as described by the Maxwell theory. Transmitting energy in the distance, it is utilized for telecommunication in a wide range of frequencies, for which intense amplitudes are required. Prompted by pressing demands since WWII, significant advances have been achieved in manipulating microwaves and coherent optical radiation, resulting in today's advanced technologies and spectroscopic studies on modern materials. Among recent advances, coherent radiation has opened up a large area of applications beyond the traditional scope of classical electromagnetism.

In Part 4, we duscuss guides waves and resonant cavities as indispensable devices in modern electronics. At optical frequencies, we particularly encounter the limit of classical theory, regarding the coherence problem. It is important for us to realized that there is a significant limit for the classical Maxwell theory.

17 Waveguide Transmission

17.1. Orthogonality Relations of Waveguide Modes

In the previous chapter, we discussed TE and TM modes of electromagnetic propagations in a rectangular waveguide, and a TEM mode in a coaxial cable. In a waveguide, axial components of $\boldsymbol{E}$ or $\boldsymbol{H}$ vectors behave like scalar variables of the phase of propagation, and in a coaxial cable the uniaxial vector potential $\boldsymbol{A}$ plays such a role as a scalar variable. Thus, transversal fields depend on the type of waveguide; on the other hand, the axial component plays essential role for propagation.

Summarizing the previous results, TE and TM modes in a rectangular waveguide are expressed as

$$\left(\Delta_{\mathrm{t}}+k_{\mathrm{c}}^{2}\right) H_{x}=0,\ \boldsymbol{H}_{t}=\frac{i k_{x}}{k_{\mathrm{c}}^{2}} \operatorname{grad}_{\mathrm{t}} H_{x}, \text{ and } \boldsymbol{E}_{\mathrm{t}}=-\frac{i k Z_{\mathrm{o}}}{k_{\mathrm{c}}^{2}}[\mathbf{i} \times \operatorname{grad}_{\mathrm{t}} H_{x}] \quad \text{for TE}$$

and

$$\left(\Delta_{\mathrm{t}}+k_{\mathrm{c}}^{2}\right) E_{x}=0,\ \boldsymbol{E}_{\mathrm{t}}=\frac{i k_{x}}{k_{\mathrm{c}}^{2}} \operatorname{grad}_{\mathrm{t}} E_{x}, \text{ and } \boldsymbol{H}_{\mathrm{t}}=\frac{i k}{Z_{\mathrm{o}} k_{\mathrm{c}}^{2}}[\mathbf{i} \times \operatorname{grad}_{\mathrm{t}} E_{x}] \quad \text{for TM,}$$

where

$$k_{\mathrm{c}}^{2}=\left(\frac{m \pi}{a}\right)^{2}+\left(\frac{n \pi}{b}\right)^{2}, \quad \text{m and n are 0 or integers.}$$

Although specified by two indexes (m, n), each waveguide mode is characterized by k_c that is a single parameter related to m and n. Therefore, in the following we distinguish different modes by the single index k_{c} when discussing orthogonality relations.

We show that any two modes, indexed by different α and β are mutually orthogonal. Such waveguide modes can be represented by an effective scalar function, and therefore we consider a function $\varphi_1 \operatorname{grad}_{\mathrm{t}} \varphi_2$ composed of two scalars φ_1 and φ_2 and its divergence, i.e.,

$$\operatorname{div}_{\mathrm{t}}(\varphi_1 \operatorname{grad}_{\mathrm{t}} \varphi_2)=(\operatorname{grad}_{\mathrm{t}} \varphi_1) \cdot (\operatorname{grad}_{\mathrm{t}} \varphi_2)-\varphi_1 \Delta_t \varphi_2.$$

Calculating the flux of two-dimensional vector $\varphi_1 \mathrm{grad}_t \varphi_2$ with the Gauss theorem, we have

$$\int_v \mathrm{div}(\varphi_1 \mathrm{grad}_t \varphi_2)\mathrm{d}v = \int_S \varphi_1 \frac{\partial \varphi_2}{\partial n}\mathrm{d}S = \int_v \{(\mathrm{grad}_t \varphi_1)\cdot(\mathrm{grad}_t \varphi_2) - \varphi_1 \Delta_t \varphi_2\}\mathrm{d}v,$$

where $\mathrm{d}v = \mathrm{d}S\mathrm{d}x$ represents the volume element taken along the waveguide, and $\mathrm{d}S$ is the differential element on the cross-section S. Therefore, from the last equation we have

$$\oint_C \varphi_1 \frac{\partial \varphi_2}{\partial n}\mathrm{d}s = \int_S \{(\mathrm{grad}_t \varphi_1)(\mathrm{grad}_t \varphi_2) - \varphi_1 \Delta_t \varphi_2\}\mathrm{d}S.$$

Taking the boundary conditions $\varphi_{1,2} = 0$ and $\dfrac{\partial \varphi_{1,2}}{\partial n} = 0$ into account, the line integral on the left side vanishes, and we obtain

$$\int_S (\mathrm{grad}_t \varphi_1)\cdot(\mathrm{grad}_t \varphi_2)\mathrm{d}S = \int_S \varphi_1 \Delta_t \varphi_2 \,\mathrm{d}S. \qquad \text{(i)}$$

We consider $\varphi_1 = H_{x1}$ and $\varphi_2 = H_{x2}$, distinguishing them by k_{c1} and k_{c2}, respectively, thereby expressing the relation (i) as

$$\frac{k_{c1}^2 k_{c2}^2}{k_{x1}k_{x2}} \int_S \boldsymbol{H}_{t1}.\boldsymbol{H}_{t2}\mathrm{d}S = -k_{c2}^2 \int_S H_{x1}H_{x2}\mathrm{d}S.$$

It is clear that this relation is correct regardless of values of these k_c, k_x, provided that the orthogonality relations

$$\int_S \boldsymbol{H}_{t1}.\boldsymbol{H}_{t2}\mathrm{d}S = 0 \quad \text{and} \quad \int_S H_{x1}H_{x2}\mathrm{d}S = 0 \qquad (17.1\text{a})$$

hold for different modes 1 and 2. Such a relation can also be derived for $\boldsymbol{E}$ vectors because $\boldsymbol{E}_t = Z_o \boldsymbol{H}_t$ with a constant phase factor, and hence we can write

$$\int_S \boldsymbol{E}_{t1}.\boldsymbol{E}_{t2}\mathrm{d}S = 0 \qquad (17.1\text{b})$$

for different modes 1 and 2. Also, from the relation $k_{x1}\mathbf{i}(\boldsymbol{E}_{t1} \times \boldsymbol{H}_{t2}) = Z_o \boldsymbol{H}_{t1} \cdot \boldsymbol{H}_{t2}$, we can show that vectors $\mathbf{i}$, $\boldsymbol{E}_{t1}$, and $\boldsymbol{H}_{t2}$ are mutually orthogonal. These relations (17.1ab) constitute orthogonality relations for TE mode.

For TM mode we can derive similar relations to (17.1ab), namely,

$$\int_S E_{x1}E_{x2}\mathrm{d}S = 0, \quad \int_S \boldsymbol{E}_{t1}.\boldsymbol{E}_{t2}\mathrm{d}S = 0 \quad \text{and} \quad \int_S \boldsymbol{H}_{t1}.\boldsymbol{H}_{t2}\mathrm{d}S = 0. \qquad (17.2)$$

Further, we can show orthogonality relations hold also between TE and TM modes. For instance,

$$\boldsymbol{E}_{t1}(TE).\boldsymbol{E}_{t2}(TM) = -\frac{Z_o k k_{x2}}{k_{c1}^2 k_{c2}^2}[(\mathbf{i} \times \mathrm{grad}_t H_{x1}).\mathrm{grad}_t E_{x2}]$$
$$= \frac{Z_o k k_{x2}}{k_{c1}^2 k_{c2}^2}\mathrm{div}_t\{E_{x2}\mathrm{curl}_t(\mathbf{i}H_{x1})\},$$

and hence

$$\int_S \boldsymbol{E}_{t1}(TE).\boldsymbol{E}_{t2}(\mathrm{TM})\mathrm{d}S = \frac{Z_o k k_{x2}}{k_{c1}^2 k_{c2}^2}\oint_C \frac{\partial}{\partial n}\{E_{t2}(\mathrm{curl}_t\mathbf{i}H_{t1})\}.\mathrm{d}s = 0,$$

because $E_{t2} = 0$ on the conducting wall. By a similar procedure, we can also show that $\boldsymbol{H}_{t1}$(TE) and $\boldsymbol{H}_{t2}$(TM) are orthogonal.

Although purely mathematical, these orthogonal relations provide necessary conditions for all waveguide modes to be linearly independent.

17.2. Impedances

So far we have discussed idealized waveguides, assuming infinite length. However, a practical waveguide is connected to an oscillator at one end, and to another object at the other far end. Inside of a waveguide, it is practical consider, effectively, waves, propagating back and forth along the guide, which can be observed as a *standing wave*. We realize that such a description is useful if we can accurately define the phase of propagation for dealing with flowing energy. For such a description it is significant that such a periodic repetition be determined within the timescale of observation τ. In other words, such a scale τ must be sufficiently long compared with the period of repetition.

Under normal circumstances, such waves are conveniently expressed by complex functions with the normalization conditions

$$\int_S |\boldsymbol{E}_{t\alpha}.\boldsymbol{E}^*_{t\alpha}|\mathrm{d}S = 1, \quad \int_S |\boldsymbol{H}_{t\alpha}.\boldsymbol{H}^*_{t\alpha}|\mathrm{d}S = \frac{k_x^2}{k^2 Z_o^2}, \text{ and } \int_S \mathbf{i}.(\boldsymbol{E}_{t\alpha} \times \boldsymbol{H}^*_{t\alpha})\mathrm{d}S = \frac{k_x}{kZ_o}, \tag{17.3}$$

where the * indicates complex conjugate. For practical application, the TE_{10} mode is employed as the *principal mode*, although the above argument can be applied to any other mode.

A guided wave consists of longitudinal and transversal components of field vectors with respect to the x-axis. It is convenient to express it as the superposition of forward and backward waves, both being characterized by longitudinal and transversal components, (E_x, H_x) and $(\boldsymbol{E}_t, \boldsymbol{H}_t)$, respectively. Therefore, for the principal mode, we write

$$\boldsymbol{E} = \boldsymbol{E}_t \{A \exp[i(k_x x - \omega t)] + B \exp[i(-k_x x - \omega t)]\} \\ + \mathbf{i} E_x \{A \exp[i(k_x x - \omega t)] - B \exp[i(-k_x x - \omega t)]\}$$

and

$$\boldsymbol{H} = \boldsymbol{H}_t \{A \exp[i(k_x x - \omega t)] - B \exp[i(-k_x x - \omega t)]\} \\ + \mathbf{i} H_x \{A \exp[i(k_x x - \omega t)] + B \exp[i(-k_x x - \omega t)]\}, \tag{17.4}$$

where the phase $-k_x x - \omega t$ signifies the backward wave, but the coefficient B is left undetermined, unless the terminal object is specified. Generally, the field is more complex than (17.4), but at a distant point from the terminals, the backward (or reflected) wave is expressed approximately by the same principal mode with a phase shift and a smaller amplitude B than A of the forward wave.

It is noted that the mode vectors $\boldsymbol{E}_t$ can usually be detected by a probe inserted perpendicular to the guide, whereas the axial component E_x is not directly measurable in practice. By this reasoning, we only need to deal with $\boldsymbol{E}_t$ in the waveguide. As discussed in Chapter 16, $\boldsymbol{E}_t$ and the corresponding $\boldsymbol{H}_t$ of the principal mode are linearly related by (16.8), so that we can define the ratio $Z_g = |\boldsymbol{E}_t|/|\boldsymbol{H}_t|$ as the effective *wave impedance*. From (16.8), we can write

$$Z_g = \frac{k}{k_x} Z_o, \tag{17.5}$$

which is specifically called the *guide impedance*.

Writing (17.4) at specific phases $\phi = k_x x - \omega t = 2\pi \times (0 \text{ or integer})$, from (17.4) we obtain

$$(A + B)^2 = \boldsymbol{E}_t.\boldsymbol{E}_t^* \quad \text{and} \quad (A - B)^2 = \boldsymbol{E}_t.\boldsymbol{H}_t^*,$$

and hence

$$Z_g = \frac{A - B}{A + B}.$$

Figure 17.1 shows two partitions installed in a waveguide, where section 2 can be regarded as a working space. These partitions should have openings of some kind, as illustrated in Figures 17.2 and 17.3, acting as couplings between separated sections. Although mathematically not so easy to represent, field-lines near a coupling can be inferred using physical intuition from the principal mode shown in Figure 16.4. For example, Figure 17.2 shows a coupling via magnetic field-lines of a rectangular waveguide in TE_{10} mode. On the other hand, Figure 17.3 shows that by extending the center conductor at the end of a coaxial

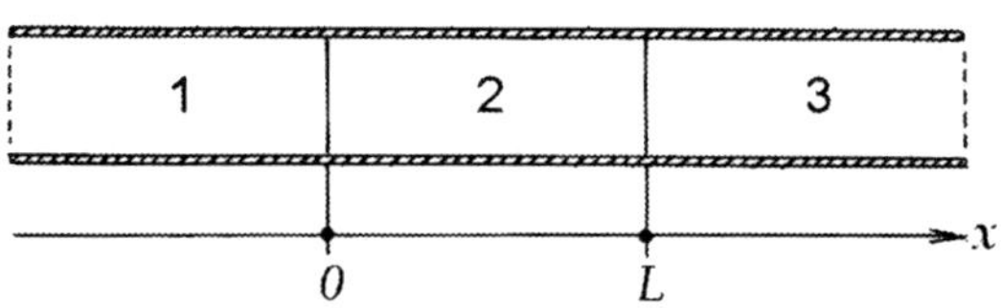

FIGURE 17.1. A uniform waveguide with two partitions at $x = 0$ and $x = L$.

FIGURE 17.2. (a) An iris coupling in general view. (b) H-field connected at an iris coupling.

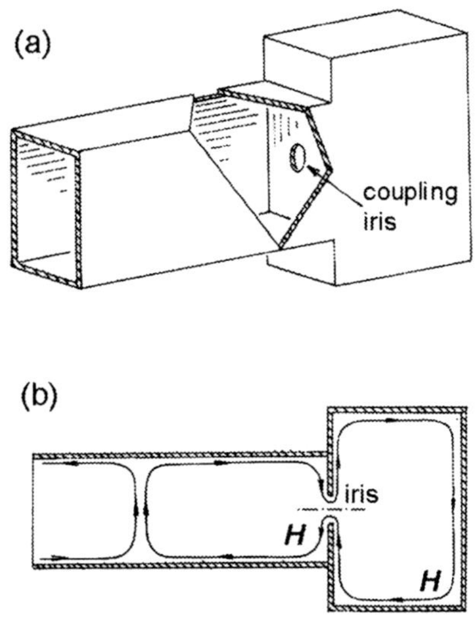

cable, an oscillating electric dipole is formed, thereby modifying field-lines at the coupling.

In these examples, at least approximately, the dominant $\boldsymbol{H}$ vector or dominant $\boldsymbol{E}$ vector at the partition is responsible for the coupling magnitude in these figures, respectively. However, depending generally on its geometric structure, such a coupling is determined by both $\boldsymbol{E}$ and $\boldsymbol{H}$ at the position. Nevertheless, the reflected wave from such a coupling can be characterized effectively by a positive or negative phase shift. In principle, such a coupling should be subjected to the boundary conditions on the conductive area $S$and open area S' of the partition. Giving the field vectors as we did in the principal mode, the coupling can be calculated as a perturbation, resulting in a phase shift φ of backward wave. Consequently,

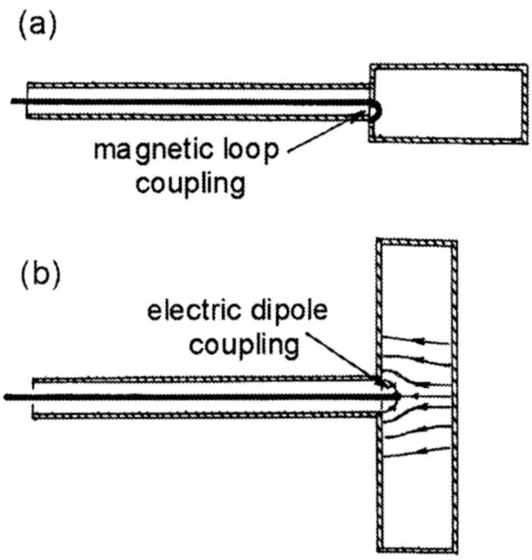

FIGURE 17.3. (a) A magnetic loop coupling between a coaxial cable and a rectangular guide. (b) An electric dipole coupling between a coaxial cable and a rectangular guide.

expressing the reflected wave with a phase shift φ with amplitude B, the guided wave can logically be written as

$$A\exp[i(k_x x-\omega t)]+B\exp[i(-k_x x-\omega t+\varphi)]$$
$$=A\exp[i(k_x x-\omega t)]\left\{1+\frac{B}{A}\exp[i(-2k_x x+\varphi)]\right\}.$$

Writing

$$R=\frac{B}{A}\exp[i(-2k_x x+\varphi)], \tag{17.6}$$

the field vectors at a given (x,t) in the waveguide are expressed by

$$\boldsymbol{E}=\boldsymbol{E}_t A(1+R)\exp[i(k_x x-\omega t)]+\mathbf{i}E_x A(1-R)\exp[i(k_x x-\omega t)]$$
$$\boldsymbol{H}=\boldsymbol{H}_t A(1-R)\exp[i(k_x x-\omega t)]+\mathbf{i}H_x A(1+R)\exp[i(k_x x-\omega t)].$$

As already mentioned, a probe is utilized for measuring transversal fields along the vectors $\boldsymbol{E}_\mathrm{t}$ or $\boldsymbol{H}_\mathrm{t}$ so that we can set aside the longitudinal components E_x and H_x in the following calculation.

Such a terminated waveguide is said to be "loaded" by an obstacle and represented by the impedance Z_x at a point x defined by the ratio

$$Z_x=\frac{|\boldsymbol{E}_\mathrm{t}|(1+R)}{|\boldsymbol{H}_\mathrm{t}|(1-R)}=Z_\mathrm{g}\frac{1+R}{1-R}. \tag{17.7}$$

At the position of an obstacle $x=0$, we write (17.7) as Z_L, which is called the *load impedance*, or the *coupling impedance*, in this case. Also we define $R_\mathrm{o}=\frac{B}{A}\exp(i\varphi)$ at $x=0$, and express (17.7) as

$$Z_\mathrm{L}=Z_\mathrm{g}\frac{1+R_\mathrm{o}}{1-R_\mathrm{o}}\quad\text{and}\quad R_\mathrm{o}=\frac{Z_\mathrm{L}-Z_\mathrm{g}}{Z_\mathrm{L}+Z_\mathrm{g}}. \tag{17.8}$$

In the above, we have assumed that the waveguide is made of perfectly conducting materials, however, the energy loss in the waveguide during transmission is inevitable. Although such a loss of electromagnetic energy on the conducting walls is not a simple mathematical problem, we only deal with the rate for diminishing amplitude per length. In this case, the propagation vector $\boldsymbol{k}$ can be expressed as a complex parameter, i.e., $k=k'+ik''$, where the imaginary part k'' describes the decay along the guide. We can also describe the propagation by a phase $\gamma x-i\omega t$ and $R_x=(B/A)\exp(-2\gamma x)$ to emphasize decaying propagation by $\gamma=ik$, in which case we can write the following relation between Z_x and Z_L:

$$\frac{Z_x}{Z_\mathrm{g}}=\frac{Z_\mathrm{L}\cosh\gamma x-Z_\mathrm{g}\sinh\gamma x}{Z_\mathrm{g}\cosh\gamma x-Z_\mathrm{L}\sinh\gamma x}=\frac{\dfrac{Z_\mathrm{L}}{Z_\mathrm{g}}-\tanh\gamma x}{1-\dfrac{Z_\mathrm{L}}{Z_\mathrm{g}}\tanh\gamma x}, \tag{17.9}$$

which is a *bilinear transformation* between complex variables $\dfrac{Z_x}{Z_\mathrm{g}}$ and $\dfrac{Z_\mathrm{L}}{Z_\mathrm{g}}$. The derivation of (17.9) is straightforward, but left to readers as an exercise. As an extreme case, we see that $Z_x=Z_\mathrm{L}$ if $Z_\mathrm{L}=\pm Z_\mathrm{g}$, indicating that $Z_x=\pm Z_\mathrm{g}$ represent independent waves of an equal amplitude propagating in the $\pm x$ directions. On the other hand, if $|Z_L| \neq Z_g$ we have $A \neq B$, giving rise to a finite *standing wave*.

17.3. Power Transmission Through a Waveguide

In the expression (17.4) of the field vectors, by amplitude factors A and B we can describe forward and backward waves in the waveguide, respectively representing incident waves to and reflected waves from the coupling iris. However, in a waveguide the word "reflection" should indicate a *group of plane waves* carrying an electromagnetic energy, as interpreted by the Poynting theorem.

As a measurable quantity in the waveguide, a flow of energy should be expressed as an average over the phase in repetition, since the timescale of usual observation is much longer than $2\pi/\omega$ of radiation. Also, the mode vectors $\boldsymbol{E}_t$ and $\boldsymbol{H}_t$ may not be exactly in phase, for which it is logical to consider the Poynting vector $\boldsymbol{E}(t) \times \boldsymbol{H}^*(t')$ that is averaged over distributed phase $\omega(t - t') = d\phi$. Therefore, for the energy transport in a waveguide, we write the Poynting theorem as

$$-\frac{dU}{dt} = \int_S dS\left[\frac{1}{2\pi}\int_0^{2\pi} \mathbf{i}.(\boldsymbol{E} \times \boldsymbol{H})\,d\phi\right]$$
$$= A^2 \exp\left(-2k_x''x\right)|1 - R_x|^2 \int_S \mathbf{i}.\left\langle \boldsymbol{E}_t \times \boldsymbol{H}_t^*\right\rangle dS, \qquad (17.10)$$

where S is the cross-sectional area, and the brackets $\langle\rangle$ signify the time average over repeated periods.

For a TE mode of a rectangular waveguide, the integrand in (17.10) can be expressed as

$$\mathbf{i}\cdot\langle \boldsymbol{E}_t \times \boldsymbol{H}_t^*\rangle = \langle(\mathbf{i} \times \boldsymbol{E}_t)\cdot \boldsymbol{H}_t^*\rangle = \frac{\omega\mu_o}{2k_x}|\boldsymbol{H}_t|^2 = \frac{\omega\mu_o}{2k_x}\frac{k_x^2}{k_c^4}\,|\mathrm{grad}_t H_x|^2,$$

and hence

$$P = \frac{\omega\mu_o}{2k_x}\int_S |H_x|^2 dS.$$

On the other hand, the electromagnetic energy is calculated from the Poynting theorem

$$\int_S \mathbf{i}.\left\langle \boldsymbol{E}_t \times \boldsymbol{H}_t^*\right\rangle dS = c\int_S u\, dS,$$

where u is the energy density

$$u = 1\varepsilon_o|\boldsymbol{E}_t|^2 + 1\mu_o|\boldsymbol{H}_t|^2 + 1\mu_o|H_x|^2$$
$$= \frac{\varepsilon}{2}\frac{k^2 Z_o^2}{k_c^4}|\mathbf{i} \times \mathrm{grad}_t H_x|^2 - \frac{\mu_o}{2}\frac{k_x^2}{k_c^4}|\mathrm{grad}_t H_x|^2 + \frac{\mu_o}{2}|H_x|^2.$$

Here, using the relation $\text{grad}_t H_x = \dfrac{k_c^2}{ik_x} H_x$, we can write $u = \dfrac{\mu_o}{2}\dfrac{k^2}{k_x^2}|H_x|^2$, and hence we have the relation

$$P = \frac{\omega}{k_x}\frac{k_x^2}{k^2} u S = c\frac{k_x}{k} U,$$

where the quantity $c(k_x/k) = v_g$ is generally known as the *group velocity*. We note that the group velocity can also be derived from the relation $k^2 = k_c^2 + k_x^2$ in Figure 16.2, where

$$v_g = \frac{d\omega}{dk_x} = c\frac{k_x}{k}. \tag{17.11}$$

As stated in (17.3), for TE mode we have the relation $\int_S \mathbf{i}.\left(\boldsymbol{E}_t \times \boldsymbol{H}_t^*\right) dS = \dfrac{1}{Z_g}$, and hence the total power flow in the waveguide can be calculated as

$$PS = \frac{A^2}{Z_g} \exp(-2k_x'' x)\{(1 - |R_x|^2) + (R_x - R_x^*)\}, \tag{17.12}$$

where the first term in the brackets represents powers flowing forward and backward in the ratio $1 : |R_x|^2$. On the other hand, the second term is pure imaginary, hence representing no real power flow, unless Z_g has an imaginary part. Ignoring the latter as a mathematical conjecture, we consider the attenuating flow of power as proportional to $1 - |R_x|^2$.

17.4. Multiple Reflections in a Waveguide

We consider now two partitions that are typically iris couplings, between which is a "working space." Such partitions divide the waveguide into three parts, marked 1, 2, and 3 in Figure 17.1. An oscillator is connected to the left end of section 1, section 3 has an opening at the left, and the right end is at infinity. In the center of section 2, the principal waves are reflected back and forth between the two couplings, whereas there are incident and reflected waves in section 1.

Each sections are characterized by wave impedances Z_1, Z_2, and Z_3 and the partitions are located at $x = 0$ and $x = L$. If the second opening is considered as the load impedance in the third guide, it should be Z_3, assuming no reflection in this section. Therefore, at a position x inside section 2, i.e., $0 \le x \le L$, the impedance of (17.9) in section 2 can be expressed

$$\frac{Z_x}{Z_2} = \frac{\dfrac{Z_3}{Z_2} - \tanh\left(\gamma_{2x} x\right)}{1 - \dfrac{Z_3}{Z_2}\tanh\left(\gamma_{2x} x\right)},$$

giving $Z_{x=L} = Z_3$ if $\tanh\left(\gamma_{2x} L\right) = 0$ at the second boundary. At the first boundary $x = 0$, $Z_{x=0} = Z_3$, which is assumed to be the same as Z_1.

At a arbitrary position $0 < x < L$ the impedance is given by the above Z_x, to which we can add that

$$\frac{Z_x}{Z_1} = \frac{1+R}{1-R}$$

should be applied. The reflection coefficient R in section 1 can be obtained by eliminating Z_x from the last two equations and setting $x = L$. That is,

$$\frac{1+R}{1-R} = \frac{Z_2}{Z_1} \frac{\frac{Z_3}{Z_2} - \tanh\left(\gamma_{2x} L\right)}{1 - \frac{Z_3}{Z_2} \tanh\left(\gamma_{2x} L\right)}.$$

If the second partition is a perfect surface without an iris, Z_3 in this expression can be set equal to zero, and we obtain

$$\frac{1+R}{1-R} = -\frac{Z_2}{Z_1} \tanh\left(\gamma_{2x} L\right). \qquad (17.13)$$

In this case, the second a box-shaped section shows a *resonant* property if L is adjustable.

Figure 17.4 shows a rectangular *cavity* where the distance L is adjustable by the attached moving plunger. If the waveguide is free from loss of electromagnetic

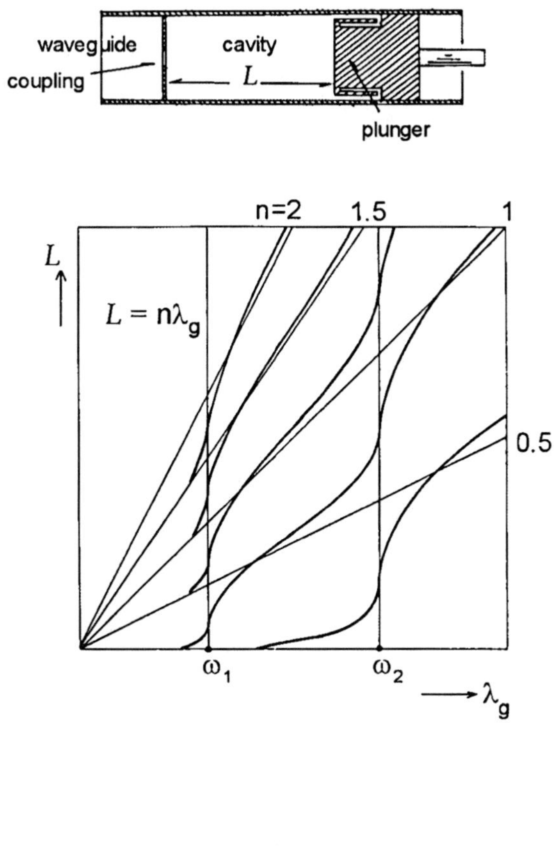

FIGURE 17.4. A tuneable rectangular cavity, and the tuning character $L = n\lambda_g$.

energy, the constant γ_{2x} is pure imaginary, i.e., $\gamma_{2x} = ik_x$, and tanh $(ik_x L) = i\ \tan(k_x L)$. Then, from (17.13), if $\tan(k_x L) = 0$, we obtain $R = 0$, signifying that there is no reflection if $k_x L = \pi \times$ integer. Using the relation $k_x = \dfrac{2\pi}{\lambda_g}$, this condition for resonance can be rephrased as

$$L = \frac{\lambda_g}{2} \times \text{integer}. \tag{17.14}$$

At resonance, electromagnetic fields in such an idealized cavity reflect back and forth between two walls, resulting in stationary waves with large amplitude, as specified by (17.14). However, a real cavity cannot be idealized as free from energy loss. Nevertheless, a cavity can be represented approximately by a loaded impedance $Z_L = -iZ_2 \tan(k_x L)$ behaving as a *reactance* for an arbitrary L.

18
Resonant Cavities

18.1. Slater's Theory of Normal Modes

In Section 17.3, we discussed the property of a resonant cavity in a rectangular TE mode. Primarily, it is a matter of analytical convenience, but in principle resonance can be found at a given frequency for a cavity of arbitrary shape. In this chapter, following Slater, we discuss resonant properties of a cavity of general shape, for which the Maxwell equations can provide oscillating modes with intense amplitudes (J. C. Slater, *Microwave Electronics*, Van Nostrand, 1950.)

Electromagnetic waves from an oscillator are transmitted through the waveguide to a cavity connected at the end. In this case the energy in the cavity exhibits a resonance behavior at a particular frequency ω_i. Slater showed that the amplitudes of field vectors $\boldsymbol{E}$ and $\boldsymbol{H}$ inside a cavity are oscillatory and described in terms of the *normal mode* function at a frequency ω_i. The oscillator frequency ω should therefore be equal to ω_i of a normal mode, i.e., $\omega = \omega_i$, which is the condition required for resonance at which the field vectors are expressed in terms of the normal mode i.

We pay attention to the fact that any vector field consists of *solenoidal* and *irrotational* components signified as free from *divergence* (div) and *rotation* (curl), respectively. For such electromagnetic fields we should obtain two solenoidal functions $\boldsymbol{e}_i$ and $\boldsymbol{h}_i$ to express the solenoidal part of $\boldsymbol{E}$ and the fully solenoidal $\boldsymbol{H}$, while we need an irrotational function $\boldsymbol{e}'_i$ for the irrotational part of $\boldsymbol{E}$. To express these $\boldsymbol{E}$ and $\boldsymbol{H}$ expanded with regard to these mode functions the orthogonality relations should be established among mode functions $\boldsymbol{e}_i$, $\boldsymbol{h}_i$ and $\boldsymbol{e}'_i$.

Owing to the relations div $\boldsymbol{e}_i = 0$ and div $\boldsymbol{h}_i = 0$, solenoidal vectors $\boldsymbol{e}_i$ and $\boldsymbol{h}_i$ can be expressed by curls of $\boldsymbol{h}_i$ and $\boldsymbol{e}_i$, respectively. We therefore assume that

$$k_i\boldsymbol{e}_i = \text{curl}\,\boldsymbol{h}_i \quad \text{and} \quad k_i\boldsymbol{h}_i = \text{curl}\,\boldsymbol{e}_i, \tag{18.1}$$

where k_i is a constant. Then, using the vector identity relation, curl(curl) = grad(div) $-\Delta$, we can obtain the wave equations characterized by the same k_i for these solenoidal vectors $\boldsymbol{e}_i$ and $\boldsymbol{h}_i$, i.e.,

$$\left(\Delta + k_i^2\right)(\boldsymbol{e}_i, \boldsymbol{h}_i) = 0, \tag{18.2}$$

indicating that these $\boldsymbol{e}_\mathrm{i}$ and $\boldsymbol{h}_\mathrm{i}$ can constitute a *normal mode* i specified by the constant k_i.

Further, using the relations in (17.1) we can show that $\boldsymbol{e}_\mathrm{i}$ and $\boldsymbol{h}_\mathrm{i}$ can satisfy the boundary conditions on the conducting walls S and opening areas S' of a cavity if the field vectors $\boldsymbol{E}$ and $\boldsymbol{H}$ can be represented by these mode vectors. Denoting the unit vector normal to inner surfaces by $\boldsymbol{n}$, we obtain

$$\boldsymbol{n} \times \boldsymbol{e}_\mathrm{i} = 0 \text{ on } S \quad \text{and} \quad \boldsymbol{n} \times \boldsymbol{h}_\mathrm{i} = 0 \text{ on } S', \tag{18.3a}$$

by applying Stokes' theorem to (18.1). Also, applying Gauss's theorem we can show that the normal components are continuous, i.e.,

$$\boldsymbol{n}.\boldsymbol{h}_\mathrm{i} = 0 \text{ on } S \quad \text{and} \quad \boldsymbol{n}.\boldsymbol{e}_\mathrm{i} = 0 \text{ on } S'. \tag{18.3b}$$

Using these boundary conditions, next we can obtain orthogonality relations of these normal modes. For that purpose we use the identity relation

$$\begin{aligned}\operatorname{div}(\boldsymbol{e}_\mathrm{j} \times \operatorname{curl} \boldsymbol{e}_\mathrm{i}) - \operatorname{div}(\boldsymbol{e}_\mathrm{i} \times \operatorname{curl} \boldsymbol{e}_\mathrm{j}) = {} & \operatorname{curl} \boldsymbol{e}_\mathrm{i}.\operatorname{curl} \boldsymbol{e}_\mathrm{j} - \boldsymbol{e}_\mathrm{j}.\operatorname{curl}(\operatorname{curl} \boldsymbol{e}_\mathrm{i}) \\ & - \operatorname{curl} \boldsymbol{e}_\mathrm{j}.\operatorname{curl} \boldsymbol{e}_\mathrm{i} + \boldsymbol{e}_\mathrm{j}.\operatorname{curl}(\operatorname{curl} \boldsymbol{e}_\mathrm{i}),\end{aligned}$$

where the left side can be simplified as $\left(k_\mathrm{j}^2 - k_\mathrm{i}^2\right) \boldsymbol{e}_\mathrm{i}.\boldsymbol{e}_\mathrm{j}$ by (18.2). Applying the Gauss theorem to the above expression in the cavity volume v, converting then to the surface integral over $S + S'$, we obtain

$$\int_{S,S'} \boldsymbol{n}.(\boldsymbol{e}_\mathrm{j} \times k_\mathrm{i}\boldsymbol{h}_\mathrm{i} - \boldsymbol{e}_\mathrm{i} \times k_\mathrm{j}\boldsymbol{h}_\mathrm{j})\mathrm{d}(S, S') = (k_\mathrm{j}^2 - k_\mathrm{i}^2) \int_v \boldsymbol{e}_\mathrm{i}.\boldsymbol{e}_\mathrm{j}\, \mathrm{d}v \tag{i}$$

The surface integrals on the left side can be written as

$$\int_S \{k_\mathrm{i}\boldsymbol{h}_\mathrm{i}.\left(\boldsymbol{n} \times \boldsymbol{e}_\mathrm{j}\right) - k_\mathrm{j}\boldsymbol{h}_\mathrm{j}.(\boldsymbol{n} \times \boldsymbol{e}_\mathrm{i})\}\, \mathrm{d}S + \int_{S'} \{k_\mathrm{j}\boldsymbol{e}_\mathrm{i}.(\boldsymbol{n} \times \boldsymbol{h}_\mathrm{j}) - k_\mathrm{i}\boldsymbol{e}_\mathrm{j}.(\boldsymbol{n} \times \boldsymbol{h}_\mathrm{i})\}\, \mathrm{d}S'$$

where all terms vanish on surfaces S and S' owing to the boundary conditions (18.3a). Hence, the volume integral in (i) vanishes for $k_\mathrm{i} \neq k_\mathrm{j}$, that is,

$$\int_v \boldsymbol{e}_\mathrm{i}.\boldsymbol{e}_\mathrm{j}\mathrm{d}v = 0 \quad \text{for} \quad \mathrm{i} \neq \mathrm{j}.$$

In addition, we consider that these $\boldsymbol{e}_\mathrm{i}$ are *normalized* to 1, i.e., $\int_v \boldsymbol{e}_\mathrm{i}^2 \mathrm{d}v$ for $\mathrm{i} = \mathrm{j}$, and hence the orthonormal condition for the vector $\boldsymbol{e}_\mathrm{i}$ is expressed as

$$\int_v \boldsymbol{e}_\mathrm{i}.\boldsymbol{e}_\mathrm{j}\, \mathrm{d}v = \delta_\mathrm{ij}, \tag{18.4a}$$

where $\delta_\mathrm{ij} = 1$ if $\mathrm{i} = \mathrm{j}$, otherwise 0 for all $\mathrm{i} \neq \mathrm{j}$, and is known as the Kronecker delta.

The orthogonalty for the vector $\boldsymbol{h}_\mathrm{i}$ can be verified in a manner similar to $\boldsymbol{e}_\mathrm{i}$. In fact, $\boldsymbol{h}_\mathrm{i}$ and $\boldsymbol{e}_\mathrm{i}$ are related by (18.1), therefore if $\boldsymbol{e}_\mathrm{i}$ is normalized, $\boldsymbol{h}_\mathrm{i}$ is also normalized. We therefore have

$$\int_v \boldsymbol{h}_i.\boldsymbol{h}_j \mathrm{d}v = \delta_{ij}. \tag{18.4b}$$

For the irrotational vector $\boldsymbol{e}'_i$, we can write

$$k_i \boldsymbol{e}'_i = \operatorname{grad} \psi_i, \tag{18.5}$$

where ψ_i is a scalar. For such a vector $\boldsymbol{e}'_i$ to be a mode function, ψ_i should also be a mode function. Therefore, we assume that ψ_i satisfies the wave equation

$$\left(\Delta + k_i^2\right) \psi_i = 0,$$

and consider the boundary conditions as given by

$$\psi_i = \text{const and} \quad \boldsymbol{n} \times \boldsymbol{e}'_i = 0.$$

Using these conditions, we can show the orthogonality relations

$$\int_v \psi_i \psi_j \mathrm{d}v = 0 \quad \text{and} \quad \int_v \boldsymbol{e}'_i.\boldsymbol{e}'_j \mathrm{d}v = 0 \quad \text{for} \quad i \neq j.$$

These functions can usually be defined as normalized to 1, and the orthonormal relations can be expressed as

$$\int_v \psi_i \psi_j \mathrm{d}v = \int_v \boldsymbol{e}'_i.\boldsymbol{e}'_j \mathrm{d}v = \delta_{ij}. \tag{18.6}$$

This relation can be proved by an identity

$$\operatorname{div}(\psi_i \operatorname{grad}\psi_j) = \psi_i \Delta \psi_j + \operatorname{grad}\psi_i.\operatorname{grad}\psi_j$$

with the equation $\left(\Delta + k_i^2\right) \psi_i = 0$, however, the proof is left to the reader as an exercise.

Finally, the orthogonality between an irrotational vector $\boldsymbol{e}'_i$ and a solenoidal one $\boldsymbol{e}_j$ should be verified, i.e.,

$$\int_v \boldsymbol{e}'_i.\boldsymbol{e}_j \mathrm{d}v = 0. \tag{18.7}$$

To prove this, we use the identity

$$\operatorname{div}(\psi_i \boldsymbol{e}_j) = \psi_i \operatorname{div} \boldsymbol{e}_j + \operatorname{grad} \psi_i.\boldsymbol{e}_j = k_i \boldsymbol{e}'_i.\boldsymbol{e}_j,$$

and we can immediately obtain (18.7) by using the Gauss theorem.

Signified by the presence of resonant frequencies, in the above theory a set of normal functions is defined to meet the boundary conditions required for the fields in a cavity. However, the completeness of normal modes is uncertain, if not duly proved. Besides this, Slater pointed out that two vectors $\boldsymbol{e}_i$ or $\boldsymbol{h}_i$ are not required because of the relation (18.1), and that the characteristic values of solenoidal and irrotational modes are not necessarily identical in his theory. Nevertheless, we may consider that these mode functions ($\boldsymbol{e}_i$, $\boldsymbol{h}_i$, $\boldsymbol{e}'_i$, and ψ_i) are sufficiently adequate to express the electromagnetic field $\boldsymbol{E}_i$, $\boldsymbol{H}_i$, and associated quantities at resonance.

18.2. The Maxwell Equations in a Cavity

We consider electromagnetic fields in a cavity, where the field vectors $\boldsymbol{E}$ and $\boldsymbol{H}$ and the source $\boldsymbol{j}$ and ρ are all expanded into series with respect to normal modes $\boldsymbol{e}_i$, $\boldsymbol{h}_i$, $\boldsymbol{e}'_i$, and ψ_i. We therefore write that

$$\boldsymbol{E} = \sum_i \left(E_i\boldsymbol{e}_i + E'_i\boldsymbol{e}'_i\right), \quad \text{where} \quad E_i = \int_v \boldsymbol{E}.\boldsymbol{e}_i\mathrm{d}v \quad \text{and} \quad E'_i = \int_v \boldsymbol{E}.\boldsymbol{e}'_i\mathrm{d}v;$$

$$\boldsymbol{H} = \sum_i H_i\boldsymbol{h}_i, \quad \text{where} \quad H_i = \int_v \boldsymbol{H}.\boldsymbol{h}_i\mathrm{d}v;$$

$$j = \sum_i \left(j_i\boldsymbol{e}_i + j'_i\boldsymbol{e}'_i\right), \quad \text{where} \quad j_i = \int_v \boldsymbol{j}.\boldsymbol{e}_i\mathrm{d}v \quad \text{and} \quad j'_i = \int_v \boldsymbol{j}.\boldsymbol{e}'_i\mathrm{d}v;$$

$$\rho = \sum_i \rho_i\psi_i, \quad \text{where} \quad \rho_i = \int_v \rho\psi_i\mathrm{d}v. \tag{18.8}$$

On the other hand, we expand curl $\boldsymbol{E}$ with respect to $\boldsymbol{h}_i$ for mathematical convenience, that is,

$$\text{curl } \boldsymbol{E} = \sum_i \boldsymbol{h}_i \int_v (\text{curl } \boldsymbol{E}).\boldsymbol{h}_i\mathrm{d}v,$$

to which we apply the identity relation

$$\begin{aligned}\text{div}(\boldsymbol{E} \times \text{curl } \boldsymbol{e}_i) &= (\text{curl } \boldsymbol{E}).(\text{curl } \boldsymbol{e}_i) - \boldsymbol{E}.(\text{curlcurl } \boldsymbol{e}_i)\\ &= k_i\boldsymbol{h}_i.(\text{curl } \boldsymbol{E}) - k_i^2\boldsymbol{E}.\boldsymbol{e}_i,\end{aligned}$$

and obtain

$$\begin{aligned}\int_v (\text{curl } \boldsymbol{E}).\boldsymbol{h}_i\mathrm{d}v &= k_i \int_v \boldsymbol{E}.\boldsymbol{e}_i\mathrm{d}v + \frac{1}{k_i}\int_v \text{div}\,(\boldsymbol{E} \times \text{curl } \boldsymbol{e}_i)\,\mathrm{d}v.\\ &= k_i \int_v \boldsymbol{E}.\boldsymbol{e}_i\mathrm{d}v + \int_{S,S'} \boldsymbol{n}.(\boldsymbol{E} \times \boldsymbol{h}_i)\mathrm{d}(S, S').\end{aligned}$$

The second surface integration on the right can be performed separately on S and S' as $\int_S \boldsymbol{h}_i.(\boldsymbol{n} \times \boldsymbol{E})\mathrm{d}S' + \int_{S'} \boldsymbol{E}.(\boldsymbol{h}_i \times \boldsymbol{n})\mathrm{d}S'$; however the boundary condition (18.3a) on S' makes the second integral zero. Therefore, curl $\boldsymbol{E}$ can be expanded as

$$\text{curl } \boldsymbol{E} = \sum_i \boldsymbol{h}_i\left[k_iE_i + \int_S (\boldsymbol{n} \times \boldsymbol{E}).\boldsymbol{h}_i\mathrm{d}S\right]. \tag{18.9a}$$

Similarly, we can write

$$\operatorname{curl} \boldsymbol{H} = \sum_{\mathrm{i}} \boldsymbol{e}_{\mathrm{i}} \left[k_{\mathrm{i}} H_{\mathrm{i}} + \int_{S'} (\boldsymbol{n} \times \boldsymbol{H}).\boldsymbol{e}_{\mathrm{i}} \mathrm{d}S' \right]. \tag{18.9b}$$

To expand div $\boldsymbol{E}$, we use the formula

$$\operatorname{div}(\psi_{\mathrm{i}} \boldsymbol{E}) = \psi_{\mathrm{i}} \operatorname{div} \boldsymbol{E} + \boldsymbol{E}.\operatorname{grad} \psi_{\mathrm{i}} = \psi_{\mathrm{i}} \operatorname{div} \boldsymbol{E} + k_{\mathrm{i}} \boldsymbol{E}.\boldsymbol{e}'_{\mathrm{i}},$$

and derive

$$\begin{aligned} \operatorname{div} \boldsymbol{E} &= \sum_{\mathrm{i}} \psi_{\mathrm{i}} \left[-k_{\mathrm{i}} \int_{v} \boldsymbol{E}.\boldsymbol{e}'_{\mathrm{i}} \mathrm{d}v + \int_{S,S'} (\boldsymbol{E}.\boldsymbol{n}) \psi_{\mathrm{i}} \mathrm{d}\left(S, S'\right) \right] \\ &= -\sum_{\mathrm{i}} k_{\mathrm{i}} \psi_{\mathrm{i}} \int_{v} \boldsymbol{E}.\boldsymbol{e}'_{\mathrm{i}} \mathrm{d}v. \end{aligned}$$

In the normal mode i, the amplitude of the field component varies as a function of time t. Accordingly, from the equation $\operatorname{curl}\boldsymbol{E} + \mu_{\mathrm{o}} \dfrac{\partial \boldsymbol{H}}{\partial t} = 0$, we can write a differential equation

$$\mu_{\mathrm{o}} \frac{\mathrm{d}H_{\mathrm{i}}}{\mathrm{d}t} + k_{\mathrm{i}} E_{\mathrm{i}} = -\int_{S} (\boldsymbol{n} \times \boldsymbol{E}).\boldsymbol{h}_{\mathrm{i}} \mathrm{d}S, \tag{18.10}$$

and for $\operatorname{curl}\boldsymbol{H} - \varepsilon_{\mathrm{o}} \dfrac{\partial \boldsymbol{E}}{\partial t} = \boldsymbol{j}$ the solenoidal amplitude can be determined by the equation

$$-\varepsilon_{\mathrm{o}} \frac{\mathrm{d}E'_{\mathrm{i}}}{\mathrm{d}t} + k_{\mathrm{i}} H_{\mathrm{i}} = j_{\mathrm{i}} - \int_{S'} (\boldsymbol{n} \times \boldsymbol{H}).\boldsymbol{e}_{\mathrm{i}} \mathrm{d}S', \tag{18.11}$$

while for the irrotational amplitude we have

$$-\varepsilon_{\mathrm{o}} \frac{\mathrm{d}E'_{\mathrm{i}}}{\mathrm{d}t} = j'_{\mathrm{i}}. \tag{18.12}$$

Further, the equation div $\boldsymbol{E} = \dfrac{\rho}{\varepsilon_{\mathrm{o}}}$ can be converted to

$$-k_{\mathrm{i}} \varepsilon_{\mathrm{o}} E'_{\mathrm{i}} = \rho_{\mathrm{i}},$$

which is however identical to (18.12), if $k_{\mathrm{i}} j'_{\mathrm{i}} = \dfrac{\mathrm{d}\rho_{\mathrm{i}}}{\mathrm{d}t}$. We can show that this relation is none other than the charge-current continuity. By (18.5),

$$k_{\mathrm{i}} j'_{\mathrm{i}} = k_{\mathrm{i}} \int_{v} \boldsymbol{j}.\boldsymbol{e}'_{\mathrm{i}} \mathrm{d}v = \int_{v} \boldsymbol{j}.\operatorname{grad} \psi_{\mathrm{i}} \mathrm{d}v = \int_{v} \{\operatorname{div}(\psi_{\mathrm{i}} \boldsymbol{j}) - \psi_{\mathrm{i}} \operatorname{div} \boldsymbol{j}\} \mathrm{d}v,$$

where the first integral on the right is equal to $\int_{S.S'} \psi_i \frac{\partial \boldsymbol{j}}{\partial \boldsymbol{n}} \mathrm{d}\left(S, S'\right) = 0$, because $\psi_i = 0$ on S and S'. Hence,

$$k_i j_i' = -\int_v \psi_i \mathrm{div}\boldsymbol{j} \mathrm{d}v = \int_v \psi_i \frac{\partial \rho}{\partial t} \mathrm{d}v = \frac{\mathrm{d}\rho_i}{\mathrm{d}t},$$

which is consistent with the continuity equation.

Equations (18.10) and (18.11) are written for amplitudes of the solenoidal fields and distinct from the irrotational fields due to space charge-current. In the absence of such a source in an empty cavity, the inside fields should be entirely solenoidal, being described by (18.10) and (18.11). From these combined we can write the following equations for the amplitudes E_i and H_i:

$$\varepsilon_o \mu_o \frac{\mathrm{d}^2 E_i}{\mathrm{d}t^2} + k_i^2 E_i = -\mu_o \frac{\mathrm{d}j_i}{\mathrm{d}t} - \mu_o \frac{\mathrm{d}}{\mathrm{d}t} \int_{S'} (\boldsymbol{n} \times \boldsymbol{H}).\boldsymbol{e}_i \mathrm{d}S' - k_i \int_S (\boldsymbol{n} \times \boldsymbol{E}).\boldsymbol{h}_i \, \mathrm{d}S \tag{18.13}$$

and

$$\varepsilon_o \mu_o \frac{\mathrm{d}^2 H_i}{\mathrm{d}t^2} + k_i^2 H_i = k_i \left\{ j_i - \int_{S'} (\boldsymbol{n} \times \boldsymbol{H}).\boldsymbol{e}_i \mathrm{d}S' \right\} - \varepsilon_o \frac{\mathrm{d}}{\mathrm{d}t} \int_S (\boldsymbol{n} \times \boldsymbol{E}).\boldsymbol{h}_i \, \mathrm{d}S. \tag{18.14}$$

Noted that (18.13) (18.14) are equations of forced harmonic oscillators, indicating the amplitudes E_i and H_i are forced to increase by the integral terms on the right side. Due to the Ohm law, the terms of j_i in these equations represent damping of oscillations, whereas the integral term $\int_{S'} \ldots \mathrm{d}S'$ in (18.13) is essential for the coupling. Nevertheless, ignoring these terms from (18.13) and (18.14), we can obtain equations of free harmonic oscillation, in which the frequency is determined by the relation $\omega_i = \frac{k_i}{\sqrt{\varepsilon_o \mu_o}} = ck_i$.

18.3. Free and Damped Oscillations

Assuming that a cavity is made of a perfect conductor, completely enclosed with no opening, in (18.13) and (18.13), terms of $\int_{S'} \ldots \mathrm{d}S'$ can be omitted, and $\boldsymbol{n} \times \boldsymbol{E} = 0$, which is the boundary condition on S. Therefore, we obtain

$$\frac{\mathrm{d}^2}{\mathrm{d}t^2} (E_i, H_i) + \omega_i^2 (E_i, H_i) = 0.$$

indicates that E_i and H_i can be expressed as proportional to $\exp[i(\omega_i t + \varphi)]$. However, (18.10) and (18.11) indicate that these functions E_i and H_i are not in phase, and therefore in such an idealized cavity, we have

$$\mu_o \frac{\mathrm{d}H_i}{\mathrm{d}t} + k_i E_i = 0 \quad \text{and} \quad -\varepsilon_o \frac{\mathrm{d}E_i}{\mathrm{d}t} + k_i H_i = 0,$$

hence

$$\frac{E_{\rm i}}{H_{\rm i}} = -i\sqrt{\frac{\mu_{\rm o}}{\varepsilon_{\rm o}}} = -iZ_{\rm o}, \tag{18.15}$$

indicating that $E_{\rm i}$ and $H_{\rm i}$ have a phase difference $\frac{\pi}{2}$.

On the other hand, in practical cavities the damping is significant. On the conducting walls energy loss is primarily due to a large conductivity, so that the last term of (18.14) is not zero but $\boldsymbol{n} \times \boldsymbol{E} = \boldsymbol{H}\sqrt{\frac{\omega\mu_{\rm o}}{2\sigma}}(1+i)$ is related with surface currents on the conducting surface S, whereas the term of $\boldsymbol{n} \times \boldsymbol{E}$ in (18.13) acts as a driving force for $E_{\rm i}$. Interpreting similarly, the tangential component $\boldsymbol{n} \times \boldsymbol{H}$ on the opening area S' is responsible for leaking field-lines from the cavity, whereas it drives $H_{\rm i}$ in (18.14). Although these terms are difficult to estimate for a given case, we may simply consider them as damping parameters and driving terms.

In the absence of $j_{\rm i}$ in the cavity the damping term can be expressed by

$$-\varepsilon_{\rm o}\frac{\rm d}{{\rm d}t}\int_S (\boldsymbol{n} \times \boldsymbol{E}).\boldsymbol{h}_{\rm i}{\rm d}S = -\gamma\frac{{\rm d}H_{\rm i}}{{\rm d}t},$$

where

$$\gamma = \varepsilon_{\rm o}\sqrt{\frac{\omega\mu_{\rm o}}{2\sigma}}(1+i)\int_S \boldsymbol{h}_{\rm i}^2{\rm d}S$$

is a complex parameter. On the other hand, the $\boldsymbol{H}$ vector on S' is responsible for driving the oscillatory fields if it is the $\boldsymbol{H}$ vector in the input waveguide. In this case we can write that

$$-\int_{S'} (\boldsymbol{n} \times \boldsymbol{H}).\boldsymbol{e}_{\rm i}{\rm d}S' \propto F\exp(i\omega t),$$

where F represents the effective amplitude of a driving magnetic field at a frequency ω.

With these γ and F defined above, the equation of motion for $H_{\rm i}$ can be written as

$$\frac{{\rm d}^2 H_{\rm i}}{{\rm d}t^2} + \gamma\frac{{\rm d}H_{\rm i}}{{\rm d}t} + \omega_{\rm o}^2 H_{\rm i} = F\exp(i\omega t),$$

which represents a *forced* harmonic oscillator, showing a large amplitude of $H_{\rm i}$ in a cavity at resonance, i.e., $\omega = \omega_{\rm i}$. Using a response function defined as $\chi_{\rm i}(\omega) = H_{\rm i}/F$, the resonant property can be expressed by the steady-state solution. That is,

$$\chi_{\rm i}(\omega) = \frac{1}{-\omega^2 + i\omega\gamma + \omega_{\rm i}^2}, \tag{18.16}$$

which may be called the *susceptibility* of the cavity. For a complex damping parameter written as $\gamma(1+i)$, (18.16) can be approximated as

$$\chi_i(\omega) = \frac{1}{-(\omega - \frac{1}{2}\gamma)^2 + i(\omega - \frac{1}{2}\gamma)\gamma + \omega_i^2}$$

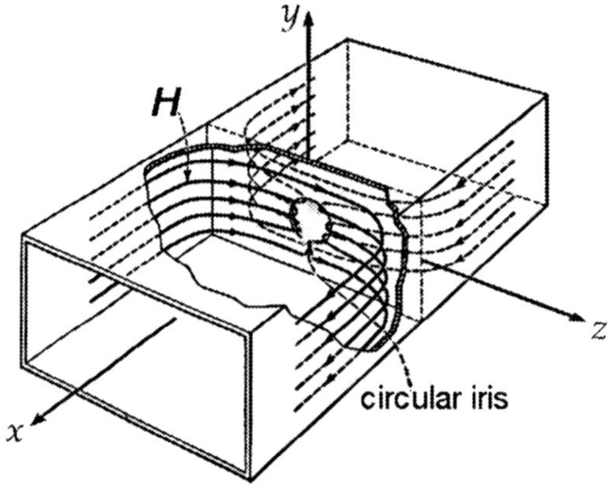

FIGURE 18.1. A view of H-lines near an iris coupling between two waveguide sections.

for a small value of γ, showing a maximum peak at $\omega = \omega_i + \frac{1}{2}\gamma$. The damping constant γ can, in fact, be evaluated from the susceptibility at resonance, i.e., $\chi_i(\omega_i) = \dfrac{1}{i\left(\omega_i + \frac{1}{2}\gamma\right)\gamma}$.

18.4. Input Impedance of a Cavity

A cavity coupled with a waveguide can be expressed by a load impedance Z_L when electromagnetic waves are transmitted from an oscillator. To maintain a steady amplitude of the field in a cavity, the *coupling* plays an important role in supplying energy continuously from the oscillator to offset the loss of energy in the system.

Figure 18.1 shows a commonly used coupling of a circular hole at the center of conducting wall of connection, which is determined by the integral $\int_{S'} (\boldsymbol{n} \times \boldsymbol{H}).\boldsymbol{e}_i dS'$ in (18.14), whereas the energy leak rate is determined by its time derivative. Here, as determined by the field $\boldsymbol{H} = \boldsymbol{H}_t$ in the waveguide coupled with the cavity mode $\boldsymbol{e}_i$ in the open area S′, the figure illustrates a *magnetic coupling*. Needless to say, such a description is useful but only approximate. Such a coupling can also be expressed in terms of a *magnetic moment* induced on the iris by $\boldsymbol{H}_t$.

We consider fields of the principal mode TE_{01} in a rectangular waveguide. However, as perturbed by the iris in the vicinity of a coupling, the mode is modified as given by a linear combination with other waveguide modes. We can therefore write the general relation between the cavity and waveguide modes as

$$\boldsymbol{e}_i = \sum_j v_{ij} \boldsymbol{E}_{tj}, \tag{18.17}$$

expressing that many transversal waveguide modes j may contribute to the effective coupling with the cavity mode i. Nevertheless, these constants v_{ij} can be considered as very small for the magnetic coupling that is activated by the $\boldsymbol{H}_t$ of the waveguide.

In this case, we also write

$$\boldsymbol{H} = \sum_{\mathrm{j}} I_{\mathrm{j}} Z_{\mathrm{o}} \boldsymbol{H}_{\mathrm{tj}}, \tag{18.18}$$

where I_{j} represents the amplitude of a TE$_{\mathrm{j}}$ mode, for which the wave impedance should be written as $Z_{\mathrm{j}} = Z_{\mathrm{o}}(k/k_x)$ if I_{j} implies a current, but in this case the factor k/k_x is included in the definition of I_{j} . For the coupling, $\boldsymbol{n} \times \boldsymbol{H}$ can be expressed as

$$\boldsymbol{n} \times \boldsymbol{H} = \sum_{\mathrm{j}} I_{\mathrm{j}} Z_{\mathrm{o}} \left(\boldsymbol{n} \times \boldsymbol{H}_{\mathrm{tj}}\right) = \sum_{\mathrm{j}} I_{\mathrm{j}} \boldsymbol{E}_{\mathrm{tj}},$$

so that

$$\int_{S'} (\boldsymbol{n} \times \boldsymbol{H}).\boldsymbol{e}_{\mathrm{i}} \mathrm{d}S' = \sum_{\mathrm{j}} I_{\mathrm{j}} \upsilon_{\mathrm{ij}}.$$

Then, the steady solution for E_{i} can be obtained from the oscillator equation (18.13), by considering $I_{\mathrm{j}} \propto \exp(i\omega t)$, that is,

$$E_{\mathrm{i}} = \sum_{\mathrm{j}} \frac{I_{\mathrm{j}} \upsilon_{\mathrm{ij}}}{i\varepsilon_{\mathrm{o}}\omega_{\mathrm{i}}} \left(\frac{\omega}{\omega_{\mathrm{i}}} - \frac{\omega_{\mathrm{i}}}{\omega}\right)^{-1}$$

if damping on S is neglected.

Corresponding to the $\boldsymbol{H}$ vector in (18.18), the $\boldsymbol{E}$ vector in the waveguide is written as

$$\boldsymbol{E} = \sum_{\mathrm{j}} V_{\mathrm{j}} \boldsymbol{E}_{\mathrm{tj}},$$

where V_{j} is the amplitude of the j mode.

On the other hand, owing to the continuity of $\boldsymbol{n} \times \boldsymbol{E}$ on S', this waveguide field $\boldsymbol{e}_{\mathrm{tj}}$ should be equal to $\boldsymbol{E} = \sum_{\mathrm{i}} E_{\mathrm{i}} \boldsymbol{e}_{\mathrm{i}}$ inside the cavity. Hence, using (18.17),

$$\boldsymbol{E} = \sum_{\mathrm{i}} \left(\sum_{\mathrm{j}} \upsilon_{\mathrm{ij}} \boldsymbol{E}_{\mathrm{tj}}\right) E_{\mathrm{i}}.$$

Accordingly, we can write the relation

$$V_{\mathrm{j}} = \sum_{\mathrm{i}} I_{\mathrm{i}} Z_{\mathrm{ij}}, \quad \text{where} \quad Z_{\mathrm{ij}} = \frac{\upsilon_{\mathrm{ij}}^2}{i\varepsilon_{\mathrm{o}}\omega_{\mathrm{i}}} \left(\frac{\omega}{\omega_{\mathrm{i}}} - \frac{\omega_{\mathrm{i}}}{\omega}\right)^{-1}. \tag{18.19a}$$

The power flow to the cavity can then be expressed as

$$\frac{1}{2} \int_{S'} \mathbf{i}.\left(\boldsymbol{E} \times \boldsymbol{H}^*\right) \mathrm{d}S' = \frac{1}{2} \sum_{\mathrm{j}} V_{\mathrm{j}} I_{\mathrm{j}}^* Z_{\mathrm{o}} \int_{S'} \mathbf{i}.\left(\boldsymbol{E}_{\mathrm{tj}} \times \boldsymbol{H}_{\mathrm{tj}}^*\right) \mathrm{d}S' = \frac{1}{2} \sum_{\mathrm{j}} V_{\mathrm{j}} I_{\mathrm{j}}^*.$$

By analogy, the transversal $\boldsymbol{E}_{\mathrm{t}}$ and $\boldsymbol{H}_{\mathrm{t}}$ can be regarded as voltage and current, and Z_{ij} represents impedances between waveguide and cavity modes.

If the cavity mode i $= 1$ is well isolated from the other, (18.19a) can be simplified for the waveguide at j $= 1$, e.g., TE_{01} as

$$V_1 = I_1 Z_{11}, \quad \text{where} \quad Z_{11} = \frac{v_{11}^2}{i\varepsilon_o\omega_1}\left(\frac{\omega}{\omega_1} - \frac{\omega_1}{\omega}\right)^{-1} + Z'_{11}, \tag{18.19b}$$

where Z'_{11} represents small contributions from other terms than the principal mode i $= 1$. Although small in magnitude, such a Z'_{11} is significant for the cavity coupling, and we can rewrite the above as

$$i\left(\frac{\omega}{\omega_1} - \frac{\omega_1}{\omega}\right) = \frac{v_{11}^2}{\varepsilon_o\omega_1}\frac{1}{Z_{11} - Z'_{11}}.$$

Further, denoting the characteristic wave impedance of TE_{01} mode by $Z_1(= kZ_o/k_x)$, we define the coupling of the cavity with TE_{01} mode as

$$\frac{1}{Q_{\text{ext},1}} = \frac{v_{11}^2}{\varepsilon_o\omega_1 Z_1}, \tag{18.20}$$

and write

$$\frac{Z_1}{Z_{11} - Z'_{11}} = g + ib. \tag{18.21}$$

With these notations, we can express

$$i\left(\frac{\omega}{\omega_1} - \frac{\omega_1}{\omega} + \frac{b}{Q_{\text{ext},1}}\right) + \frac{g}{Q_{\text{ext},1}} = 0,$$

from which we obtain the relations

$$\frac{Z_{11}}{Z_1} = \frac{\dfrac{1}{Q_{\text{ext},1}}}{i\left(\dfrac{\omega}{\omega'_1} - \dfrac{\omega'_1}{\omega}\right) + \dfrac{g}{Q_{\text{ext},1}}}, \quad \text{where} \quad \omega'_1 = \omega_1\left(1 - \frac{b}{2Q_{\text{ext},1}}\right). \tag{18.22}$$

Here, the factor $Q_{\text{ext},1}$ defined by (18.20) signifies the ratio between energy stored in the cavity vs. energy loss due to the opening area in this case per cycle of oscillation. It is realized that the stored energy in a cavity is limited not only by a finite conductivity on the wall, but also by leaking through the opening. Hence, these mechanisms should be combined for the description of resonance performance.

Now that the oscillatory cavity fields can be described by equations of forced oscillation (18.13) and (18.14), the cavity resonance can be discussed on the analogy of *LCR* circuits. We can convert the damping constant γ to a loss factor denoted as Q_{wall}. We define

$$Q_{\text{wall}} = \omega'_i\frac{\text{stored energy}}{\text{conductung loss}} = \frac{2\omega'_i}{\gamma} \quad \text{or} \quad \frac{1}{Q_{\text{wall}}} = \frac{\delta}{2}\int_S h_i^2 \mathrm{d}S,$$

where $\delta g = \sqrt{2/\omega_i \sigma \mu_o}$ is the skin depth. Combining the damping effect with the loss factor due to an iris (18.20), we can express the total Q-factor as

$$\frac{1}{Q} = \frac{1}{Q_{\text{wall}}} + \frac{g}{Q_{\text{ext},1}}, \tag{18.23a}$$

and (18.21) can be revised as

$$\frac{Z_{11}}{Z_1} = \frac{\dfrac{1}{Q_{\text{ext},1}}}{i\left(\dfrac{\omega}{\omega_1'} - \dfrac{\omega_1'}{\omega}\right) + \dfrac{1}{Q}}, \quad \text{where} \quad \omega_1' = \omega_1\left(1 - \frac{1}{Q_{\text{wall}}} - \frac{b}{2Q_{\text{ext},1}}\right). \tag{18.23b}$$

In the above the coupling is represented by the *reduced* admittance $g + ib$, while effects of non-resonant modes are packaged into the parameter Z_{11}'. Calculations of these parameters are complicated and therefore left for empirical evaluation. We consider that the equations (18.22a) and (18.22b) are sufficient for analyzing a cavity as a circuit element. As indicated by (18.22b), the characteristic frequency ω_1' shifts from the idealized case, although close to ω_1, and the input admittance $g + ib$ is adjustable in a practical circuit. The impedance Z_{11} is called the *input impedance of a cavity* when the transmission line is loaded in TE_{01} mode.

18.5. Example of a Resonant Cavity

By analogy with an acoustic resonator, we say that a cavity can be electromagnetically *tuned* when the applied frequency ω is adjusted to be equal to the characteristic frequency ω_1 of a cavity. Such electromagnetic resonators are not only significant in high frequency circuits, but also inside fields are large in amplitude and hence a useful device for applications where the intensity matters.

In practice the resonant condition $\omega = \omega_i$ can be achieved either by a mechanical tuning of a cavity or by a scanning frequency of the oscillator. Here, we discuss a cavity of rectangular TE_{01} mode again as an example of the impedance formula (18.22). Figure 18.2 shows such a cavity that can be tuned by adjusting the distance L. In the previous discussion, we showed that if there is no reflection, the impedance of the cavity is given by

$$Z_{11} = -iZ_1 \tan\frac{2\pi L}{\lambda_g}, \tag{18.24}$$

where $\lambda_g = \dfrac{\omega_1}{2\pi c}$ is related with $\lambda_g = \dfrac{\omega}{2\pi c}$ as determined by

$$\frac{1}{\lambda_g^2} = \frac{1}{\lambda^2} + \frac{1}{\lambda_c^2},$$

and $Z_1 = \dfrac{\lambda_g Z_o}{\lambda}$ is the constant modified for TE_{01} mode.

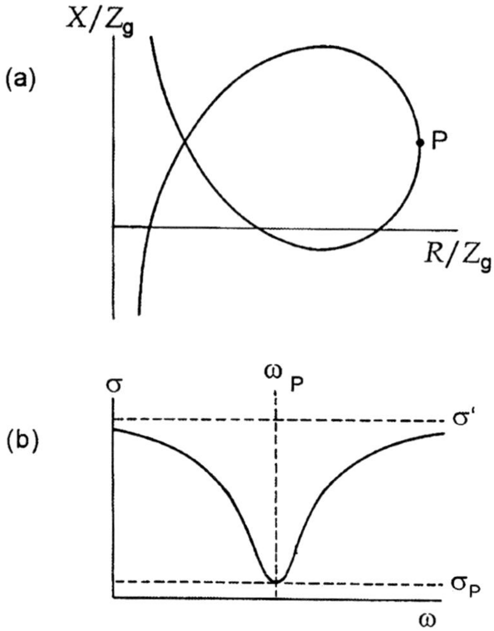

FIGURE 18.2. (a) The tuning character in terms of the cavity impedance $R + iX$ in the vicinity of a resonance ω_P. (b) The resonance monitored by SWR σ. $\sigma = \sigma_P$ at resonance, and $\sigma = \sigma'$ off resonance.

If the energy loss can be totally ignored the impedance Z_{11} is equal to that described by (18.22), and hence we write

$$\tan\left(\frac{2\pi L}{\lambda_g}\right) = \frac{\dfrac{1}{Q_{\text{ext},1}}}{\dfrac{\omega}{\omega_1'} - \dfrac{\omega_1'}{\omega}}$$

or

$$L = \frac{\lambda_g}{2\pi}\tan^{-1}\frac{\dfrac{1}{Q_{\text{ext},1}}}{\dfrac{\omega}{\omega_1'} - \dfrac{\omega_1'}{\omega}} + \frac{\lambda_g}{2}\mathrm{n},$$

where n is an integer. The frequency shift occurs as $\omega_1' = \omega_1 \pm \Delta\omega$ by a small $\Delta\omega$, and hence

$$\frac{\omega}{\omega'} - \frac{\omega_1'}{\omega} \approx \mp\frac{2\Delta\omega}{\omega_1},$$

approximately. Corresponding to such small $\Delta\omega$, an angle θ can be defined by the relation

$$\tan\left(\tfrac{1}{2}\pi \pm \theta\right) = \pm\frac{\omega_1}{2Q_{\text{ext},1}\Delta\omega},$$

thereby writing

$$L = \frac{\lambda_g}{2}\left(n \pm \frac{\theta}{\pi}\right), \quad \text{where} \quad \theta = 2Q_{ext,1}\Delta(\ln\omega), \tag{18.25}$$

and

$$\frac{dL}{d\lambda_g} = \frac{n}{2} + \frac{Q_{ext,1}}{\pi}\frac{d(\ln\omega)}{d(\ln\lambda_g)} = \frac{n}{2} + \frac{Q_{ext,1}}{\pi}\frac{v_g}{c}.$$

In Figure 17.4, shown are curves for L vs. λ_g sketched primarily as straight lines $L = \frac{1}{2}\lambda_g \times n$, which, however, shift at resonances by increasing n by 1. Such transitions are shown in the figure by smooth deviations by $\pm\theta/\pi$, which are also contributed by Q_{wall} in practical cases. For illustration, only two resonances are marked, ω_1 and ω_2 in the figure.

18.6. Measurements of a Cavity Resonance

In usual practice an $\boldsymbol{E}_t$ vector can be monitored by means of a small antenna attached to a semi-conductor diode, which acts as a *detector*. A cavity resonance can be detected by measuring the reflection, which is signified by the impedance.

The input to a cavity can generally be expressed by the impedance formula:

$$\frac{Z(\omega)}{Z_g} = \frac{\dfrac{1}{Q_{ext,1}}}{i\left(\dfrac{\omega}{\omega_1} - \dfrac{\omega_1}{\omega}\right) + \dfrac{1}{Q}} + \frac{Z'}{Z_g}, \tag{18.26}$$

where ω_1, Q, and $Q_{ext,1}$ are the resonant frequency, the overall quality factor, and the coupling of the cavity, respectively; Z_g is the effective wave impedance of the connecting waveguide. Here, the additional term of Z' represents contribution from all other modes other than the principal mode 1. At resonance $\omega = \omega_1$, we obtain from (18.26)

$$\frac{Z(\omega_1)}{Z_g} = \frac{Q}{Q_{ext,1}} + \frac{Z'}{Z_g}. \tag{18.27}$$

Ignoring Z', $Z(\omega_1)/Z_g$ can be larger or smaller than 1, depending on $Q > Q_{ext,1}$ or $Q < Q_{ext,1}$. These cases are called *overcoupled* or *undercoupled*, respectively, referring to the standing wave measurements, where VSWR $= V_{max}/V_{min} = |Z(\omega_1)|/Z_g$ at resonance. Denoting VSWR as σ, (18.27) can be expressed as

$$\sigma = \frac{Q}{Q_{ext,1}} + \sigma',$$

where $\sigma = Z(\omega_1)/Z_g$ and $\sigma' = Z'/Z_g$ are VSWR *on*- and *off* resonance, respectively. Writing further that

$$\delta = Q_{ext,1}\left(\frac{\omega}{\omega_1} - \frac{\omega_1}{\omega}\right) \sim 2Q_{ext,1}\frac{\omega - \omega_1}{\omega_1} \tag{18.28a}$$

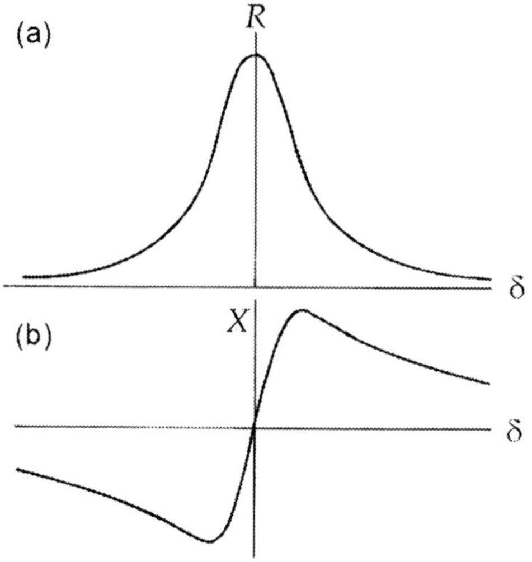

FIGURE 18.3. Resonant behaviors in the vicinity of $\omega = \omega_P$. (a) For $R(\delta)$ and (b) for $X(\delta)$ where $\delta = (\omega - \omega_P)/\omega_P$.

for a small deviation from ω_1, (18.26) can be expressed as

$$\frac{Z(\omega)}{Z_g} = \frac{1}{i\delta + \dfrac{1}{\sigma - \sigma'}} + \sigma'. \tag{18.28b}$$

Using the complex expression $Z(\omega) = R(\omega) + iX(\omega)$, the behavior of $Z(\omega)$ can be studied in the complex plane, as shown in Figure 18.2(a), where the reduced resistance and reactance, $R(\omega)/Z_g$ and $X(\omega)/Z_g$, are considered as the real and imaginary axes, respectively. The point P for $\omega = \omega_P$ corresponds to maximum σ_P, as determined by $R(\omega_P)$ and $X(\omega_P) = 0$, whereas the minimum is σ' at $\omega = \pm\infty$. Figure 18.2(b) shows that the VSWR varies as a function of ω between σ_P and σ'.

Ignoring Z', from (18.26) $R(\omega)$ and $X(\omega)$ can be expressed as a function of δ that is defined by (18.28a). Namely,

$$\frac{R(\delta)}{Z_g} = \frac{\dfrac{Q_{ext,1}}{Q}}{\left(1 + \dfrac{Q_{ext,1}}{Q}\right)^2 + \delta^2} \quad \text{and} \quad \frac{X(\delta)}{Z_g} = \frac{\delta}{\left(1 + \dfrac{Q_{ext,1}}{Q}\right)^2 + \delta^2}. \tag{18.29}$$

The real part $R(\delta)$ exhibits a curve in dumbbell shape in the vicinity of $\delta = 0$, as shown in Figure 18.3(a). Although approximate, such a curve is typical for power *absorption* by a cavity in this case, where the width between *half peak points*, i.e.,

$$\delta_+ - \delta_- = 1 + \frac{Q_{ext,1}}{Q}$$

or the equivalent difference in frequencies,

$$\frac{2(\omega - \omega_1)}{\omega_1} = \frac{1}{Q_{ext,1}} + \frac{1}{Q_{wall}} = \frac{1}{Q}, \tag{18.30}$$

can specify the sharpness of resonance.

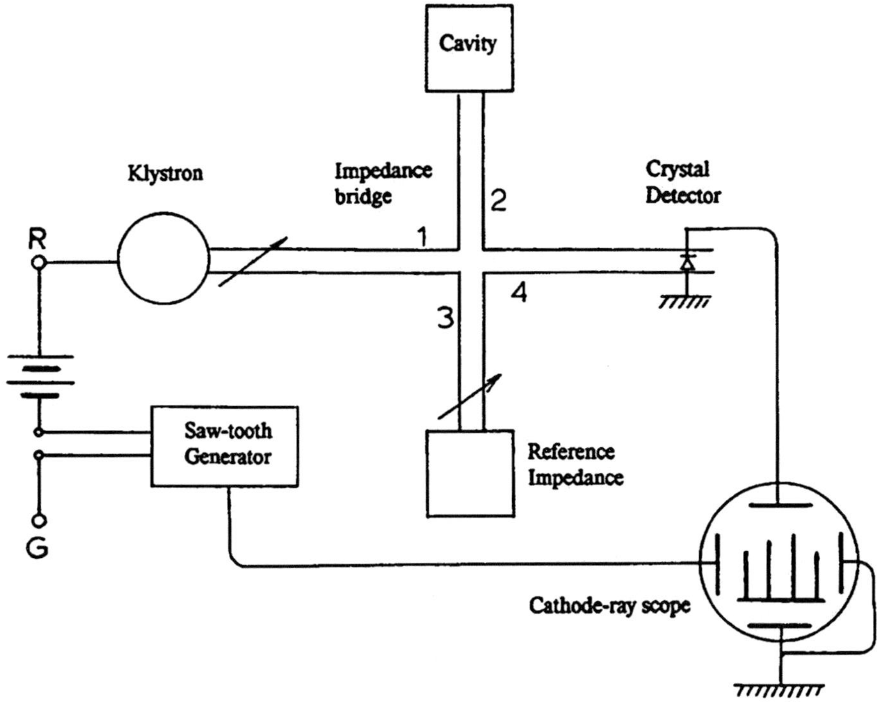

FIGURE 18.4. Waveguide circuit for impedance measurements.

In contrast, the imaginary part $X(\delta)$ exhibits a curve as shown in Figure 18.3(b), indicating a steep slope symmetrically on both sides of $\delta = 0$, which is referred to as a *dispersion* curve of resonance. It is sufficient to measure either $R(\omega)$ or $X(\omega)$ for detecting the cavity resonance, as these are related, nevertheless providing a typical absorption or dispersion curve if observed with *modulated* frequencies.

The technical method for impedance measurements is beyond the scope of this book, however it is worth looking at some details. Figure 18.4 shows a block diagram for an impedance *bridge*, which can be adjusted for measuring $R(\omega)$ or $X(\omega)$. The bridge consists of four waveguide arms, 1-2-3-4, where the oscillator is connected to arm 1, and the detector is in arm 4. A cavity and a reference impedance $Z(\omega)$ and Z_0 are respectively on arms 2 and 3 for comparison, and the reflected waves are mixed by adjusting phases so that either R or X can dominate at the detector in arm 4. Here, the oscillator is frequency-modulated by a *sawtooth* voltage so that the detected signal can be displayed on the CRO scope.

19
Electronic Excitation of Cavity Oscillations

19.1. Electronic Admittance

A resonator is an essential device for sustained oscillations, such as a triode oscillator, in which the oscillating current is controlled by a resonant LC load. In microwave oscillators a cavity resonator controls the electronic current to sustain oscillation in the feedback mechanism. The electronic current density $\boldsymbol{j}$ can interact with a cavity when passing through an area dominated by the $\boldsymbol{E}$ vector. Either from electrons traveling to the cavity or in the opposite direction, the transferred energy operates the device like an oscillator or an electron accelerator. Sharing a similar principle, a *betatron* accelerator is operated at low frequencies for accelerating electrons by an alternating dynamic $\boldsymbol{E}$ field that arises from a sinusoidal magnetic flux.

In the equations (18.13) and (18.14) for amplitudes of the normal mode i, the terms of $\mathrm{d}j_\mathrm{i}/\mathrm{d}t$ and j_i can act as the driving force for forced oscillation if $\boldsymbol{j}_\mathrm{i}$ at the opening area S' interacts with the fields of the cavity mode. Since electronic currents are technically feasible at high intensity, such interactions can be utilized for a variety of applications in modern technologies, such as high-power oscillators and particle accelerators. We define the *electronic admittance* for such interactions, as discussed in the following.

A coupling between the waveguide and cavity mode was previously expressed by the electric field

$$\boldsymbol{E} = \sum_\mathrm{i} E_\mathrm{i}\boldsymbol{e}_\mathrm{i} = \sum_\mathrm{j} V_\mathrm{j}\boldsymbol{E}_\mathrm{tj} \quad \text{and} \quad V_\mathrm{j} = \sum_\mathrm{j} v_\mathrm{ij}E_\mathrm{j},$$

together with the corresponding $\boldsymbol{H}$ given by

$$\boldsymbol{H} = \sum_\mathrm{j} I_\mathrm{j} Z_\mathrm{o} \boldsymbol{H}_\mathrm{tj}$$

at the iris. The input impedance can be defined with coefficients V_j and I_j as

$$Z_\mathrm{input} = \sum_\mathrm{j} \frac{V_\mathrm{j}}{I_\mathrm{j}} = \sum_\mathrm{j} \frac{v_\mathrm{ij}}{I_\mathrm{j}} E_\mathrm{j} = \frac{v_{11}}{I_1} E_1 + Z',$$

where

$$Z' = \sum_{j\neq 1} \frac{v_{1j}}{I_j} E'_j,$$

and hence

$$I_1 = \frac{v_{11}E_1}{Z_{\text{input}} - Z'}.$$

The integral for coupling on the area S' in (18.13) and (18.14) can be written as

$$\int_{S'} (\boldsymbol{n} \times \boldsymbol{H}).\boldsymbol{e}_1 \mathrm{d}S' = I_1 v_{11} + \sum_{j\neq 1} I_j v_{1j} = I_1 v_{11} + \sum_{j\neq 1} \frac{v_{1j}^2 E_1}{Z_{\text{input}} - Z'},$$

whereas the integral for damping on S can be expressed as

$$\int_S (\boldsymbol{n} \times \boldsymbol{E}).\boldsymbol{h}_1 \mathrm{d}S = E_1 \sqrt{\frac{\omega\mu_o}{2\sigma}}(1+j) \int_S \boldsymbol{h}_1^2 \mathrm{d}S.$$

By an oscillating $I_1 \propto \exp(i\omega t)$, the cavity mode E_1 is forced to oscillate, for which equation (18.13) has a steady solution

$$\begin{aligned}&(-\varepsilon_o\mu_o\omega^2 + k_1^2)E_1\\ &\quad= -i\omega\mu_o j_1 + \varepsilon_o I_1 v_{11} + \varepsilon_o E_1 \sum_j \frac{v_{1j}^2}{Z_{\text{input}} - Z'} - k_1 E_1 \sqrt{\frac{\omega\mu_o}{2\sigma}}(1+i)\int_{S'} \boldsymbol{h}_1^2 \mathrm{d}S'\end{aligned}$$

or

$$\begin{aligned}&\left[i\left(\frac{\omega}{\omega_1} - \frac{\omega_1}{\omega}\right) + \frac{(1+i)\delta}{2}\int_S \boldsymbol{h}_1^2 \mathrm{d}S + \frac{1}{\varepsilon_o\omega_1}\sum_{j\neq 1} \frac{v_{j1}^2}{Z_{\text{input}} - Z'}\right]\\ &\quad\times E_1 + \frac{1}{\varepsilon_o\omega_1} j_1 = \frac{I_1 v_{11}}{\varepsilon_o\omega_1},\end{aligned}$$

where δ is the skin depth. As defined in Section 18.4, we write

$$\frac{\delta}{2}\int_S \boldsymbol{h}_1^2 \mathrm{d}S = \frac{1}{Q_{\text{wall}}}, \quad \frac{1}{Z_1}\sum_{j\neq 1} \frac{v_{1j}^2}{\varepsilon_o\omega_1} \frac{Z_1}{Z_{\text{input}} - Z'} = \frac{1}{Q_{\text{ext}}}$$

$$\text{and} \quad \frac{Z_1}{Z_{\text{input}} - Z'} = g_1 + ib_1,$$

and obtain the expression

$$E_1 = \frac{\dfrac{I_1 v_{11}}{\varepsilon_o\omega_1}}{i\left(\dfrac{\omega}{\omega'_1} - \dfrac{\omega'_1}{\omega}\right) + \dfrac{1}{Q_{\text{wall}}} + \dfrac{g_1}{Q_{\text{ext}}} + \dfrac{j_1}{\varepsilon_o\omega_1 E_1}},$$

where

$$\omega_1' = \omega_1\left(1 - \frac{1}{Q_{\text{wall}}} - \frac{b_1}{Q_{\text{ext}}}\right).$$

Here, in the expression of E_1, the last term in the denominator can be interpreted as an energy loss due to the current j_1 in the electric field E_1, which is therefore expressed as

$$\frac{1}{Q_{\text{el}}} = \frac{\frac{1}{2}j_1 E_1^*}{\left(\frac{1}{2}\varepsilon_o E_1 E_1^*\right)\omega_1}. \tag{19.1}$$

It is noted that this electronic Q_{el} can be *negative* if the current j_1 and the field E_1 are *out of phase*; but Q_{el} is *positive* if they are *in phase*. Such a negative Q_{el} corresponds to negative feedback in a triode oscillator at low frequencies, where the grid current and anode voltage are out of phase in oscillating condition. On the other hand, a positive Q_{el} gives a condition for acceleration if the resonator is used as an accelerator.

Although unspecified in the above, for efficient interactions electronic currents should pass through a part of the cavity field where E_1 is predominant with a high density of field-lines. Considering that such a part of E_1 is signified by a high voltage difference ΔV across a narrow gap d, analogous to a static case, we can write $E_1 = \Delta V/d$, and

$$\frac{1}{Q_{\text{el}}} \sim \frac{I_1}{A}\frac{d}{\varepsilon_o\omega_1\Delta V} = \frac{I_1}{C\omega_1\Delta V} = \frac{g_e + ib_e}{C\omega_1}, \tag{19.2a}$$

where A is the effective area for the current I_1, $C = \dfrac{\varepsilon_o A}{d}$ is a capacity by static analogy, and $\dfrac{I_1}{\Delta V} = g_e + ib_e$ may be called the *electronic admittance*. Thus, for a cavity interacted with a current, we have the formulae

$$\frac{1}{Q} = \frac{1}{Q_{\text{wall}}} + \frac{g_1}{Q_{\text{ext}}} + \frac{g_e}{Q_{\text{el}}} \tag{19.2b}$$

and

$$\omega_1' = \omega_1\left(1 - \frac{1}{Q_{\text{wall}}} - \frac{b_1}{2Q_{\text{ext}}} - \frac{b_e}{2Q_{\text{el}}}\right). \tag{19.2c}$$

19.2. A Klystron Cavity

In a cavity of a *klystron* oscillator, $\boldsymbol{E}$-lines are distributed in high density in a particular region, shown in Figure 19.1. The cavity is basically a short cylindrical coaxial cavity with both ends closed, and the central conductor is a hollow cylinder with a narrow gap of parallel grids, as illustrated in Figure 19.1(a). The field-lines of $\boldsymbol{E}$ are radial with regard to the x-axis in most parts of the coaxial space, except for in the gap area, where they are almost parallel to the axis. The hollow central cylinder is designed for electrons to pass inside through the grids, where they

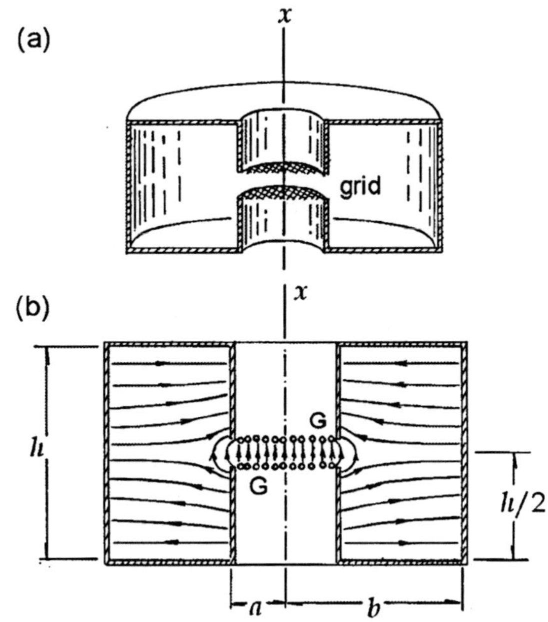

FIGURE 19.1. A klystron cavity. (a) A sectional view. (b) E-lines near the grids.

interact with the E-field. In fact, it is not so easy to find the mathematical solution to the Maxwell equations in such a cavity; however, the important feature of the gap is intuitively clear.

We can consider that the cavity is in a continuous coaxial structure, in which a gap in the center conductor can be considered as a perturbation. Slater discussed such a perturbation, one that modifies the conducting boundary of the cavity, resulting in a shift of the resonance frequency. Consider that the conducting boundary S is displaced inward to S^* by making a small dent on S, as illustrated in Figure 19.2. In this case, in a small volume v^* between S and S^*, the field vectors $\boldsymbol{E}$ and $\boldsymbol{H}$ should become zero by displacing S to S^*, hence creating a discontinuity of the tangential component of $\boldsymbol{H}$ on S^*. Consequently, the equations of the normal oscillation are forced by an additional perturbation $-\int_{S^*}(\boldsymbol{n}\times\boldsymbol{H}).\boldsymbol{e}_1\mathrm{d}S^*$. Assuming $\boldsymbol{H}\approx H_1\boldsymbol{h}_1$, this perturbation can be written as

$$-H_1\int_{S^*}(\boldsymbol{n}\times\boldsymbol{h}_1).\boldsymbol{e}_1\mathrm{d}S^* = -H_1\int_{S^*}(\boldsymbol{e}_1\times\boldsymbol{h}_1).\mathrm{d}\boldsymbol{S}^*,$$

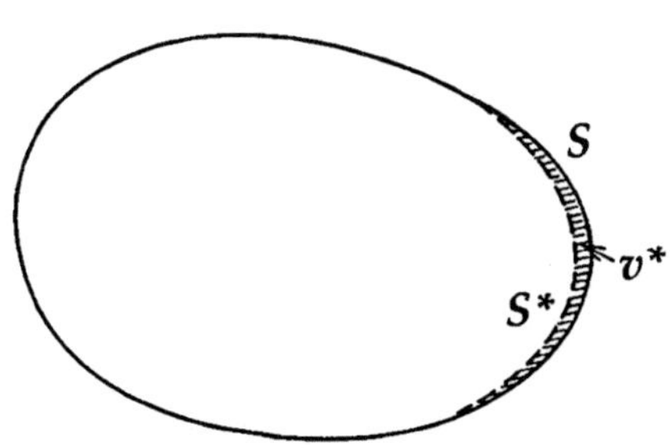

FIGURE 19.2. A boundary perturbation for Slater's theorem.

whereas

$$\int_S (\boldsymbol{e}_1 \times \boldsymbol{h}_1).\mathrm{d}\boldsymbol{S} = 0.$$

Applying Gauss's theorem to the volume v^*, we can write

$$H_1 \left\{ \int_{S+S'} (\boldsymbol{e}_1 \times \boldsymbol{h}_1).\mathrm{d}(S + S') \right\} = H_1 \int_{v^*} \mathrm{div}(\boldsymbol{e}_1 \times \boldsymbol{h}_1)\,\mathrm{d}\boldsymbol{v}^*,$$

where

$$\mathrm{div}\,(\boldsymbol{e}_1 \times \boldsymbol{h}_1) = \boldsymbol{h}_1.\mathrm{curl}\,\boldsymbol{e}_1 - \boldsymbol{e}_1.\mathrm{curl}\,\boldsymbol{h}_1 = k_1 \left(\boldsymbol{h}_1^2 - \boldsymbol{e}_1^2\right).$$

Therefore, resulting from such a perturbation the steady solution of (18.14) can be shifted as given by

$$-\varepsilon_o \mu_o \omega^2 + k_1^2 = -k_1^2 \int_{v^*} \left(\boldsymbol{h}_1^2 - \boldsymbol{e}_1^2\right) \mathrm{d}v^*,$$

or

$$\omega^2 = \omega_1^2 \left\{ 1 + \int_{v^*} \left(\boldsymbol{h}_1^2 - \boldsymbol{e}_1^2\right) \mathrm{d}v^* \right\}, \quad \text{where} \quad \omega_1 = \frac{k_1}{\sqrt{\varepsilon_o \mu_o}}. \tag{19.3}$$

As shown in Figure 19.1, between the two grids, the field is predominantly electric, so that at least qualitatively, $\omega < \omega_1$, giving a resonant frequency lower than the normal mode in the unperturbed cavity. In contrast, we can obtain $\omega > \omega_1$ if we are considering a similar magnetic perturbation.

Such a frequency shift due to a capacitive perturbation can also be estimated by the following calculation. Using cylindrical coordinates r, θ, x, the unperturbed coaxial mode at the lowest normal mode can be expressed by

$$E_r(r, x, t) = \frac{E_o}{r} \sin\left(\frac{2\pi x}{\lambda} - \omega t\right)$$

and

$$H_\theta(r, x, t) = -i\sqrt{\frac{\varepsilon_o}{\mu_o}} \frac{E_o}{r} \cos\left(\frac{2\pi x}{\lambda} - \omega t\right),$$

representing a standing wave along the x direction. This is in fact the TEM mode of propagation, and $k = 2\pi/\lambda$, for which we can define the voltage and current functions as

$$V(x, t) = -\int_a^b E_r \mathrm{d}r = -E_o \ln\frac{b}{a} \sin\left(\frac{2\pi x}{\lambda} - \omega t\right)$$

and

$$\begin{aligned} I(x, t) &= \oint H_\theta r \mathrm{d}\theta = -i\,(2\pi r)\sqrt{\frac{\varepsilon_o}{\mu_o}} \frac{E_o}{r} \cos\left(\frac{2\pi x}{\lambda} - \omega t\right) \\ &= -2\pi i \sqrt{\frac{\varepsilon_o}{\mu_o}} E_o \cos\left(\frac{2\pi x}{\lambda} - \omega t\right). \end{aligned}$$

Assuming that the grids of gap d are located exactly at the center as in the figure, we can calculate the potential difference as

$$\Delta V(t) = V\left(\frac{h+d}{2}, t\right) - V\left(\frac{h-d}{2}, t\right)$$
$$\approx -2E_o \ln\frac{b}{a}\sin\left(\frac{\pi h}{\lambda} - \omega t\right), \quad \text{if } d \ll h.$$

On the other hand, the current is

$$I(1h, t) = -2\pi i\sqrt{\frac{\varepsilon_o}{\mu_o}}E_o\cos\left(\frac{\pi h}{\lambda} - \omega t\right)$$

under the same condition. Therefore

$$\frac{1}{i\omega C} = \frac{V\left(\frac{1}{2}h, t\right)}{I\left(\frac{1}{2}h, t\right)} = \frac{1}{\pi i}\sqrt{\frac{\mu_o}{\varepsilon_o}}\ln\frac{b}{a}\tan\left(\frac{\pi h}{\lambda} - \omega t\right),$$

from which we can define a capacity $C = \dfrac{\varepsilon_o \pi a^2}{d}$, and write the relation

$$\frac{\lambda}{a} = \sqrt{\frac{2\pi^2 h}{d}\ln\frac{b}{a}},$$

justifying that $\lambda > a$, if $d < h$. Therefore, the wavelength λ for the resonance can be increased significantly by reducing the gap d, making the cavity substantially small. Such a small cavity can be easily installed in a vacuum tube of convenient size.

Further, we can verify that such a cavity has an important feature, required for an efficient interaction with electrons. In an unperturbed coaxial cavity the field energy distributed over the whole volume can be calculated as

$$\int_{\text{cavity}} E^2 \mathrm{d}v = E_o^2 \int_a^b \frac{2\pi r \mathrm{d}r}{r^2}\int_0^\lambda \sin^2\frac{2\pi x}{\lambda}\mathrm{d}x = \frac{\pi E_o^2}{6}\left(\frac{2\pi}{\lambda}\right)^2 h^3 \ln\frac{b}{a},$$

while the energy in the capacitive volume in the perturbed cavity can be estimated as

$$\int_{\text{gap}} E^2 \mathrm{d}v = \frac{\Delta V\left(\frac{1}{2}h\right)^2}{C} = \pi a^2 E_o^2\left(\frac{2\pi}{\lambda}\right)^2\frac{h^2}{d}\ln\frac{b}{a}.$$

The ratio between them can then given by

$$\int_{\text{gap}} E^2 \mathrm{d}v \Big/ \int_{\text{cavity}} E^2 \mathrm{d}v = \frac{\pi^2}{3}\left(\frac{h}{\lambda}\right)^2,$$

which is large, if $\lambda \gg h$.

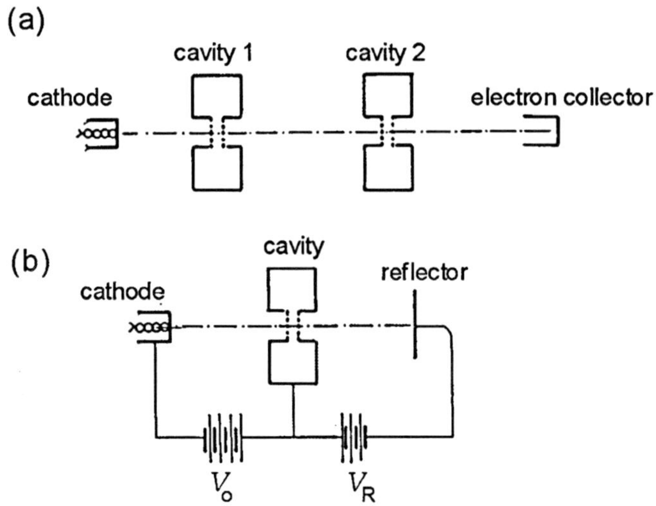

FIGURE 19.3. (a) Two-cavity klystron. (b) Reflex klystron.

19.3. Velocity Modulation

We consider a continuous flow of electrons in one dimension from the cathode, which is described by the continuity equation of charge and current. Such a current as caused by a static voltage is primarily a steady current, as characterized by a constant charge density, i.e., $\partial\rho/\partial t = 0$, at any point in the path of electrons moving at a constant density, i.e., $\mathrm{div}\boldsymbol{j} = 0$ at any time. If this condition is violated, ρ and $\boldsymbol{j}$ should be changed with respect to the continuity law, i.e., $\mathrm{div}\boldsymbol{j} + \partial\rho/\partial t = 0$, where $\boldsymbol{j} = \rho v$. In a one-dimensional flow with $v =$ constant, we have a solution $\rho = \rho(x - vt)$, expressing that all electrons are moving at the same constant speed v.

Figure 19.3 shows a practical arrangement for such a current in a vacuum tube. Electrons emitted by the cathode are accelerated by the potential V_o between the cathode and cavity, as shown in Figure 19.3(b). Denoting the mass and charge of an electron by m and e, respectively, we have the energy relation, i.e., $1mv_o^2 = eV_o$, which is valid before electrons enter the cavity. Needless to say, V_o must be negative to accelerate electrons to a speed v_o. Between the two grids there is a potential difference $\Delta V\,\cos(\omega_1 t)$, where the speed changes from v_o to v_1 according to the relation

$$1\,mv_1^2 - 1\,mv_o^2 = -e\Delta V\cos\omega_1 t,$$

hence we can write that

$$1\,mv_1^2 = e[V_o + \Delta V\cos(\omega_1 t)],$$

or

$$v_1 = v_o\left[1 - \frac{\Delta V}{2V_o}\cos(\omega_1 t)\right] \quad \text{for} \quad \Delta V \ll V_o. \tag{19.4}$$

Thus, electrons are further accelerated or decelerated, depending on the sign of the potential $\Delta V\ \cos(\omega_1 t)$. Owing to the periodic potential, accelerated electrons catch up with those that are slowed down after passing through the cavity, and consequently the electron density is periodically modulated. Electrons are thus said to be *velocity modulated*, resulting in so-called electron *bunching*. Generally, such a modulated current is rich in harmonic frequencies of ω_1, hence it is utilized for generating intense coherent radiation at ω_1 or other applications.

It is noted that after it is modulated at t_o in the cavity (called a *buncher*), the bunched density arrives at a position x at a later time $t = t_o + x/v_1$, which can be written as

$$t = t_o + \frac{x}{v_o\sqrt{1 + \dfrac{\Delta V}{V_o}\cos\omega_1 t_o}}. \tag{19.5}$$

Therefore, the modulated beam current can be expressed as

$$I(x,t) = e\frac{\mathrm{dN}}{\mathrm{d}t} = e\frac{\mathrm{dN}}{\mathrm{d}t_o}\frac{\mathrm{d}t_o}{\mathrm{d}t} = I_o\frac{\mathrm{d}t_o}{\mathrm{d}t}, \tag{19.6}$$

where $I_o = e\dfrac{\mathrm{dN}}{\mathrm{d}t_o}$. Here, $\mathrm{d}t_o/\mathrm{d}t$ can be obtained from (19.6), i.e.,

$$\frac{\mathrm{d}t_o}{\mathrm{d}t} = \frac{1}{1 + \alpha\sin\omega_1 t_o},$$

where the factor $\alpha = \dfrac{1}{2}\dfrac{x\omega_1}{v_o}\dfrac{\Delta V}{V_o}$ is called the *bunching parameter*. For a small value of α we also obtain the relation $\omega_1(t - \dfrac{x}{v_o}) = \omega_1 t_o - \alpha\cos(\omega_1 t_o)$, which can be used as the phase variable for the Fourier expansion of the current $I(x,t)$.

Considering that the electron density at $x = 0$ is modulated by the potential ΔV in a period of time $0 \le t_o \le 2\pi/\omega_1$, we can express $I(x,t)$ at an arbitrary point x in the drift space by the Fourier series

$$I = I_1 + \sum_n \left\{A_n \cos n\omega_1\left(t - \frac{x}{v_o}\right) + B_n \sin n\omega_1\left(t - \frac{x}{v_o}\right)\right\},$$

where

$$I_1 = \frac{\omega_1}{2\pi}\int_0^{2\pi/\omega_1} I(x,t)\,\mathrm{d}t = I_o,$$

$$A_n = \frac{\omega_1}{\pi}\int_0^{2\pi/\omega_1} I(x,t)\cos n\omega_1\left(t - \frac{x}{v_o}\right)\mathrm{d}t,$$

and

$$B_n = \frac{\omega_1}{\pi} \int_0^{2\pi/\omega_1} I(x, t) \cos n\omega_1 \left(t - \frac{x}{v_o}\right) dt.$$

We can rewrite these coefficients by using an angular variable $\omega_1 t = \varphi$, that is,

$$A_n = \frac{I_o}{\pi} \int_0^{2\pi} \cos n(\varphi - \alpha \cos\varphi)\, d\varphi$$
$$= (-1)^{n/2}\, 2I_o J_n(n\alpha) \quad \text{for } n \text{ even, } 0 \text{ for } n \text{ odd,}$$

and

$$B_n = \frac{I_o}{\pi} \int_0^{2\pi} \sin n(\varphi - \alpha \cos\varphi)\, d\varphi$$
$$= (-1)^{\frac{1}{2}(n-1)}\, 2I_o J_n(n\alpha) \quad \text{for } n \text{ odd, } 0 \text{ for } n \text{ even,}$$

where $J_n(n\alpha)$ is the Bessel function in integral form.

The velocity modulated current I_{bunch} is therefore expressed as a series of component waves with amplitudes changing periodically as described by Bessel's functions:

$$\begin{aligned} I_{\text{bunch}} = {} & I_o + 2I_o J_1(\alpha) \sin\left[\omega_1\left(t - \frac{x}{v_o}\right)\right] \\ & - 2I_o J_2(2\alpha)\cos\left[2\omega_1\left(t - \frac{x}{v_o}\right)\right] - 2I_o J_3(3\alpha)\sin\left[3\omega_1\left(t - \frac{x}{v_o}\right)\right] \\ & + 2I_o J_4(4\alpha)\cos\left[4\omega_1\left(t - \frac{x}{v_o}\right)\right] + \cdots . \end{aligned} \tag{19.7}$$

Figure 16.5 shows curves of some Bessel's functions, varying periodically as characterized by the parameter α, each of which indicates a maximum at a specific phase of propagation. Electrons and the cavity field can interact significantly if they are in phase at the same frequency ω_1. Obviously, a bunch for $n = 1$, i.e., the second term $2I_o J_1 \sin\left[\omega_1\left(t - \frac{x}{v_o}\right)\right]$, is the most important component among others, interacting in phase with an oscillating field at ω_1. From Table 16.1, shown in Appendix, the curve of $J_1(\alpha)$ is maximum at $\alpha_{\max} = 1.841$, where $J_1(\alpha_{\max}) = 0.5819$, giving a significantly large amplitude.

19.4. A Reflex Oscillator

Figures 19.3(a) and 19.3(b) show a schematic layout in two-cavity and reflex klystrons, respectively, which are vacuum tube devices commonly used as microwave oscillators. The former is generally suitable for high-power generation, whereas the latter is designed as a low-power oscillator in other applications. As shown in Figure 19.3(a), a second cavity is installed for interactions with bunched currents from the first cavity, whereas in the reflex oscillator of Figure 19.3(b) the bunched current is reflected back by an additional electrode, called the *reflector*, to interact with the cavity. For whichever operation—either oscillator or accelerator—the transit time of the bunched current is essential, and it is adjustable by a negative potential difference V_R applied between the two-cavity klystron in (a), and between the cavity and reflector in (b). Here, we discuss a reflex klystron, in which the reflector voltage V_R is essential for selecting mode of operation.

Letting d be the distance between the cavity and the reflector and t_o the time for a bunch to start from the cavity, the speed should change to

$$v = v_o - \frac{eV_R}{md}(t - t_o).$$

Considering that the speed is $-v_o$ when electrons are returning to the cavity, the time t can be expressed as

$$t = t_o + \frac{2md}{eV_R}v = t_o + \frac{2mv_o d}{eV_R}\left(1 - \frac{\Delta V}{2V_o}\cos\omega_1 t_o\right).$$

In terms of the phase, the current phase is shifted from the starting phase $\omega_1 t_o$ as

$$\omega_1 t = \omega_1 t_o + \theta + \frac{1}{2}\theta\frac{\Delta V}{V_o}\cos\omega_1 t_o,$$

where $\theta = \omega_1 \dfrac{2mv_o d}{eV_R}$ signifies a phase difference arising from the transit process, and in the last term $\beta = \dfrac{1}{2}\theta\dfrac{\Delta V}{V_o}$ represents the bunching effect, similar to the parameter α. It is noted, however, that the parameter β, as related to θ, is not identical to α.

When returned from the reflector to $x = 0$, the phase of the reflected electron bunch is considered as $\omega_1 t - \theta = \omega_1 t_o - \beta\cos\omega_1 t_o$. Therefore, by the Fourier expansion, the reflected current $I_{ref}(t)$ can be expressed as

$$I_{ref}(t) = I_1 + \sum_n \{A_n\cos[n(\omega_1 t - \theta)] + B_n\sin[n(\omega_1 t - \theta)]\},$$

we have

$$I_1 = I_o,\ A_n = \frac{\omega_1}{2\pi}\int_0^{2\pi/\omega_1} I_{ref}(t)\cos[n(\omega_1 t - \theta)]\,dt$$

$$= \frac{I_o}{\pi}\int_0^{2\pi}\cos[n(\varphi - \beta\cos\varphi)]\,d\varphi$$

and

$$B_n = \frac{\omega_1}{\pi} \int_0^{2\pi/\omega_1} I_{\text{ref}}(t) \sin[n(\omega_1 t - \theta)]\, dt = \frac{I_o}{\pi} \int_0^{2\pi} \sin[n(\varphi - \beta \cos\varphi)]\, d\varphi,$$

where the variable $\varphi = \omega_1 t_o$ is considered a continuous variable in the range $0 \leq \varphi \leq 2\pi$. For bunching, we have already discussed that the last integrals in the above are Bessel's functions of the nth order, i.e.,

$$A_n = (-1)^{n/2}\, 2I_o J_n(n\beta) \quad \text{for} \quad n \text{ even}$$

and

$$B_n = (-1)^{n+1/2}\, 2I_o J_n(n\beta) \quad \text{for} \quad n \text{ odd.}$$

Therefore, the reflected current at $x = 0$ can be expressed as

$$I_{\text{ref}}(t) = I_o - 2I_o J_1(\beta) \sin(\omega_1 t - \theta) + 2I_o J_2(2\beta) \cos 2(\omega_1 t - \theta) + \cdots . \quad (19.8)$$

For a reflex klystron, we are only interested in the component $2I_o J_1(\beta)\sin(\omega_1 t - \theta)$ against the voltage $\Delta V \cos(\omega_1 t)$ across the grids. Due to the interaction energy $\boldsymbol{j}.\boldsymbol{E}$, the negative sign is required for sustaining oscillation in the cavity, for which the phase difference θ plays an essential role. In fact, such an oscillatory condition can be met at any angle of $\theta - (3\pi/2) \times \text{m}$, where m is an integer, which can be achieved by adjusting the reflector voltage V_R, since θ is defined as it relates to V_R. In a practical reflex klystron the oscillator power exhibits a series of peaks as V_R is varied.

Using complex forms for this particular component of $I_{\text{ref}}(t)$ and grid voltage ΔV, we can write the electronic admittance as

$$g_e + ib_e = \frac{2I_o J_1(\beta) \exp\left\{i\left(\omega_1 t - \theta + \frac{3\pi}{2}\right)\right\}}{\Delta V \exp\{i(\omega_1 t)\}},$$

where $\Delta V = \frac{2\beta V_o}{\theta}$ by definition. Accordingly, we have

$$g_e + ib_e = \frac{I_o}{V_o}\theta \exp\left\{-i\left(\theta - \frac{3\pi}{2}\right)\right\} \frac{J_1(\beta)}{\beta},$$

and these, the electronic conductance and susceptance, are related as

$$\frac{b_e}{g_e} = -\tan\left(\theta - \frac{3\pi}{2}\right).$$

The power for exciting the cavity field can then be expressed as

$$P_e = -2I_o J_1(\beta) \cos\left(\theta - \frac{3\pi}{2}\right) \times \frac{2\beta V_o}{\theta} = -\frac{1}{2} g_e V_o^2 \beta J_1(\beta)$$

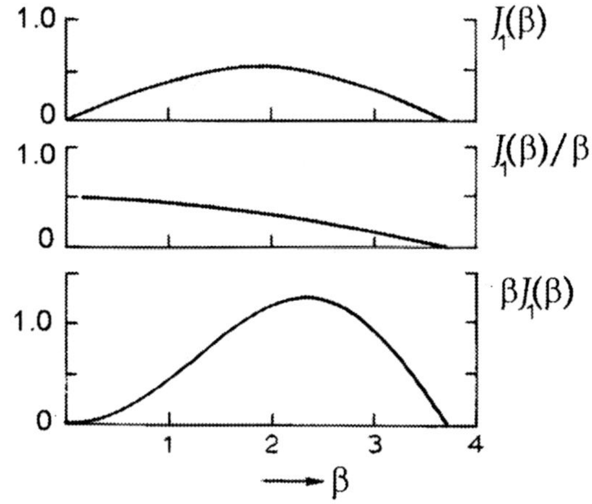

FIGURE 19.4. Bessel's function $J_1(\beta)$, and $J_1(\beta)/\beta$, $\beta J_1(\beta)$ vs. β.

where

$$g_e = -\frac{I_o}{V_o}\theta \cos\left(\theta - \frac{3\pi}{2}\right)\frac{J_1(\beta)}{\beta},$$

which is directly related to the reflector voltage V_R, expressed as

$$V_R \propto \frac{1}{\theta}\cos\left(\theta - \frac{3\pi}{2}m\right).$$

where m is an integer. Figure 19.4 shows sketches of $J_1(\beta)$, $J_1(\beta)/\beta$ and $\beta J_1(\beta)$ as functions of β, which can be used for evaluating the oscillation spectrum and intensities.

20 Dielectric and Magnetic Responses in Resonant Electromagnetic Fields

20.1. Introduction

Regarding conducting properties, materials were traditionally classified into categories of "conductor" or "insulator." Today, many types of semi-conductors as well as magnetic materials are used for a variety of applications. Modern materials can be discussed properly with respect to structure, but such a discussion is beyond the scope of this book. Nevertheless, modern empirical knowledge of electronic charge and spin, among other properties, must be taken for granted in order to understand classical electromagnetic theory. In addition, the presence of electric and magnetic moments in constituent species is empirical.

We consider that these moments are located in solid insulating materials, ignoring their interactions. With this assumption, the moments can interact primarily with applied electric or magnetic field. Macroscopically, such a response to an applied field can be interpreted as arising from the macroscopic moment. Referring to as the linear response, it is meaningful only in the limit of weak applied field. It is primarily an adiabatic process, but necessarily involved in a thermal relaxation process. In practice, the response from an individual moment is too weak to detect with conventional instruments. For example, the dipole moment of a NH_3 molecule is of the order of 0.5×10^{-29} C-m, and therefore a group of a sufficiently large number of identical moments is required for dealing with the molecular dipole. On the other hand, in most condensed states such a response can be measured through the use of standing electromagnetic waves of a sizable intensity at a wavelength longer than the material size, where the field is regarded as uniform over the sample. We therefore require an *LC* circuit resonator at low frequencies or a cavity resonator at higher frequencies for response experiments, where the field has sufficiently large amplitude to interact with a group of identical moments. We can study the time-dependence of microscopic dipoles under such circumstances.

20.2. The Kramers–Krönig Formula

We consider a system of microscopic dipole moments embedded in a material that is assumed as isotropic and non-conducting. For a response from the system of dipoles to a standing wave of a time-dependent field, microscopic moments $p(t)$ are considered as flexible and rotatable at their locations, but not as migratory in motion. Hence, the applied field $X(t)$ can be assumed only as a function of time, and expressed by a Fourier integral

$$X(t) = \frac{1}{2\pi} \int_{-\infty}^{+\infty} \overline{X}(\omega) \exp(i\omega t)\, d\omega,$$

where

$$\overline{X}(\omega) = \int_{-\infty}^{+\infty} X(t) \exp(-i\omega t)\, dt$$

is the Fourier amplitude of $X(t)$. Considering that a moment $p(t)$ is driven by such a field $X(t)$, the response function $\chi(\omega)$ can be defined from the relation

$$p(t) = \frac{1}{2\pi} \int_{-\infty}^{+\infty} \chi(\omega) \overline{X}(\omega) \exp(i\omega t)\, d\omega. \tag{20.1a}$$

Assuming that $X(t) = X_o \delta(t)$, where the delta function $\delta(t) = 1$ only if $t = 0$, otherwise zero at all $t \neq 0$, we can interpret that the field $\overline{X}(\omega) = X_o \delta(\omega) = X_o$ is switched on at $t = 0$ for infinitely short duration. Writing $p(t)/X_o = \alpha(t)$ for simplicity, (20.1a) can be expressed as

$$\alpha(t) = \frac{1}{2\pi} \int_{-\infty}^{+\infty} \chi(\omega) \exp(i\omega t)\, d\omega, \tag{20.1b}$$

where $\alpha(t)$ is the Fourier transform of the response function $\chi(\omega)$. Writing the function in a complex form, i.e., $\chi(\omega) = \chi'(\omega) - i\chi''(\omega)$, (20.1b) can be re-expressed as

$$\alpha(t) = \frac{1}{2\pi} \int_{-\infty}^{+\infty} \left[\chi'(\omega)\cos(\omega t) + \chi''(\omega)\sin(\omega t)\right] d\omega$$
$$+ \frac{i}{2\pi} \int_{-\infty}^{+\infty} \left[\chi'(\omega)\sin(\omega t) - \chi''(\omega)\cos(\omega t)\right] d\omega.$$

For a real $\alpha(t)$ the integral of the second term should vanish, so that

$$\int_{-\infty}^{+\infty} \chi'(\omega)\sin(\omega t)\,d\omega = \int_{-\infty}^{+\infty} \chi''(\omega)\cos(\omega t)\,d\omega,$$

and

$$\alpha(t) = \frac{1}{2\pi}\int_{-\infty}^{+\infty} \left[\chi'(\omega)\cos(\omega t) + \chi''(\omega)\sin(\omega t)\right]d\omega. \tag{20.1c}$$

Further, since the field is turned on at $t = 0$, we consider $\delta(t) = 0$ for all $t < 0$, and hence $\alpha(t) = 0$ in (20.1c) at all negative $t < 0$. Accordingly,

$$\frac{1}{\pi}\int_0^{\infty} \chi'(\omega)\cos(\omega t)d\omega = -\frac{1}{\pi}\int_0^{\infty} \chi''(\omega)\sin(\omega t)d\omega \quad \text{for } t < 0.$$

Therefore, for positive time $t \geq 0$, (20.1c) can be expressed as

$$\alpha(t) = \frac{1}{\pi}\int_0^{\infty} \chi'(\omega)\cos(\omega t)d\omega \quad \text{or} \quad = \frac{1}{\pi}\int_0^{-\infty} \chi''(\omega)\sin(\omega t)d\omega. \tag{20.2}$$

In the complex form, the susceptibility for $t \geq 0$ can be written as

$$\chi(\omega) = \int_0^{\infty} \alpha(t)\exp(-i\omega t)\,dt,$$

and for $t < 0$

$$\chi(\omega) = 0.$$

Therefore, the real and imaginary parts are given by

$$\chi'(\omega) = \int_0^{\infty} \alpha(t)\cos(\omega t)dt \quad \text{and} \quad \chi''(\omega) = -\int_0^{\infty} \alpha(t)\sin(\omega t)dt. \tag{20.3}$$

Combining (20.2) and $\chi''(\omega)$ in (20.3), we have

$$\chi''(\omega) = -\frac{1}{\pi}\int_0^{\infty}\int_0^{\infty} \chi'(\omega')\cos(\omega' t)\sin(\omega t)\,d\omega' dt.$$

Integrating over the time t, the result can be expressed as

$$\chi''(\omega) = -\frac{2\omega}{\pi}\mathrm{P}\int_0^{\infty} \frac{\chi'(\omega')\,d\omega'}{\omega'^2 - \omega^2}. \tag{20.4a}$$

Similarly, using (20.2) in $\chi'(\omega)$ in (20.3), we can obtain

$$\chi'(\omega) = \frac{2}{\pi} \mathrm{P} \int_0^\infty \frac{\omega' \chi''(\omega')\, d\omega'}{\omega'^2 - \omega^2}, \tag{20.4b}$$

where P in (20.4a) and (20.4b) indicates the *principal value* of these integrals that is calculated with the *Cauchy theorem* in the theory of complex functions. Equations (20.4a) and (20.4b) are together known as *Kramers–Krönig formula*, implying that the real and imaginary parts are not independent, but related to each other. On the other hand, these parts of $\chi(\omega)$ are measurable by separate experiments, resulting in the relation consistent with the Kramers–Krönig formula.

20.3. Dielectric Relaxation

An electric dipole moment $\boldsymbol{p}$ takes a stable direction in a stationary electric field $\boldsymbol{E}$ since the potential $U = -\boldsymbol{p} \cdot \boldsymbol{E}$ becomes minimum when $\boldsymbol{p} \parallel \boldsymbol{E}$. However, as it is rotatable dynamically, it takes a certain time for the dipole to settle in the equilibrium direction, for which time a *friction* can be considered responsible, analogous to what is seen in macroscopic media. Also, taking a kinetic energy of rotation, the dipole may fluctuate around the equilibrium direction. Due to inevitable energy loss in this case, such a description as above is adequate for these processes called *thermal relaxation* and *fluctuation*.

Denoting equilibrium values of $\boldsymbol{p}$ and U by $\boldsymbol{p}_o$ and U_o, the relaxation can be determined by

$$\frac{d\boldsymbol{p}}{dt} + \frac{\boldsymbol{p} - \boldsymbol{p}_o}{\tau} = 0 \quad \text{and} \quad \frac{dU}{dt} + \frac{U - U_o}{\tau} = 0, \tag{20.5a}$$

where the parameter τ is called the *relaxation time*, and the solutions show an exponential decay

$$\boldsymbol{p} = \boldsymbol{p}_o \exp\left(-\frac{t}{\tau}\right) \quad \text{and} \quad U = U_o \exp\left(-\frac{t}{\tau}\right). \tag{20.5b}$$

The macroscopic polarization $\boldsymbol{P}$ in the presence of an external field $\boldsymbol{E}$ can generally be expressed as

$$P(t) = \varepsilon_o \int_0^\infty \alpha(t - t') E(t')\, dt',$$

where

$$\alpha(t - t') = \exp\left(-\frac{t - t'}{\tau}\right)$$

represents the response from the polarization per unit volume due to microscopic correlations between t and t'.

If $t - t' \to \infty$, we may consider that directions of these dipole moments become random, making no contribution to the polarization $P(\infty)$. Writing however the dielectric constant at such a condition as $\varepsilon(\infty)$, the electric flux density can be expressed as

$$D(t) = \varepsilon(\infty) E(t) + \varepsilon_o \int_0^\infty \alpha(t - t') E(t') \, dt'.$$

In a sinusoidal field $E(t) = E_o \cos(\omega t)$, writing $t - t' = t''$, we have

$$D(t) = E_o \cos(\omega t) \left\{ \varepsilon(\infty) + \varepsilon_o \int_0^\infty \alpha(t'') \cos(\omega t'') dt'' \right\} + E_o \sin(\omega t) \left\{ \varepsilon_o \int_0^\infty \alpha(t'') \sin(\omega t'') dt'' \right\},$$

which can be re-expressed as

$$D(t) = \varepsilon'(\omega) E_o \cos(\omega t) + \varepsilon''(\omega) E_o \sin(\omega t),$$

where

$$\varepsilon'(\omega) = \varepsilon(\infty) + \varepsilon_o \int_0^\infty \alpha(t'') \cos(\omega t'') dt'' \tag{20.6a}$$

and

$$\varepsilon''(\omega) = \varepsilon_o \int_0^\infty \alpha(t'') \sin(\omega t'') dt''. \tag{20.6b}$$

Here, the correlation function $\alpha(t'')$ is the Fourier transform of the dielectric function $\varepsilon(\omega)$, i.e.,

$$\alpha(t'') = \frac{2}{\pi} \int_0^\infty \left[\varepsilon'(\omega') - \varepsilon(\infty)\right] \cos(\omega' t'') d\omega' = \frac{2}{\pi} \int_0^\infty \varepsilon''(\omega') \sin(\omega' t'') d\omega'.$$

Equations (20.6a) and (20.6b) can be combined in a complex form by defining the complex dielectric constant $\varepsilon(\omega) = \varepsilon'(\omega) - i\varepsilon''(\omega)$ i.e.,

$$\varepsilon(\omega) = \varepsilon(\infty) + \varepsilon_o \int_0^\infty \alpha(t'') \exp(-i\omega t'') \, dt''. \tag{20.6c}$$

Putting the expressions of $\alpha(t'')$ back into (20.5a) and (20.6b), we obtain the Kramers–Krönig relations:

$$\varepsilon'(\omega) - \varepsilon(\infty) = \frac{2}{\pi}\int_0^\infty \frac{\varepsilon''(\omega')}{\omega'^2 - \omega^2}\mathrm{d}\omega'$$

and

$$\varepsilon''(\omega) = \frac{2}{\pi}\int_0^\infty \frac{\varepsilon'(\omega') - \varepsilon(\infty)}{\omega'^2 - \omega^2}\mathrm{d}\omega',$$

indicating that the respective real and imaginary components $\varepsilon'(\omega)$ and $\varepsilon''(\omega)$ are not independent.

In a static case signified by $\omega = 0$ from (20.6.a) and (20.6.b) we have

$$\varepsilon'(0) = \varepsilon(\infty) + \varepsilon_o \int_0^\infty \alpha(t'')\mathrm{d}t'' \quad \text{and} \quad \varepsilon''(0) = 0. \tag{20.6d}$$

Using $\alpha(t'') = \exp(-\frac{t''}{\tau})$ in (20.6c) and (20.6d), the integrated results are

$$\varepsilon(\omega) - \varepsilon(\infty) = \frac{\varepsilon_o\tau}{1 + i\omega\tau} \quad \text{and} \quad \varepsilon'(0) - \varepsilon(\infty) = \varepsilon_o\tau,$$

hence we can write the relation for the complex $\varepsilon(\omega)$gnd $\varepsilon(0)$, i.e.,

$$\varepsilon(\omega) - \varepsilon(\infty) = \frac{\varepsilon(0) - \varepsilon(\infty)}{1 + i\omega\tau}. \tag{20.7}$$

For the real and imaginary parts, we have

$$\varepsilon'(\omega) = \varepsilon(\infty) + \frac{\varepsilon(0) - \varepsilon(\infty)}{1 + \omega^2\tau^2}$$

and

$$\varepsilon''(\omega) = [\varepsilon(0) - \varepsilon(\infty)]\frac{\omega\tau}{1 + \omega^2\tau^2},$$

respectively. Eliminating $\omega\tau$ from these expressions for $\varepsilon'(\omega)$ and $\varepsilon''(\omega)$, we obtain the formula

$$\left\{\varepsilon'(\omega) - \frac{\varepsilon(0) + \varepsilon(\infty)}{2}\right\}^2 + \varepsilon''(\omega)^2 = \left\{\frac{\varepsilon(0) - \varepsilon(\infty)}{2}\right\}^2, \tag{20.8}$$

indicating that the curve of $\varepsilon'(\omega)$ vs. $\varepsilon''(\omega)$, if plotted as a function of frequency ω, is a semi-circle of radius $r = \frac{\varepsilon(0) - \varepsilon(\infty)}{2}$ centered at a point $A(\frac{1}{2}\{\varepsilon(0) + \varepsilon(\infty)\}, 0)$ on the real ε'-axis, as shown in Figure 20.1. Such a diagram, called the *Cole-Cole plot*, is typical for a dielectric relaxation, which is also referred to as relaxation of *Debye's type*, a useful model for evaluating the relaxation time τ.

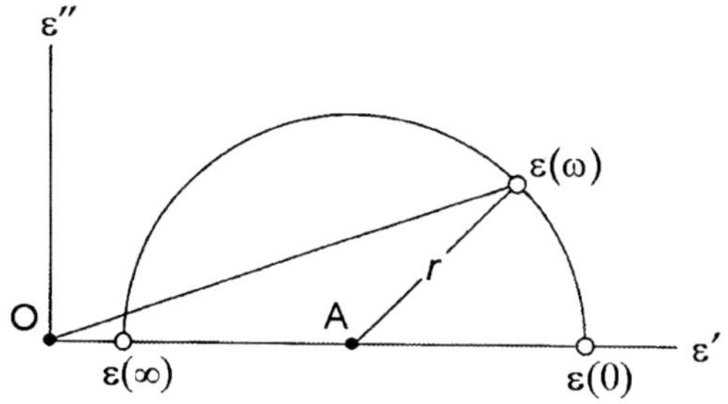

FIGURE 20.1. Cole-Cole plot for a complex dielectric function $\varepsilon(\omega) = \varepsilon'(\omega) + \varepsilon''(\omega)$.

The dielectric response is not always of Debye's type, for the kinetic energy of dipoles can often exceed the frictional energy for relaxation. Typically, the motion is oscillatory around the equilibrium position, as described by the equation of motion

$$m\frac{\mathrm{d}^2x}{\mathrm{d}t^2} + \gamma\frac{\mathrm{d}x}{\mathrm{d}t} + kx = eE_o\exp(i\omega t)$$

if in one-dimension, where m, γ, k, and e are mass, damping constant, restoring parameter, and charge, respectively. Writing this equation as

$$\frac{\mathrm{d}^2x}{\mathrm{d}t^2} + \gamma\frac{\mathrm{d}x}{\mathrm{d}t} + \omega_o^2x = \frac{eE_o}{m}\exp(i\omega t),$$

where γ/m that would result in the second term is redefined as γ, and where $\omega_o = \sqrt{k/m}$ is the characteristic frequency, the response function can be obtained from the stationary solution as

$$\alpha(\omega) \propto \frac{e/m}{\omega_o^2 - \omega^2 + i\omega\gamma} \propto \frac{\omega_o^2 - \omega^2 - i\omega\gamma}{\left(\omega_o^2 - \omega^2\right)^2 + \omega^2\gamma^2}.$$

Using this $\alpha(\omega)$, the flux density can be expressed as

$$\begin{aligned} D(\omega, t) &= \varepsilon(\infty)E_o\exp(i\omega t) + \varepsilon_o\alpha(\omega)E_o\exp(i\omega t) \\ &= \left\{\varepsilon(\infty) + \frac{\omega_o^2 - \omega^2}{\left(\omega_o^2 - \omega^2\right)^2 + \omega^2\gamma^2}\right\}E_o\cos(\omega t) \\ &\quad - \frac{i\omega\gamma E_o\sin(\omega t)}{\left(\omega_o^2 - \omega^2\right)^2 + \omega^2\gamma^2}. \end{aligned}$$

Therefore the complex dielectric constant can be given in complex form $\varepsilon(\omega) = \varepsilon'(\omega) - i\varepsilon''(\omega)$, where $\varepsilon'(\omega)$ and $\varepsilon''(\omega)$ are expressed by

$$\varepsilon'(\omega) = \varepsilon(\infty) + \varepsilon_o\frac{\omega_o^2 - \omega^2}{\left(\omega_o^2 - \omega^2\right)^2 + \omega^2\gamma^2} \tag{20.9a}$$

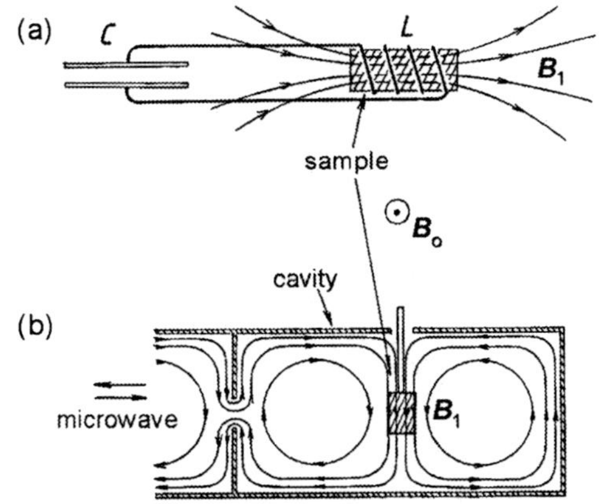

FIGURE 20.2. Resonators for magnetic resonance. (a) An LC resonator for a nuclear magnetic resonance. (b) A cavity resonator for a paramagnetic resonance.

and

$$\varepsilon''(\omega) = \varepsilon_o \frac{\omega\gamma}{\left(\omega_o^2 - \omega^2\right)^2 + \omega^2\gamma^2}. \tag{20.9b}$$

To measure $\varepsilon(\omega)$, a sample is placed in a capacitor that is a part of a resonant LC_o circuit, as shown in Figure 20.2(a). This is similar to a magnetic measurement. Here, the empty capacitor C_o can accommodate a sample, and the capacity changes to $C = (\varepsilon' - i\varepsilon'')C_o$, assuming the space is fully occupied by the dielectric material. The impedance of the circuit is given by

$$Z(\omega) = R + i\omega L + \frac{1}{i\omega C} = R + \frac{1}{\omega\varepsilon'' C_o} + i\omega L - \frac{1}{i\omega\varepsilon'' C_o},$$

and the resonance frequency is determined by $\mathrm{Im}\,Z(\omega) = 0$, i.e., $\omega = \dfrac{1}{\sqrt{\varepsilon' L C_o}}$. Hence, at resonance with the sample, the resonance frequency shifts from the value $\omega_o = \dfrac{1}{\sqrt{LC_o}}$. Namely,

$$\Delta\omega = \omega - \omega_o = \left(\frac{1}{\sqrt{\varepsilon'}} - 1\right)\omega_o. \tag{20.10a}$$

The energy loss represented by the resistance R accompanies an additional term $\dfrac{1}{\varepsilon''\,\omega C_o}$ so that the quality factor should shift by

$$\frac{\Delta Q}{Q} = \frac{Q_o - Q}{Q} = \frac{\varepsilon''}{\varepsilon_o} Q_o. \tag{20.10b}$$

Thus, ε' and ε'' can be determined by measuring the resonance shift and change in Q-factor.

At microwave frequencies, a sample should be placed in a volume at a location in the resonant cavity where the field is sufficiently uniform over the sample volume.

In this case, the volume ratio, i.e., sample/cavity, must be calibrated for the formula (20.10a) and (20.10b) to provide reliable values of ε' and ε''.

20.4. Magnetic Resonance

Figure 20.2 illustrates resonators for magnetic studies, similar to what are used for dielectric measurements. In such a resonator, a sample is placed at a location where the magnetic flux density is high: inside a coil at low frequencies, as shown in Figure 20.2(a), and at the location indicated in Figure 20.2(b) for microwave measurements. At these locations the flux density is also assumed to be sufficiently uniform over the sample volume.

Magnetic moments originate from unpaired electrons of ions in specific groups of elements. The electronic motion in ions is much faster than the dipole moments, as signified by a distinctive difference in their relaxation times. It is also known that various atomic nuclei are characterized by their magnetic moments. Virtually independent of the host material, the properties of these ions and nuclei can be investigated under an adiabatic condition with applied magnetic fields.

Hence, a magnetic moment in non-magnetic media can be regarded as primarily isolated from the rest of the material and can be set in precession when a static magnetic field is applied. In the presence of a uniform magnetic field $\boldsymbol{B}_o$, there is a *torque* $\mu \times \boldsymbol{B}_o$ exerted on such a magnetic moment μ as described by the equation of motion

$$\frac{\mathrm{d}\boldsymbol{J}}{\mathrm{d}t} = \mu \times \boldsymbol{B}_o,$$

where $\boldsymbol{J}$ is the angular momentum of a carrier of μ. Writing $\mu = \gamma\, \boldsymbol{J}$, where γ is called the *gyromagnetic ratio*, we have

$$\frac{\mathrm{d}\mu}{\mathrm{d}t} = \gamma\,(\mu \times \boldsymbol{B}_o)\,. \tag{20.11a}$$

Using rectangular coordinates x, y, z with respect to the direction of $\boldsymbol{B}_o$, which is taken as the z-axis as shown in Figure 20.3, equation (20.11) can be expressed as

$$\frac{\mathrm{d}\mu_x}{\mathrm{d}t} = \gamma\,\mu_y B_o, \quad \frac{\mathrm{d}\mu_y}{\mathrm{d}t} = -\gamma\,\mu_x B_o \quad \text{and} \quad \frac{\mathrm{d}\mu_z}{\mathrm{d}t} = 0. \tag{20.11b}$$

From the last equation, μ_z can be a constant of motion, i.e.,

$$\mu_z = \text{const}, \tag{20.12a}$$

whereas for perpendicular components μ_x and μ_y, combining the first and second equations in (20.11b) in complex form gives

$$\frac{\mathrm{d}\left(\mu_x + i\mu_y\right)}{\mathrm{d}t} = \gamma\, B_o\left(\mu_y - i\mu_x\right) = -i\gamma\, B_o\left(\mu_x + i\mu_y\right),$$

FIGURE 20.3. Principle of magnetic resonance.

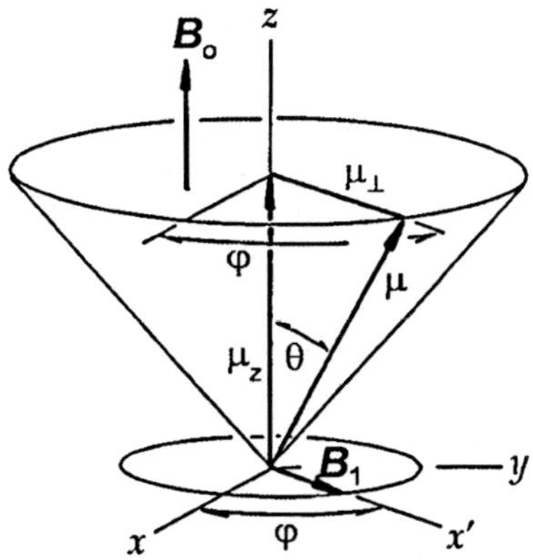

where the component $\mu_x + i\mu_y = \mu_\perp$ perpendicular to $\boldsymbol{B}_o$ has a stationary solution that can be expressed as

$$\mu_x + i\mu_y = \left(\mu_x + i\mu_y\right)_{t=0} \exp\left(-i\omega_L t\right), \quad \omega_L = \gamma B_o. \tag{20.12b}$$

As illustrated in Figure 20.3, the motion of μ is a precession around the z-axis, as described by (20.12a) and (20.12b). Such a motion is known as the *Larmor precession*, where ω_L, called the Larmor frequency, is characterized by γ, depending on B_o.

For *magnetic resonance* we apply an additional oscillating magnetic field $\boldsymbol{B}_1 \exp(-i\omega t + \varphi)$ in the xy-plane perpendicular to $\boldsymbol{B}_o$. If the condition

$$\omega = \omega_L \tag{20.13}$$

is met, $\mu_\perp$ and B_1 are synchronized in phase, and accordingly, the torque $\mu_\perp \times \boldsymbol{B}_1$ is responsible for changing the direction of $\mu_\perp$ in a rotating plane $x'z$, as shown in the figure. In this case the angle φ is forced to become 0 by steady rotation of field $\boldsymbol{B}_1$. It is therefore convenient to use the *rotating coordinate system*, x', y', and z, that is, rotating with $\boldsymbol{B}_1$ for $\mu_\perp \times \boldsymbol{B}_1$ to drive the precession. At $\omega = \omega_L$, the energy of oscillating field is transferred to the magnetic moment in precession, which is called the *magnetic resonance*.

It is noted that magnetic moments of atomic nuclei are about $1/10^3$ of an electronic moment. Hence, the Larmor frequency of a proton is lower by that factor than the electronic Larmor frequency in the same $\boldsymbol{B}_o$. Accordingly, the magnetic resonance frequencies are in ranges of $1 \sim 100$ MHz and $1{,}000 \sim 10{,}000$ MHz, respectively, for which resonators of an *LC* and microwave cavity are commonly used. Typical resonators for nuclear and electronic magnetic resonance are as illustrated in Figures 20.2(a) and (b).

In the above is given only the principle; the *resonance condition* (20.13) should also be expressed for a macroscopic medium, as described next by the Bloch theory.

20.5. The Bloch Theory

The Bloch theory primarily applies to nuclear resonance, but it describes conditionally electronic magnetic resonance as well. Although written for an idealized case where interactions are predominantly with external fields $\boldsymbol{B}_o$ and $\boldsymbol{B}_1$, the theory covers the resonance phenomenon in general, providing the basic approach to more complex cases.

We consider a system of magnetic nuclei μ embedded in an isotropic non-magnetic media, where nuclei are not mobile diffusively, but rotate independently at their positions. The whole system is assumed to be in thermal equilibrium at a given temperature T.

The system in a static field $\boldsymbol{B}_o || z$ is characterized by the macroscopic magnetization $M_z = M_z(t)$ due to nuclear magnetic moments μ in precession around the z-axis. In this case we can write that M_x and $M_y = 0$, implying that individual phases φ of $\mu_\perp$ are randomly distributed in the system. Therefore, in equilibrium state the components M_z are considered as related to the thermal average of μ_z, whereas $M_\perp$ vanishes, as given by the random-phase average. Bloch introduced two separate relaxation times, τ_1 and τ_2, for the time-dependent processes for M_z and $M_\perp$, respectively, by writing

$$\frac{\mathrm{d}M_z}{\mathrm{d}t} + \frac{M_z - M_o}{\tau_1} = 0 \tag{20.14a}$$

and

$$\frac{\mathrm{d}M_\perp}{\mathrm{d}t} + \frac{M_\perp}{\tau_2} = 0. \tag{20.14b}$$

Here, M_o represents the magnetization of the system of μ in the field $\boldsymbol{B}_o$, whose energy $-M_oB_o$ should be thermally associated with Gibbs thermodynamic potential. In this context he called τ_1 the *spin-lattice relation time*, indicating the rate at which the magnetization energy reaches equilibrium with the rest of the system. On the other hand, $-M_1B_1$ is related to the energy associated with the phase distribution of $\mu_\perp$ and is referred to as the *spin-spin relaxation time*. In fact, the dipole-dipole interaction in the system is responsible for random phases, which determines adiabatically the rate for individual phases of $\mu_\perp$ to become in-phase with $\boldsymbol{B}_1$. Figure 20.4 shows schematically how random phases can be forced to be in-phase with the rotating field $\boldsymbol{B}_1$. Denoting such a dipolar field of random phases

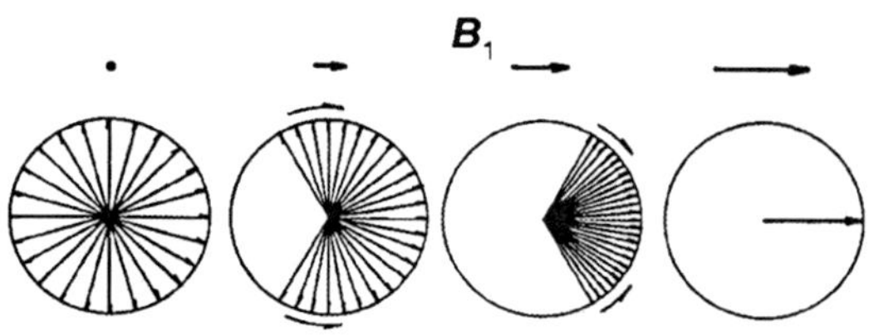

FIGURE 20.4. Adiabatic application of a linearly polarized field $\boldsymbol{B}_1$.

by ΔB, it is clear that the applied B_1 should be greater than ΔB in order to unify distributed phases. Writing $\Delta B \sim 1/\gamma\tau_2$, the condition $B_1 \gg \Delta B$ can be expressed as

$$\gamma B_1 \gg 1/\tau_2, \tag{20.15}$$

which is called the *slow passage* condition, signifying that an external field B_1 is applied adiabatically. In fact, (20.14b) is a valid assumption under the slow passage condition. Such an adiabatic phasing as shown in Figure 20.4 is reversible, and hence *dephasing* should take place when external B_1 is switched off.

Taking these relaxation mechanisms into account, Bloch wrote the equations of motion in the presence of $\boldsymbol{B}_o$ and $\boldsymbol{B}_1\exp(\pm i\omega t)$ as

$$\frac{\mathrm{d}M_\pm}{\mathrm{d}t} \pm i\gamma B_o M_\pm + \frac{M_\pm}{\tau_2} = -iM_z B_o \exp(\pm i\omega t) \tag{20.16a}$$

and

$$\frac{\mathrm{d}M_z}{\mathrm{d}t} + \frac{M_z - M_o}{\tau_1} = \frac{1}{2} i\gamma B_1 \left[M_+ \exp(-i\omega t) + M_- \exp(i\omega t)\right]. \tag{20.16b}$$

It is noted that (20.16a) and (20.16b), known as the *Bloch equations*, have stationary solutions under a slow passage condition if one assumes that $\frac{\mathrm{d}M_z}{\mathrm{d}t} = 0$ and $M_\perp = M_\pm \propto \exp(\pm i\omega t)$. With respect to the rotating coordinate system, M_z and $M_\pm$ can be related as

$$M_\pm(t) = \frac{\gamma B_1 M_z \exp(\pm i\omega t)}{\omega + \omega_L \mp \frac{i}{\tau_2}}, \quad \text{where} \quad \omega_L = \gamma B_o.$$

Accordingly,

$$\frac{M_z}{M_o} = \frac{1 + (\omega - \omega_L)^2 \tau_2^2}{1 + (\omega - \omega_L)^2 \tau_2^2 + \gamma^2 B_1^2 \tau_1 \tau_2} \tag{20.17a}$$

and

$$\frac{M_\pm}{M_o} = \frac{[(\omega - \omega_L)\tau_2 \pm i]\gamma B_1 \tau_2 \exp(\pm i\omega t)}{1 + (\omega - \omega_L)^2 \tau_2^2 + \gamma^2 B_1^2 \tau_1 \tau_2}. \tag{20.17b}$$

Equation (20.17a) indicates that $M_z \approx M_o$ at resonance, i.e., $\omega = \omega_L$, provided that $\gamma B_1^2 \tau_1 \tau_2 \ll 1$. In this case we have

$$\tan\theta = \frac{M_\pm}{M_z} \approx \frac{\gamma B_1 \tau_2}{1 + (\omega - \omega_L)^2 \tau_2^2}.$$

The complex magnetic response function $\chi(\omega)$ can be defined from the relation

$$M_\pm = \chi(\omega) B_1 \exp(\pm i\omega t) \quad \text{and} \quad M_o = \chi_o B_o,$$

and we have

$$\frac{\chi_\pm(\omega)}{\chi_o} = \frac{\gamma B_o \tau_2 \{(\omega - \omega_L)\tau_2 + i\}}{1 + (\omega - \omega_L)^2 \tau_2^2 + \gamma^2 B_1^2 \tau_1 \tau_2}.$$

Therefore, writing $\delta\omega = \dfrac{1}{\tau_2}$, the real and imaginary parts of $\chi_\pm(\omega)$ are expressed as

$$\frac{\chi'(\omega)}{\chi_o} = \frac{\omega_L(\omega - \omega_L)}{(\omega - \omega_L)^2 + \delta\omega^2 + \gamma^2 B_1^2 \tau_1 \delta\omega} \tag{20.18a}$$

and

$$\frac{\chi''(\omega)}{\chi_o} = \frac{(\omega - \omega_L)\,\delta\omega}{(\omega - \omega_L)^2 + \delta\omega^2 + \gamma^2 B_1^2 \tau_1 \delta\omega}. \tag{20.18b}$$

At resonance, $\omega = \omega_L$, from (20.18b) we obtain

$$\frac{\chi''(\omega_L)}{\chi_o} \approx \frac{\omega_L}{\delta\omega}, \tag{20.18c}$$

representing maximum energy loss. Although $\chi'(\omega_L) = 0$, the slope $d\chi'/d\omega \approx \omega_L/(\omega - \omega_L)$ is determined by a steep tangent at $\omega = \omega_L$, exhibiting a characteristic *dispersion.*

20.6. Magnetic Susceptibility Measured by Resonance Experiments

As shown in Figures 20.2(a) and 20.2(b), a sample is placed in a coil or at a place of a cavity where B_1 is maximum. When magnetic resonance occurs in the sample the inductance L changes from the off-resonance inductance L_o according to

$$L = L_o(1 + \chi) = (1 + \chi')\,L_o - i\chi'' L_o,$$

where $\chi = \chi' - \chi''$. If the resonator is not fully occupied by a sample, as in a microwave cavity, the response should be reduced by the volume ratio, i.e., $\alpha = V_{\text{sample}} / V_{\text{cavity}}$, called a *filling factor.* Assuming $\alpha = 1$, for simplicity, the impedance of the resonator can be written as

$$\begin{aligned} Z(\omega) &= R + i\omega L + \frac{1}{i\omega C} = (R + \omega\chi'' L_o) + i\omega\,(1 + \chi')\,L_o - \frac{i}{\omega C} \\ &= (R + \omega\chi'' L_o) + i\omega_o L_o\,(1 + \chi')\left(\frac{\omega}{\omega_o} - \frac{\omega_o}{\omega}\right), \end{aligned} \tag{20.19}$$

where $\omega_o = \dfrac{1}{\sqrt{(1 + \chi')\,L_o C}}$ is the resonant frequency with a sample, but with the magnetic resonance off. Hence, near the magnetic resonance, i.e., $\omega = \omega_o \pm \Delta\omega$,

$$\text{Im}Z(\omega) = 2\,(1 + \chi')\,L_o\Delta\omega \quad \text{and} \quad \text{Re}Z(\omega) = R + \omega_o\chi'' L_o, \tag{20.20}$$

indicating that χ' and χ'' can be obtained independently from the imaginary and real parts of the resonator impedance $Z(\omega)$.

It is a usual practice to vary the static field B_o, rather than the frequency ω. As related by the resonance condition $\omega = \gamma B_o$, the resonance can be observed

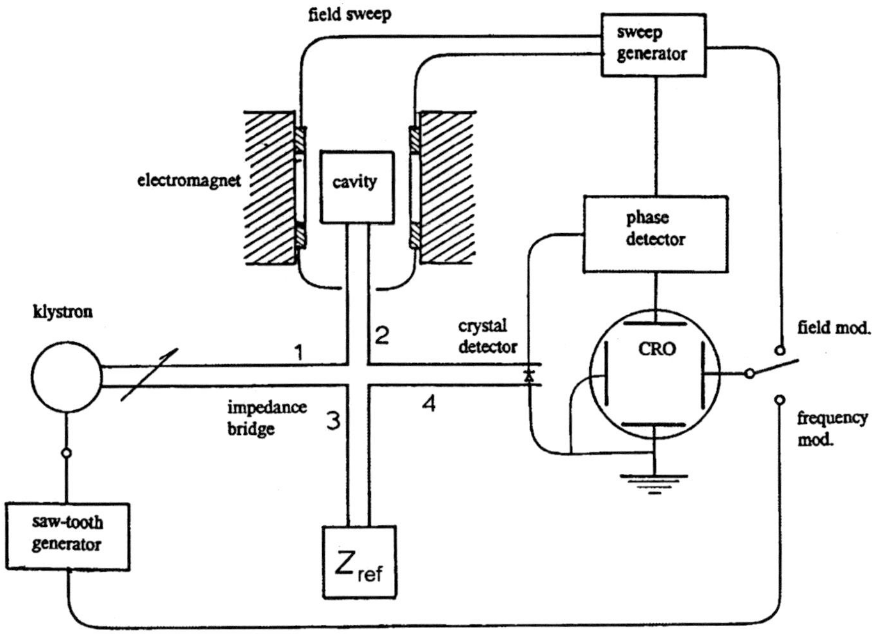

FIGURE 20.5. A typical magnetic resonance spectrometer.

by either changing B_o at a constant ω, or varying ω at a constant B_o. However, the former approach is technically easier than the latter, and commonly used in practical experiments. Figure 20.5 shows a magnetic resonance spectrometer where B_o can be swept in the vicinity of the resonance field while the oscillator frequency ω is kept constant. Here, the impedance bridge can be adjusted for measuring either real or imaginary parts of the cavity impedance by balancing with a complex reference impedance Z_{ref}. For such an experiment with sweeping B_o, $\Delta\omega$ in (20.20) can be replaced by $\Delta B_o = \Delta\omega/\gamma$, and $\Delta\,\{\mathrm{Re}Z(B_o)\} \propto \omega_o\chi''(B_o)L_o \approx \delta\left(\frac{1}{Q}\right)_{res}$, and

$$\Delta\,\{\mathrm{Im}Z(B_o)\} \propto 2\chi'(B_o)\Delta B_o,$$

exhibiting curves of magnetic absorption and dispersion that are similar in shape to those shown in Figure 18.3.

21 Laser Oscillations, Phase Coherence, and Photons

21.1. Optical Resonators

Phase-coherent radiation at high frequencies is a useful *signal carrier* for telecommunication and guidance. While wave propagation at high frequencies depends on many factors; frequency range is usually selected for specific purposes. For example, u.h.f. waves ($10^8 \sim 10^9$ Hz) and microwaves ($10^9 \sim 10^{11}$ Hz) at moderate power levels are indispensable for practical applications because of the high modulating capabilities and directivity; yet, energy dissipation in air space is a serious problem at all frequencies. Optical waves could also be used in telecommunications, in principle, but their use for such purposes was not imaginable before 1958, when Schawlow, Townes, and Prokorov demonstrated that radiation from excited atoms and molecules could be incorporated with optical resonators. Originating from quantum transitions, optical radiation is normally *incoherent*, hence, it is not suitable for any purpose requiring coherent radiation. However, the successful experiments by Schawlow et al. on simultaneous emission from excited atoms and molecules opened an entirely new era of telecommunication.

As it relates to quantum transitions, coherent optical radiation is a subject beyond the scope of classical physics; but the concept is simple enough to accept by today's standards, unless deeper insight is sought in understanding its origin. Therefore, we can be free to discuss quantum transitions for optical waves—even using a semi-classical approach to this recent application.

The Maxwell theory is well established for classical electromagnetism; however, as applied to higher frequency phenomena in the infrared and visible wave ranges, we encounter a limitation that arises from radiation sources that is predominantly quantum in nature. It is important to realize that classical theory has a limitation due to microscopic interactions. Thus "laser" oscillators pose a technical problem for incoherency, related to uncertainity in observing the phase of radiation, a sign of quantum character of electromagnetic phenomena.

In this chapter we conclude our treatment of classical electromagnetism with a brief discussion of optical oscillators. Like a microwave cavity, an optical resonator could be fabricated to accommodate many wavelengths, in principle. For example, in an optical cavity of size 1 cm^3 there should be as many as 10^8 normal modes

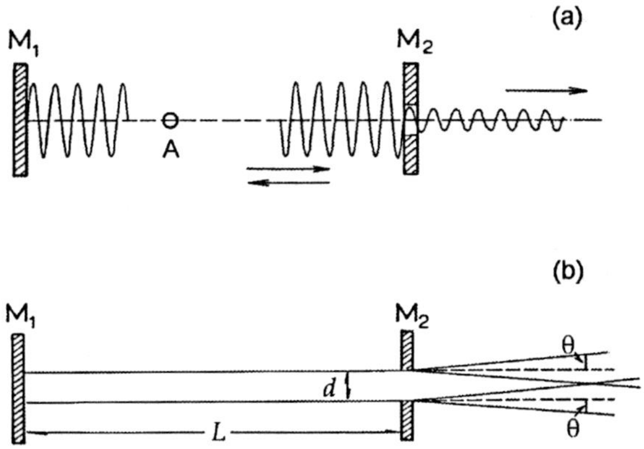

FIGURE 21.1. (a) A Fabry–Perot interferometer as an optical cavity. (b) An optical cavity with a radiation output.

for waves of $\lambda = 1\ \mu$m within a frequency width of $\Delta\omega \approx 1$ GHz. However, since the mode spectrum is virtually continuous, such a cavity cannot be tuned at a single frequency. On the other hand, for a light wave propagating strictly along one direction x, a standing wave signified by the wavevector $\pm k_x = \pm\, 2\pi/\lambda$ can be set up by placing two parallel mirrors in the $x$direction of propagation, which constitutes a specific normal mode with a large amplitude resulting from repeated reflections between the mirrors. Such a device is basically the same as a *Fabry–Perot interferometer* familiar in optics, and Schawlow and Townes used it as a feedback resonator in their first successful optical oscillator.

Figure 21.1(a) shows a typical Fabry-Perot resonator, where M_1 and M_2 are parallel mirrors that are fabricated in precision for an adjustable distance L. Between the mirrors a light-emitting gas in a glass container is installed, where atoms can be excited by electric discharge (not shown in the figure). The mirror M_2 has a semi-transparent window, through which radiation at resonance, i.e., $L = n\lambda/2$, emits with an intense amplitude.

Unlike a microwave cavity where the coupling window is approximately the size of λ, the optical wavelength is much shorter than the window size. Therefore, we apply *Huygens'principle* to deal with light *diffraction* at the optical output, leaving the detail to the theory of optics. Figure 21.1(b) describes the diffraction at the window in the resonator sketched in Figure 21.1(a), showing the angle of diffraction θ is given by the order of magnitude

$$\theta \sim \frac{\lambda}{d},$$

where d is the diameter of the window, assuming it is circular in shape. Numerically, for $d \sim 1$ cm and $\lambda \sim 1\ \mu$, such an angle is of the order of $\theta \sim 10^{-4}$ rad, which is not quite negligible but sufficiently small for practical use. Hence, the output light is regarded as nearly coherent and in good directivity, ignoring such a diffraction effect.

21.2. Quantum Transitions

Light emitting atoms and molecules must be treated quantum mechanically, although some related phenomena can be discussed within the classical theory. The most serious difficulty in the classical theory is that the electron orbit cannot remain stable, as energy is continuously lost by emitting radiation. In quantum theory, on the other hand, stability is assured by discrete energy levels of stationary orbits, while transitions between levels are restricted by so-called "selection rules." Therefore, such microscopic systems can release their radiation energy if they are in excited levels above the ground state; such natural radiation of a finite life is referred to as *spontaneous radiation*. On the other hand, those in the ground state can in no way lose energy, so remain stable.

If permitted by the selection rule, a transition occurs spontaneously between atomic levels E_j and E_i, where $E_j > E_i$, emitting radiation at a frequency ω_{ji} determined by

$$\omega_{ji} = \frac{1}{\hbar}\left(E_j - E_i\right), \tag{21.1}$$

where $\hbar = h/2\pi = 1.0 \times 10^{-34}$ J-sec is the Planck constant h divided by 2π. Spontaneous radiation can be interpreted classically as related to an oscillating electric dipole moment

$$p(t) = \frac{I_o}{i\omega_{ji}} \exp(i\omega_{ji}t), \quad \text{where} \quad p_o = \frac{I_o}{i\omega_{ji}},$$

representing an effective circular current in a loop antenna, as discussed in Section 13.3. We can calculate such radiation energy emitted during one complete circular motion as

$$\Delta E = \frac{cp_o{}^2k^4}{12\pi\varepsilon_o}\frac{2\pi}{\omega_{ji}} = \frac{p_o{}^2\omega_{ji}{}^3}{6\varepsilon_o c^3} \propto \left(\frac{\omega_{ji}}{c}\right)^3, \tag{21.2}$$

which is called the *spontaneous emission*, and $\tau = \dfrac{\hbar}{\Delta E}$ is the *lifetime* of the excited state E_j. Hence, the transition is not sharply observed, but broadened to $\Delta\omega \sim h/\tau$. It is noted that the probability for the transition (21.2) is proportional to $(\omega_{ji}/c)^3$, which is significant if ω_{ji} are optical frequencies or higher, whereas at radio- and microwave frequencies, such spontaneous emission probability is negligibly small.

In practice, those excited atoms cannot be quite independent in gaseous states, but interact with each other, so that spontaneous transitions do not occur normally at a sharp frequency, but are broadened at ω_{ji}. In addition to the *natural broadening* $\Delta\omega \sim h/\tau$ due to lifetime of the excite state, a line broadening arises not only from collisions with other atoms, known as the *pressure broadening*, but also from relative motion with respect to a point of observation, called *Doppler broadening*. Typically, the overall lifetime of light-emitting atoms is of the order of 10^{-9} s or longer, and the corresponding broadening 10^9 Hz, which is not quite negligible at optical frequency $\omega_{ji} = 10^{14} \sim 10^{15}$ Hz, but sufficiently small.

In addition to spontaneous radiation, transitions between energy levels can be induced by an electromagnetic field; these transitions are known as *induced* or *stimulated* transitions. In the presence of an electromagnetic field, the Hamiltonian of an orbiting electron in an H atom can be written as

$$\mathcal{H} = \frac{(\boldsymbol{p} + e\boldsymbol{A})^2}{2m} - \frac{e^2}{4\pi\varepsilon_o r} = H_o + \frac{e}{m}\boldsymbol{p}.\boldsymbol{A} + \frac{e^2\boldsymbol{A}^2}{2m}, \tag{21.3a}$$

where

$$\mathcal{H}_o = \frac{\boldsymbol{p}^2}{2m} - \frac{e^2}{4\pi\varepsilon_o r} \tag{21.3b}$$

describes the unperturbed orbit. Here, the vector potential $\boldsymbol{A}$ represents standing electromagnetic waves in an optical resonator and is expressed as

$$A(\boldsymbol{r}, t, \varphi) = 2A_o \cos(\boldsymbol{k} \cdot \boldsymbol{r} - \omega t + \varphi),$$

where $\boldsymbol{A}_o$ is the amplitude and φ the phase determined by the location of the interaction.

The Hamiltonian $\mathcal{H}_o$ is signified by characteristic energy levels E_i determined by the equation $\mathcal{H}_o\psi_i = E_i\psi_i$, where ψ_i is called the *eigenfunction* that describes motion of the electron in each state i. The last two terms in (21.3a) are perturbations due to the field $\boldsymbol{A}$, the last one giving rise to a diamagnetic induction effect, which was already discussed in Chapter 8. On the other hand, the other term of perturbation proportional to $\boldsymbol{p}\,.\boldsymbol{A}$ is essential for induced transitions between levels of H_o, giving rise to the selection rule.

Although properly unexplained in the framework of classical physics, the transition process may be interpreted intuitively. By definition, the momentum of the electron is given by $\boldsymbol{p} = m\mathrm{d}r/\mathrm{d}t$ where $\boldsymbol{r}$ is a radius of the orbit, and hence for the closed orbit the time average of $\boldsymbol{p}$ is zero, i.e., $\langle \boldsymbol{p} \rangle = 0$, where the $\langle\ \rangle$ are used to express the average. In quantum theory we consider that the perturbed state can be expressed by a perturbed function $\psi = \psi_i + \alpha\psi_j$, where α represents a small correction factor due to the interaction with another state j. And, the perturbation energy is expressed as

$$\int_v \frac{e}{m}\psi\,(\boldsymbol{p}\,.\boldsymbol{A})\,\psi^*\mathrm{d}v = \frac{\alpha e}{m}\int_v \psi_i(\boldsymbol{p}\,.\boldsymbol{A})\psi_j^*\mathrm{d}v$$

because $\psi_i(\boldsymbol{r}\,.\boldsymbol{A})\psi_i = \psi_j(\boldsymbol{r}\,.\boldsymbol{A})\psi_j = 0$, if we assume that orbits in states i and j are both centrally symmetric. Therefore, the interaction term $\psi_i(\boldsymbol{p}\,.\boldsymbol{A})\psi_j$, if non-zero, should play a significant role between the states E_i and E_j. Such a term, called the *off-diagonal element* ($i \neq j$) of an operator $\boldsymbol{p}\,.\,\boldsymbol{A}$, is considered to be responsible for transitions that are induced by the field $\boldsymbol{A}$.

With such a quantum concept, the transition probabilities can be calculated as follows: For the fast moving electron we express the momentum $\boldsymbol{p}$ by the time average $\langle \boldsymbol{p} \rangle$, and write

$$\frac{e}{m}\langle \boldsymbol{p}\rangle.\boldsymbol{A} = e\frac{\mathrm{d}\langle \boldsymbol{r}\rangle}{\mathrm{d}t}.\boldsymbol{A} = \frac{\partial(e\langle \boldsymbol{r}\rangle.\boldsymbol{A})}{\partial t} - e\langle \boldsymbol{r}\rangle.\frac{\partial \boldsymbol{A}}{\partial t},$$

from which we obtain the non-vanishing elements

$$\frac{e}{m}\langle \boldsymbol{p}\rangle_{\mathrm{ij}}.\boldsymbol{A} = -e\langle \boldsymbol{r}\rangle_{\mathrm{ij}}.\frac{\partial \boldsymbol{A}}{\partial \boldsymbol{t}} = -\boldsymbol{\mu}_{\mathrm{ij}}.\boldsymbol{E},$$

where $\boldsymbol{\mu}_{\mathrm{ij}} = -e\langle \boldsymbol{r}\rangle_{\mathrm{ij}}$ is the off-diagonal electric dipole element, which can be expressed as

$$\boldsymbol{\mu}_{\mathrm{ij}} = \boldsymbol{\mu}_{\mathrm{o}}\exp(i\omega_{\mathrm{ij}}t) \quad \text{where} \quad \omega_{\mathrm{ij}} = \frac{1}{\hbar}(E_{\mathrm{j}} - E_{\mathrm{i}}).$$

Writing that

$$\boldsymbol{A}(\boldsymbol{r}, t', \varphi) = \boldsymbol{A}_{\mathrm{o}}\exp\left[i(\boldsymbol{k}.\boldsymbol{r} - \omega t' + \varphi)\right] + \boldsymbol{A}_{\mathrm{o}}\exp\left[-i\left(\boldsymbol{k}.\boldsymbol{r} - \omega t' + \varphi\right)\right],$$

we have

$$\begin{aligned}-\mu_{\mathrm{ij}}(\boldsymbol{r}, t')E(\boldsymbol{r}, t', \varphi) = &-i\omega\mu_{\mathrm{o}}A_{\mathrm{o}}\exp\left[i(\boldsymbol{k}.\boldsymbol{r} + \omega t + \varphi)\right]\exp\left[i(\omega_{\mathrm{ij}} - \omega)(t - t')\right]\\ &-i\omega\mu_{\mathrm{o}}A_{\mathrm{o}}\exp\left[-i(\boldsymbol{k}.\boldsymbol{r} + \omega t + \varphi)\right]\exp\left[i(-\omega_{\mathrm{ij}} - \omega)(t - t')\right].\end{aligned}$$

Transitions $\mathrm{i} \rightarrow \mathrm{j}$ and $\mathrm{j} \rightarrow \mathrm{i}$ can take place at a time t after the perturbing field is switched on at t', whose probabilities for $t > t'$ can be determined by a square of the integral:

$$\begin{aligned}&-\left\langle \mu_{\mathrm{ij}}(\boldsymbol{r}, t)E(\boldsymbol{r}, t', \varphi)\right\rangle_{\mathrm{t}}\\ &= -i\omega\left\langle \mu_{\mathrm{o}}A_{\mathrm{o}}\exp\left[i(\boldsymbol{k}.\boldsymbol{r} + \omega t + \varphi)\right]\right\rangle_{\mathrm{ij}}\frac{2\pi}{\omega}\int_0^{t-t'}\left\{\exp\left[i(\omega_{\mathrm{ij}} - \omega)(t - t')\right]\right\}\mathrm{d}(t - t')\\ &\quad -i\omega\left\langle \mu_{\mathrm{o}}A_{\mathrm{o}}\exp\left[-i(\boldsymbol{k}.\boldsymbol{r} + \omega t + \varphi)\right]\right\rangle_{\mathrm{ji}}\frac{2\pi}{\omega}\int_0^{t-t'}\left\{\exp\left[i(-\omega_{\mathrm{ji}} - \omega)(t - t')\right]\right\}\mathrm{d}(t - t')\\ &= -2\pi i\left\langle \mu_{\mathrm{o}}A_{\mathrm{o}}\exp\left[i(\boldsymbol{k}.\boldsymbol{r} + \varphi)\right]\right\rangle_{\mathrm{ij}}\frac{\exp\left[i(\omega_{\mathrm{ij}} - \omega)t\right] - 1}{\omega_{\mathrm{ij}} - \omega}\\ &\quad -2\pi i\left\langle \mu_{\mathrm{o}}A_{\mathrm{o}}\exp\left[-i(\boldsymbol{k}.\boldsymbol{r} + \varphi)\right]\right\rangle_{\mathrm{ji}}\frac{\exp\left[i(-\omega_{\mathrm{ji}} - \omega)t\right] - 1}{-\omega_{\mathrm{ji}} - \omega}.\end{aligned}$$

It is noted that the above quantities indicate maxima at $\omega = \omega_{\mathrm{ij}}$ and $\omega = -\omega_{\mathrm{ji}}$, corresponding to induced *absorption* and *emission*, for which the probabilities are given as proportional to

$$\langle \mu_{\mathrm{o}}\rangle_{\mathrm{ij}}{}^{2}A_{\mathrm{o}}{}^{2}\frac{\sin^2\dfrac{(\omega_{\mathrm{ij}} \mp \omega)t}{2}}{(\omega_{\mathrm{ij}} \mp \omega)^2},$$

which are equal for both absorption and emission at $\omega = \mp\omega_{ij}$, and proportional to $A_o{}^2$ and $\langle\mu_o\rangle_{ij}{}^2$. In such an interpretation the electromagnetic field can be assumed as consisting of energy quanta $\hbar\omega_{ij}$ that are absorbed or emitted between atomic energy levels, as $E_i \rightleftarrows E_j$, obeying the energy conservation law.

21.3. Inverted Population and the Negative Temperature

Einstein (1917) proposed the relation between probabilities for induced transitions and spontaneous emission, giving the thermodynamic foundation to Planck's theory of blackbody radiation. His formula can deal with the light emission and absorption in a body in thermal equilibrium with its surroundings.

Considering a system of two energies E_i and E_j in a body in equilibrium at temperature T, these levels are occupied by N_i and N_j atoms determined by the Boltzmann statistics, i.e.,

$$\frac{N_i}{N_j} = \exp\left(-\frac{E_i - E_j}{k_B T}\right), \tag{21.4}$$

where $k_B = 1.38066 \times 10^{-23}$J-K^{-1} is the Boltzmann constant, signifying that always $N_i < N_j$ in equilibrium if $E_i > E_j$. However, if the atoms can absorb light quanta $\hbar\omega_{ij}$ at a rate sufficiently faster than the relaxation rate, a condition for $N_i > N_j$ can be achieved at which the temperature takes a negative value according to (21.4), i.e., $T < 0$.

When absorbing radiation quanta, the population N_i at the upper level increases, while the population N_j at the lower level decreases. For such a *pumping* process that may be in equilibrium with the radiation field, Einstein gave the formula for the radiation power,

$$P(\omega_{ij}) = \hbar\omega_{ij}\left[N_i A_{ij} + (N_i - N_j)B_{ij}\,\rho(\omega_{ij})\right], \tag{21.5}$$

where A_{ij} and B_{ij} are called the *Einstein coefficients* for spontaneous and induced emissions, respectively, and $\rho(\omega_{ij})$ the radiation energy density. Denoting the excited and ground states by i and j as in the above, we have $N_i \ll N_j$ at a normal temperature, so that the second negative term dominates the power $P(\omega_{ij})$ in (21.5). Therefore, such a system can absorb radiation quanta under normal circumstances, where the first term for emission is significant if the value of $\rho(\omega_{ij})$ is small. On the other hand, when $\rho(\omega_{ij})$ is large the second term for induced transitions becomes significant. At radio- and microwave frequencies, the spontaneous transition is negligible, and only induced transitions are predominant for the *pumping* process. However, at optical frequencies the factor $\rho(\omega_{ij})$ is not negligible, and hence the pumping rate A_{ij} should be sufficiently high to achieve a condition for $N_i > N_j$.

In any case, the condition for *inverted population* is necessary for simultaneous emissions from the excited states, from which coherent radiation can be stimulated by an oscillating electric field of a resonator that is incorporated with an atomic system. Referring to the Boltzmann formula (21.4), such a state of inverted populations can be characterized by a *negative temperature* $-T$. Nevertheless, as the pumping is an *adiabatic* process, such an equilibrium with radiation is not thermodynamically isothermal equilibrium. Although not a thermodynamic temperature, the negative temperature is often used as a convenient parameter to specify an inverted population.

21.4. Ammonium Maser

Gordon, Zeiger, and Townes (1954) performed the first ammonium *maser* experiment, which demonstrated inverted populations that occurred between two levels of NH_3 molecules at microwave frequencies. Here, "maser" is an acronym used by these authors for "microwave amplification by stimulated emission of radiation."

In vibrational spectra of NH_3 molecules there is an absorption line at about 24 GHz known as the *inversion line*, as explained in the following. An NH_3 molecule of a *pyramidal structure* is characterized by inversion symmetry with respect to the position of the N atom related to the triangular plane of H_3, as illustrated in Figure 21.2(a). The motion between inverted structures is quantum-mechanical tunnelling between the two minima of a potential $V(x)$ as shown in Figure 21.2(b), where x is a position of the N atom relative to the center of the H_3 plane. According to the quantum theory, such degenerate inversion energies are lifted by the tunnelling motion, resulting in two levels separated by an energy gap Δ, characterized by symmetric and anti-symmetric modes. Under a normal circumstance in NH_3 gas, these levels, denoted by 1 and 2, are populated as in (21.4) by the Boltzmann statistics, i.e., $\frac{N_1}{N_2} = \exp(-\frac{\Delta}{k_B T})$, hence the transition between them can be observed in the absorption spectrum.

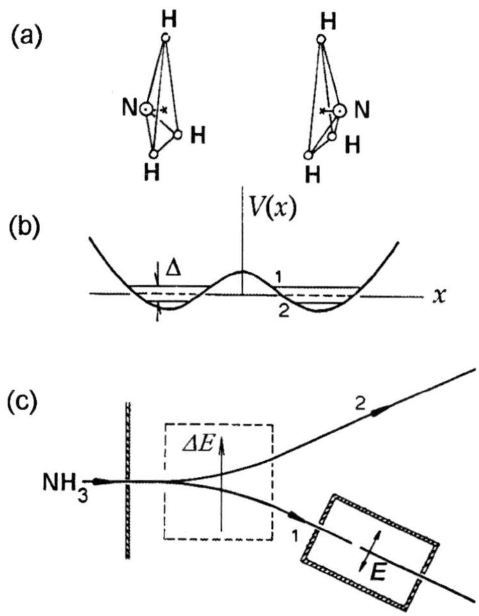

FIGURE 21.2. (a) Inversion of NH_3 molecules. (b) Inversion splitting of the degenerate NH_3 energy level. D is the splitting between two inversion levels 1 and 2.

Gordon et al. were able to create a condition for inverted population between the inversion levels, resulting in a stimulated emission $N_1 \rightarrow N_2$ in a microwave cavity. Figure 21.2(c) sketches their apparatus for observing emission line. A collimated beam of NH_3 is injected into the area indicated by dotted lines, where the molecules of different inversion energies are separated by Δ. Therefore, the molecules of energies at $\pm\frac{1}{2}\Delta$ are separated into two beams by an interaction between the dipole moment μ and the field gradient $\mathrm{d}E$in the square region in Figure 21.2(c), where the beam at the higher level 1 is shown as separated from those at the lower level 2. Thus, ammonium molecules in beam 1 have inverted populations, and consequently the *inversion line* is dominated as an emission line induced coherently by the microwave field $\boldsymbol{E}$ in the cavity.

21.5. Coherent Light Emission from a Gas Laser

Similar to maser, the word "laser" is an acronym for "light amplification by stimulated emission of radiation." Processes for population inversion depend on materials, occurring not only in gases but in solid materials as well, both of which are in practical use for laser oscillators constructed for a variety of purposes.

The prediction for inverted populations and the first successful laser were made on mixed gases. A laser action can be characterized in practice in systems characterized by three energy levels, as illustrated schematically in Figure 21.3(a). Here, the population N_3 at level 3 can be enhanced at a frequency ω_{31} by pumping up atoms from N_1; on the other hand, N_3 will then be reduced continuously via a stimulated emission to level 2 when $N_3 > N_2$, although level 2 is initially unpopulated. Such a level as 2 should essentially be of *metastable* character, where $N_2$is kept close to an unpopulated condition; the transition $2 \rightarrow 1$ should be either spectroscopically *forbidden*, or there may be another mechanism in the system to remove N_2 down to the ground level.

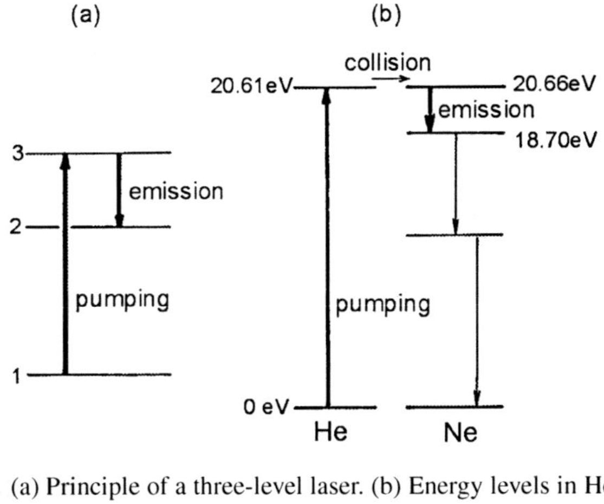

FIGURE 21.3. (a) Principle of a three-level laser. (b) Energy levels in He-Ne gas laser.

Figure 21.3(b) shows the energy level diagram for a He-Ne gas laser, consisting of these atoms in 90:10 percentage ratio. By applying a high frequency discharge, He atoms are pumped up primarily to the excited level at 20.61 eV above the ground state. Excited He atoms collide then with Ne atoms, resulting in excited Ne ($2p^5 5s^1$) at a metastable state. These excited Ne atoms emit red light by induced transitions, falling down to a lower level at 18.70 eV, as shown in the diagram. As induced transitions are practically simultaneous, such a red emission light is nearly coherent at the wavelength $\lambda = 623.8$ nm. Stepwise spontaneous transitions from the level at 18.70 eV can maintain the inverted population ratio between the 20.66 eV and 18.70 eV levels for Ne atoms.

By simultaneous pumping by high frequency discharge the subsequent induced emission can be made practically coherent, although diffraction effects are inevitable at the output of the instrument in use.

21.6. Phase Coherence and Radiation Quanta

Dynamical properties of plane waves were demonstrated by radiation pressure experiments. As they exert a force on a charged particle, electromagnetic waves can be equivalent to n particles of zero mass and momentum p. Here p is linearly related to the kinetic energy K, i.e., $K = cp$, according to the special theory of relativity. Quantum theoretically, electromagnetic waves confined to a finite volume consist of n radiation quanta $\hbar\omega$, hence their energy per volume is given by $K = n\hbar\omega$, where n is the density of quanta. On the other hand, according to the Maxwell theory each wave is characterized by a phase of propagation.

For electromagnetic waves originating from an oscillating charge-current, we generally consider that the wavelength is long compared with the size of the source, so no significant diffracted waves are considered for the radiation. In contrast, for waves at optical or higher frequencies, radiation is characteristically incoherent, as it originates from randomly distributed emitters. Optical radiation from a laser oscillator is practically coherent with a well-tuned resonator; however at the present level of technology it is not an easy task to construct lasers at ultraviolet and even shorter wavelengths since phase incoherency becomes a more serious problem.

Nevertheless, noting that the classical Maxwell theory has limitations at higher frequencies, we should realize that in radiation from microscopic sources, phases are essentially at random, and hence the Poynting vector should be expressed as

$$\boldsymbol{P} = c\boldsymbol{u}\langle U\rangle, \quad \text{where} \quad U = \varepsilon_o\left(\sum_i \boldsymbol{E}_i\right)\cdot\left(\sum_j \boldsymbol{E}_j\right)^*. \tag{21.6}$$

Here

$$\boldsymbol{E}_i = \boldsymbol{E}_{oi}\exp(i\phi_i),$$

where $\phi_i = \boldsymbol{k}\cdot\boldsymbol{r}_i - \omega t_i + \varphi_\alpha$ is the phase of radiation, in which the shift φ_α signifies the phase of emission from a microscopic source α in random distribution. Here, in reality, $\boldsymbol{P}$ and $\langle U \rangle$ should be subjected not only to the timescale of observation but also to distributed light sources. At any rate, the energy U in (21.6) can be written as

$$U = \varepsilon_o \sum_i \boldsymbol{E}_{oi} \cdot \boldsymbol{E}_{oi}{}^* + \varepsilon_o \sum_{i \neq j} \boldsymbol{E}_{oi} \cdot \boldsymbol{E}_{oj}{}^* \exp(i\Delta\phi_{ij}), \tag{21.7}$$

where $\Delta\phi_{ij} = \phi_i - \phi_j$ are phase differences between waves from different emitters, signifying incoherence of radiation. The first term on the right of (21.7) represents quantum energy $\langle U_o \rangle = n_o \hbar\omega$ for $\Delta\phi_{ij} = 0$, where n_o is the number of coherent photons, and the second term is the energy deviation $\langle U \rangle - \langle U_o \rangle$ due to $\Delta\phi_{ij} \neq 0$, indicating incoherence. Hence, we can write

$$\langle U \rangle - \langle U_o \rangle = (n - n_o)\hbar\omega = \varepsilon_o \sum_{i \neq j} \boldsymbol{E}_{oi}{}^2 \langle \exp(i\Delta\phi_{ij}) \rangle. \tag{21.8}$$

Assuming that such $\Delta\phi_{ij}$ are continuously distributed in a range between $-\Delta\phi$ and $+\Delta\phi$, the average $\langle \exp(i\Delta\phi_{ij}) \rangle$ in (21.8) can be calculated as

$$\langle \exp(i\Delta\phi_{ij}) \rangle = \frac{1}{2\Delta\phi} \int_{-\Delta\phi}^{+\Delta\phi} \exp(i\Delta\phi) d(\Delta\phi) = \frac{\cos\Delta\phi}{\Delta\phi} \approx \frac{1}{\Delta\phi}$$

for a small $\Delta\phi$. Therefore, we have

$$\varepsilon_o \langle \boldsymbol{E}_i \cdot \boldsymbol{E}_j \rangle_{i \neq j} = \frac{n\hbar\omega}{\Delta\phi}. \tag{21.9}$$

Writing $n - n_o = \Delta n$ to indicate *uncertainty* of the photon number, from (21.8) and (21.9) we obtain the relation

$$\Delta n = n/\Delta\phi \qquad \text{or} \qquad \frac{\Delta n}{n} \Delta\phi = 1. \tag{21.10}$$

This is an *uncertainty relation* between the phase and the number of photons. Photons are a legitimate quantum-theoretical model of electromagnetic radiation, which is particularly useful at very high frequencies, where the intensity can be measured by *counting photons*, though there is accompanying unavoidable uncertainty of Δn related with incoherent phases. In contrast, at low frequency, electromagnetic waves are coherent with a well-defined phase, i.e., $\Delta\phi \sim 0$, over which a large uncertainty of $\Delta n/n$ prevails.

Appendix

Mathematical Notes

A.1. Orthogonal Vector Space

In a rectangular coordinate system O-xyz, a vector $\boldsymbol{A} = (A_x, A_y, A_z)$ can be expressed as

$$\boldsymbol{A} = A_x\,\mathbf{i} + A_y\,\mathbf{j} + A_z\,\mathbf{k},$$

where the unit vectors $\mathbf{i}, \mathbf{j}$, and $\mathbf{k}$ along their respective x, y, and z axes are orthogonal to each other, as related by

$$\mathbf{i.i} = \mathbf{j.j} = \mathbf{k.k} = 1 \quad \text{and} \quad \mathbf{j.k} = \mathbf{k.i} = \mathbf{i.j} = 0.$$

In Chapter 14 such orthonormal relations were extended to the four dimensional space-time system by adding the fourth coordinate $x_4 = ict$ to ordinary three-dimensional space coordinates $x_1 = x$, $x_2 = y$, and $x_3 = z$. Thus, a position in such a four-dimensional "space" can be specified by a four-vector, which can be represented by a row matrix $\langle \boldsymbol{x}| = (x_1, x_2, x_3, x_4)$ or a column matrix

$$|\boldsymbol{x}\rangle = \begin{pmatrix} x_1 \\ x_2 \\ x_3 \\ x_4 \end{pmatrix}.$$

Defining unit vectors by

$$\langle \boldsymbol{u}_1| = (1, 0, 0, 0), \quad \langle \boldsymbol{u}_2| = (0, 1, 0, 0), \quad \langle \boldsymbol{u}_3| = (0, 0, 1, 0), \quad \langle \boldsymbol{u}_4| = (0, 0, 0, 1)$$

and

$$|\boldsymbol{u}_1\rangle = \begin{pmatrix} 1 \\ 0 \\ 0 \\ 0 \end{pmatrix}, |\boldsymbol{u}_2\rangle = \begin{pmatrix} 0 \\ 1 \\ 0 \\ 0 \end{pmatrix}, |\boldsymbol{u}_3\rangle = \begin{pmatrix} 0 \\ 0 \\ 1 \\ 0 \end{pmatrix}, |\boldsymbol{u}_4\rangle = \begin{pmatrix} 0 \\ 0 \\ 0 \\ 1 \end{pmatrix},$$

we can write that

$$\langle \boldsymbol{x}| = \sum_{\mathrm{i}} x_{\mathrm{i}}\,\langle \boldsymbol{u}_i| \text{ and } |\boldsymbol{x}\rangle = \sum_{\mathrm{i}} x_{\mathrm{i}}\,|\boldsymbol{u}_{\mathrm{i}}\rangle \quad \text{and} \quad \langle \boldsymbol{u}_{\mathrm{i}}|\,\boldsymbol{u}_{\mathrm{j}}\rangle = \delta_{\mathrm{ij}},$$

where $\delta_{ij} = 1$ if $i = j$, but 0 if $i \neq j$, representing orthonormality in four-dimensional space.

These relations can be extended formally to an n-dimensional mathematical space, or even to infinite dimensions. If obeying a linear differential equation, the physical quantity can often be required to expand in a multi-dimensional space composed of linearly independent solutions. For example, for a static problem a potential function can generally be expanded in a multi-dimensional space of the Legendre functions $P_n(\cos\theta)$ that are solutions of the Laplace equation. A superposition of elementary waves is another feature of a linear wave equation whose independent solutions are $\exp(in\phi)$, where ϕ is the phase variable, and n an integer. For such expansions it is essential that these elementary solutions be signified by their orthonormal relations.

A.2. Orthogonality of Legendre's Polynomials

A potential function $V(\boldsymbol{r})$ is determined by solutions of the Laplace equation $\nabla^2 V = 0$. In axial symmetry along the z-axis, the potential can be expressed as a function of radius r and polar angle θ; that is,

$$V(r,\theta) = \sum_n \left(A_n r^n + \frac{B_n}{r^{n+1}}\right) P_n(\cos\theta),$$

where the coefficients A_n and B_n are to be determined by given boundary conditions. Applying this expression to the potential problem of a sphere of radius a in a uniform field, we consider that

$$V_<(r,\theta) = \sum_n A_n r^n P_n(\cos\theta) \quad \text{and} \quad V_>(r,\theta) = \sum_n \frac{B_n P_n(\cos\theta)}{r^{n+1}}$$

represent potentials inside the sphere ($r < a$) and outside the sphere ($r > a$), respectively. On the spherical boundary ($r = a$) the potential should take a unique value (per the uniqueness theorem), so that

$$V_<(a,\theta) = V_>(a,\theta), \quad \text{and} \quad \varepsilon_1 \left(\frac{\partial V_<}{\partial r}\right)_{r=a} = \varepsilon_2 \left(\frac{\partial V_>}{\partial r}\right)_{r=a},$$

from which A_n and B_n can be determined for each index n.

At a distant point ($r \gg a$) the angular function $P_n(\cos\theta)$ satisfies the Legendre equation

$$\frac{d}{d\mu}\left\{\left(1-\mu^2\right)\frac{dP_n(\mu)}{d\mu}\right\} + n(n+1)P_n(\mu) = 0, \tag{i}$$

where $\mu = \cos\theta$. Replacing n by $-(n+1)$, we can obtain the same equation for $P_{-(n+1)}$, so that,

$$P_n(\mu) = P_{-n-1}(\mu). \tag{ii}$$

For these Legendre functions we show here the orthogonality relation

$$\int_{-1}^{+1} P_n(\mu)\, P_{n'}(\mu)\, d\mu = 0 \quad \text{for} \quad n \neq n'. \tag{iii}$$

Proof: Writing the Legendre equation for $P_{n'}(\mu)$, we have

$$\frac{d}{d\mu}\left\{(1-\mu^2)\frac{dP_{n'}(\mu)}{d\mu}\right\} + n'(n'+1)\, P_{n'}(\mu) = 0. \tag{iv}$$

Calculating (i) $\times P_{n'}-$ (iv) $\times P_n$,

$$\frac{d}{d\mu}(1-\mu^2)\left(P_{n'}\frac{dP_n}{dm} - P_n\frac{dP_{n'}}{d\mu}\right) = \left[n'(n'+1) - n(n+1)\right] P_n P_{n'}.$$

Integrating this with respect to μ from -1 to $+1$,

$$(1-\mu^2)\left(P_{n'}\frac{dP_n}{d\mu} - P_n\frac{dP_{n'}}{d\mu}\right) = \left[n'(n'+1) - n(n+1)\right]\int_{-1}^{+1} P_n P_{n'} d\mu.$$

In order for this expression to be independent of μ, the left side should vanish. Therefore, the integral on the right should be zero if $n \neq n'$. Q.E.D.

The function $P_n(\mu)$ can be normalized as follows. We derived the formula (5.7),

$$(2n+1)\,\mu P_n(\mu) = (n+1)\, P_{n+1}(\mu) + n P_{n-1}(\mu).$$

Replacing n by n – 1,

$$(2n-1)\,\mu P_{n-1}(\mu) = n P_n(\mu) + (n-1)\, P_{n-2}(\mu).$$

Hence,

$$\int_{-1}^{+1} \{P_n(\mu)\}^2 d\mu = \int_{-1}^{+1} P_n(\mu)\frac{(2n-1)\,\mu P_{n-1}(\mu) - (n-1)\, P_{n-2}(\mu)}{n} d\mu$$

$$= \frac{2n-1}{n}\int_{-1}^{+1} \mu P_n(\mu)\, P_{n-1}(\mu)\, d\mu.$$

Using the recurrence formula once more, we can replace $\mu P_n(\mu)$ in the last integrand by

$$\mu P_n(\mu) = \frac{n P_{n-1}(\mu) + (n+1)\, P_{n+1}(\mu)}{2n+1}$$

and obtain

$$\int_{-1}^{+1} \{P_n(\mu)\}^2 d\mu = \frac{2n-1}{2n+1}\int_{-1}^{+1} \{P_{n-1}(\mu)\}^2 d\mu.$$

Repeating the same procedure, reducing n stepwise down to n = 0, we have

$$\int_{-1}^{+1} \{P_n(\mu)\}^2\, d\mu = \frac{2n-1}{2n+1} \times \frac{2n-3}{2n-1} \times \cdots \times \frac{3}{5} \times \frac{1}{3} \times \int_{-1}^{+1} \{P_o(\mu)\}^2\, d\mu = \frac{2}{2n+1}.$$

Thus, the orthonormal relation for $P_n(\mu)$ can be expressed as

$$\int_{-1}^{+1} P_n(\mu)\, P_{n'}(\mu)\, d\mu = \frac{2}{2n+1}\delta_{n,n'}. \qquad \text{(v)}$$

A.3. Associated Legendre Polynomials

In the absence of uniaxial symmetry, the potential function can be expressed with polar coordinates r, θ, φ. Expressing that $V(r, \theta, \varphi = R(r)\Theta(\theta)\Phi(\varphi)$, the Laplace equation can be separated into ordinary differential equations for these factors $R(r)$, $\Theta(\theta)$, and $\Phi(\varphi)$. Among these, the equation for $\Phi(\varphi)$ for a simple harmonic oscillator is

$$\frac{d^2\Phi}{d\varphi^2} + m^2\Phi = 0,$$

where m^2 is a constant for variable separation, and $m = 0, 1, \pm 2, \ldots$.

Specifically, m = 0 corresponds to an axial symmetric case, and all other $m \neq 0$ express non-axial symmetry. In this case, the function $\Theta(\theta)$ is given as solutions of the equation

$$\frac{d}{d\mu}\left\{\left(1-\mu^2\right)\frac{d\Theta}{d\mu}\right\} + \left\{n(n+1) - \frac{m^2}{1-\mu^2}\right\}\Theta = 0.$$

However, to avoid singularities at $\mu = \pm 1$ or at $\theta = 0, \pi$, it is known that $n \geq |m|$ for integers n and m, for which the function Θ can be expressed by associated Legendre functions defined by

$$P_n^m(\mu) = (-1)^m \left(1-\mu^2\right)^{\frac{m}{2}} \frac{d^m P_n(\mu)}{d\mu^m} \quad \text{for} \quad m \geq 0. \qquad \text{(vi)}$$

We therefore differentiate the Legendre equation

$$(1-\mu^2)\frac{d^2 P_n(\mu)}{d\mu^2} - 2\mu\frac{dP_n(\mu)}{d\mu} + n(n+1)\,P_n(\mu) = 0$$

m times, yielding that

$$\left(1-\mu^2\right)\frac{d^{m+2} P_n(\mu)}{d\mu^{m+2}} - 2\mu(m+1)\frac{d^{m+1} P_n(\mu)}{d\mu^{m+1}} - m(m+1)\frac{d^m P_n(\mu)}{d\mu^m} + n(n+1)\frac{d^m P_n(\mu)}{d\mu^m} = 0.$$

From (vi) we have $\dfrac{d^m P_n(\mu)}{d\mu^m} = \dfrac{(-1)^m}{\left(1-\mu^2\right)^{\frac{m}{2}}} P_n^m(\mu)$, which can be used in the above to obtain the equation for $P_n^m(\mu)$, that is,

$$\left(1-\mu^2\right)\frac{d^2}{d\mu^2}\left[\frac{P_n^m(\mu)}{\left(1-\mu^2\right)^{\frac{m}{2}}}\right] - 2(m+1)\mu\frac{d}{d\mu}\left[\frac{P_n^m(\mu)}{\left(1-\mu^2\right)^{\frac{m}{2}}}\right] + [n(n+1) - m(m+1)]\frac{P_n^m(\mu)}{\left(1-\mu^2\right)^{\frac{m}{2}}} = 0.$$

After differentiations, we obtain the equation for $P_n^m(\mu)$

$$\left(1-\mu^2\right)\frac{d^2 P_n^m(\mu)}{d\mu^2} - 2\mu\frac{dP_n^m(\mu)}{d\mu} + \left\{n(n+1) - \frac{m^2}{1-\mu^2}\right\} P_n^m(\mu) = 0. \tag{vii}$$

The orthogonality relation for associated Legendre polynomials can be derived in a process similar to the way it was done for Legendre polynomials $P_n(\mu)$. We obtain the orthogonality

$$\int_{-1}^{+1} P_n^m(\mu)\, P_{n'}^m(\mu)\, d\mu = 0 \quad \text{for} \quad n \neq n'. \tag{viii}$$

The functions $P_n^m(\mu)$ can be normalized as follows. First, we differentiate (iv) and obtain the recurrence relation

$$\left(1-\mu^2\right)^{\frac{1}{2}}\frac{dP_n^m(\mu)}{d\mu} = -P_n^{m+1}(\mu) - m\mu\left(1-\mu^2\right)^{-\frac{1}{2}} P_n^m(\mu).$$

Using this formula, we then calculate

$$\int_{-1}^{+1}\left\{P_n^{m+1}(\mu)\right\}^2 d\mu = \int_{-1}^{+1}\left\{\left(1-\mu^2\right)^{\frac{1}{2}}\frac{dP_n^m(\mu)}{d\mu} + m\mu\left(1-\mu^2\right)^{-\frac{1}{2}} P_n^m(\mu)\right\}^2 d\mu,$$

which is expanded and integrated to obtain the relation

$$\int_{-1}^{+1}\left\{P_n^{m+1}(\mu)\right\}^2 d\mu = (n+m+1)(n-m)\int_{-1}^{+1}\left\{P_n^m(\mu)\right\}^2 d\mu.$$

Using this result for reducing m to 0 in succession, we can finally express that

$$\int_{-1}^{+1}\left\{P_n^m(\mu)\right\}^2 d\mu = (n+m)(n-m+1)(n+m-1)(n-m+2)\ldots(n+1) \times n\int_{-1}^{+1}\left\{P_n^0(\mu)\right\}^2 d\mu = \frac{2}{2n+1}\frac{(n+m)!}{(n-m)!}.$$

Thus, the orthonormality of associated Legendre functions can be written as

$$\int_{-1}^{+1} P_n^m(\mu)\, P_{n'}^m(\mu)\, d\mu = \frac{2}{2n+1}\frac{(n+m)!}{(n-m)!}\delta_{n,n'}. \tag{ix}$$

The *spherical harmonics* are defined as $\Theta(\theta)\Phi(\varphi)$; i.e.,

$$Y_n^m(\theta,\varphi) = \sqrt{\frac{2n+1}{4\pi}\frac{(n-m)!}{(n+m)!}}\, P_n^m(\cos\theta)\exp(im\varphi), \tag{x}$$

for which the orthogonality relation is

$$\int_0^{2\pi} d\varphi \int_0^{\pi} Y_n^m(\theta,\varphi)\, Y_{n'}^{m'}(\theta,\varphi)^* \sin\theta d\theta = \delta_{n\cdot n'}\delta_{m\cdot m'}, \tag{xi}$$

where

$$Y_{n'}^{m'}(\theta,\varphi)^* = (-1)^{m'}\, Y_{n'}^{-m'}(\theta,\varphi). \tag{xii}$$

A.4. Fourier Expansion and Wave Equations

A periodic function repetitive in space-time coordinates can be specified by a phase variable $\phi = kx - \omega t$ in a one-dimensional case, where $|x| < \infty$, $|t| < \infty$, and the constant $\omega/k = v$ represents the speed of propagation. Such a function $f(\phi)$ is given as a solution of the wave equation

$$\frac{\partial^2 f}{\partial t^2} - v^2\frac{\partial^2 f}{\partial x^2} = 0. \tag{xiii}$$

Depending on properties of the medium, if characterized by fixed values of k and ω, or v, the solutions are given by a sinusoidal function $\exp(i\phi)$ and its power $[\exp(i\phi)]^n = \exp(in\phi)$, or their linear combinations. Functions $\varphi_n(\phi)$, such as $\exp(in\phi)$, constitute a mathematical space of n dimensions, or infinite dimensions, if the functions converge as $n \to \infty$, and we can express that

$$f(x,t) = \sum_n c_n\varphi_n(\phi) = \sum_n c_n \exp(in\phi),$$

which is called the *Fourier series*. It is noted that such an expansion is characterized by the orthogonal relation, thereby allowing us to calculate the coefficient c_m, for example, as

$$c_m = \int_0^{2\pi} \exp(-im\phi)\, f(\phi)\, d\phi = \sum_n c_n \int_0^{2\pi} \exp[i(n-m)\phi]\, d\phi,$$

which is equal to c_m if n = m; otherwise, $c_m = 0$ for n $\neq$ m. In this case the coefficients are normalized as

$$\sum_n c_n c_n^* = 2\pi.$$

On the other hand, if the medium is *dispersive*, as characterized by distributed k and ω, the Fourier expansion can generally be written as

$$f(x,t) = \int_\omega \int_k g(k,\omega) \exp[i(kx - \omega t)]\, dk\, d\omega,$$

where

$$g(k,\omega) = \int_t \int_x f(x,t) \exp[-i(kx - \omega t)]\, dx\, dt$$

is called the *Fourier transform* of $f(x,t)$. If values of k and ω are limited in narrow ranges around k_o and ω_o, i.e., $k_o - \Delta k < k < k_o + \Delta k$ and $\omega_o - \Delta\omega < \omega < \omega_o + \Delta\omega$, such integrals as the ones above are said to be *wave packets*.

Example 1. As a typical example of the Fourier method, we briefly discuss the problem of transversal vibration of an elastic string of length L supported at both ends. In this case the standing wave can be expressed by

$$f(x,t) = \sum_n \sin(k_n x)\{A_n \cos(\omega_n t) + B_n \sin(\omega_n t)\},$$

where the coefficients A_n and B_n can be determined by the initial condition for oscillation at a given point x_o, e.g., values of f and $\partial f/\partial t$ at $t = 0$ at this point. The boundary condition at the fixed ends $x = 0$ and L are already taken into account for the spatial variation sin($k_n x$), where $k_n L = n\pi$. Hence, the initial shape of the string is given by

$$f(x,0) = \sum_n A_n \sin(k_n x) \quad \text{and} \quad \frac{\partial f(x,0)}{\partial t} = \sum_n B_n \omega_n \sin(k_n x),$$

for which we can confirm the orthonormal relations

$$\int_0^L \sin\frac{m\pi x}{L} \sin\frac{n\pi x}{L} dx = 0 \quad \text{for} \quad m \neq n, \quad \text{and} \quad \int_0^L \sin^2\frac{n\pi x}{L} dx = \frac{L}{2}.$$

Hence, the Fourier coefficients are

$$A_n = \int_0^L f(x,0) \sin\frac{n\pi x}{L} dx \quad \text{and} \quad B_n = 0.$$

Further, to illustrate the method, we consider that the string is initially triangular in shape, where the center point $x = L/2$ is displaced by a. In this case,

$$f(x,0) = \frac{2a}{L}x \quad \text{for} \quad 0 \le x \le \frac{L}{2},$$
$$f(x,0) = \frac{2a}{L}(L-x) \quad \text{for} \quad \frac{L}{2} \le x \le L.$$

We can calculate the Fourier coefficients as

$$A_{\mathrm{n}} = \frac{4a}{L}\left\{\int_0^{L/2} \frac{x}{L}\sin\frac{\mathrm{n}\pi x}{L}\mathrm{d}x + \int_{L/2}^{L}\left(1-\frac{x}{L}\right)\sin\frac{\mathrm{n}\pi x}{L}\mathrm{d}x\right\} = \frac{8a}{L^2}\int_0^{L/2} x\sin\frac{\mathrm{n}\pi x}{L}\mathrm{d}x$$
$$= \frac{8a}{\pi^2}\frac{(-1)^{\frac{n-1}{2}}}{\mathrm{n}^2}, \text{ where n} = 1, 3, 5, \ldots.$$

Hence, the triangular string can be expressed in series as

$$f(x,0) = \frac{8a}{\pi^2}\left(\frac{1}{1^2}\sin\frac{\pi x}{L} + \frac{1}{3^2}\sin\frac{3\pi x}{L} + \cdots\right).$$

Since this is identical to $f\left(\frac{L}{2}, 0\right) = a$, we obtain the formula

$$\frac{\pi^2}{8} = \frac{1}{1^2} + \frac{1}{3^2} + \frac{1}{5^2} + \cdots$$

A.5. Bessel's Functions

In Chapter 19 we discussed a velocity modulation resulting in a bunched electronic current in a klystron oscillator. A modulated current by a resonator's frequency ω_1 was expressed by

$$I(t) = I_o + \sum_n \{A_n \cos[n(\omega_1 t - \theta)] + B_n \sin[n(\omega_1 t - \theta)]\},$$

and here we express these amplitudes A_n and B_n in a complex form. Defining a parameter $\beta = \frac{\theta}{2}\frac{\Delta V}{V_o}$, we showed that $\omega_1 t - \theta = \omega_1 t_o - \beta\cos(\omega_1 t_o)$. Therefore, writing $\omega_1 t_o = \varphi$, we can now define

$$A_n + iB_n = \frac{I_o}{\pi}\int_0^{2\pi} \exp[in(\varphi - \beta\cos\varphi)]\,\mathrm{d}\varphi.$$

Here, we define the function

$$i^n J_n(n\beta) = \frac{1}{2\pi}\int_0^{2\pi} \exp\{in(\varphi - \beta\cos\varphi)\}\,\mathrm{d}\varphi,$$

and

$$\exp(-in\beta\cos\varphi) = \sum_{m=n} J_m(m\beta)\, i^m \exp(-im\varphi) = \sum_{m=n} J_m(m\beta)\,\zeta^m,$$

where $\zeta = i\exp(-i\varphi)$. Using ζ and letting $n\beta = \xi$, the left side of the above can be expressed as $\exp\left\{-\frac{\xi}{2}\left(\zeta - \frac{1}{\zeta}\right)\right\}$, and hence we obtain

$$\exp\left\{-\frac{\xi}{2}\left(\zeta - \frac{1}{\zeta}\right)\right\} = \sum_n J_n(\xi)\,\zeta^n. \tag{xiv}$$

The functions $J_n(\xi)$ in the above are defined with a modulated phase, but it is shown in the following that these functions satisfy the Bessel equation, and hence are called the *Bessel functions*.

Proof: Differentiating (xiv) with respect to ζ, we obtain

$$\left(-\frac{\xi}{2}\right)\left(1 + \frac{1}{\zeta^2}\right)\sum_n J_n(\xi)\,\zeta^n = \sum_n nJ_n(\xi)\,\zeta^{n-1}. \tag{xv}$$

Differentiating with respect to ξ, we have

$$-\frac{1}{2}\left(\zeta - \frac{1}{\zeta}\right)\sum_n J_n(\xi)\,\zeta^n = \sum_n J_n'(\xi)\,\zeta^n. \tag{xvi}$$

Combining (xv) and (xvi),

$$-\xi\sum_n J_n(\xi)\,\zeta^n = \sum_n \left\{nJ_n(\xi) + \xi J_n'(\xi)\right\}\zeta^{n-1}$$

and

$$-\frac{\xi}{2}\sum_n J_n(\xi)\,\zeta^{n-2} = \sum_n \left\{nJ_n(\xi) - \xi J_n'(\xi)\right\}\zeta^{n-1}.$$

Comparing terms of the same power on both sides, from these relations we obtain

$$-\xi J_{n-1}(\xi) = nJ_n(\xi) + \xi J_n'(\xi) \quad \text{and} \quad -\xi J_{n+1}(\xi) = nJ_n(\xi) - \xi J_n'(\xi). \tag{xvii}$$

Next, we differentiate the first relation in (xvii), then replacing $J_n'(\xi)$ by using the second,

$$\begin{aligned}\frac{d}{d\xi}\left[\xi J_n'(\xi)\right] &= nJ_n'(\xi) + \frac{d}{d\xi}\left[\xi J_{n+1}(\xi)\right]\\ &= \frac{n}{\xi}\{nJ_n(\xi) + \xi J_{n+1}(\xi)\} + J_{n+1}(\xi) + \xi J_{n+1}'(\xi)\\ &= \frac{n^2}{\xi}J_n(\xi) + (n+1)\,J_{n+1}(\xi) + \xi J_{n+1}'(\xi).\end{aligned}$$

According to (xvii), the last two terms in the above can be replaced by $-\xi J_n(\xi)$, and we obtain

$$\frac{\mathrm{d}}{\mathrm{d}\xi}\left\{\xi J_n'(\xi)\right\} = \frac{n^2}{\xi} J_n(\xi) - \xi J_n(\xi),$$

which can be written in the standard form of the Bessel equation, i.e.,

$$\xi \frac{\mathrm{d}}{\mathrm{d}\xi}\left\{\xi \frac{\mathrm{d}J_n(\xi)}{\mathrm{d}\xi}\right\} + \left(\xi^2 - n^2\right) J_n(\xi) = 0. \qquad \text{(xviii)}$$

References

Corson, D.R., and Lorrain, P. (1970). *Introduction to electromagnetic fields and waves.* San Francisco: Freeman.

Flügge, S. (1986). *Rechenmethoden der Elektrodynamik.* Berlin-Heidelberg: Springer-Verlag.

Halliday, D., and Resnick, R. (1974). *Fundamentals of physics.* New York: John Wiley & Sons.

Hauser, W. (1971). *Introduction to the principles of electromagnetism.* Reading, MA: Addison-Wesley.

Jackson, J.D. (1962). *Classical electrodynamics.* New York: John Wiley & Sons.

Landau, L.D., and Lifshitz, E.M. (1962). *The classical theory of fields.* Reading, MA: Addison-Wesley.

Møller, C. (1952). *The theory of relativity.* London: Oxford University Press.

Panofsky, W.K.H., and Phillips, M. (1962). *Classical electricity and magnetism.* Reading, MA: Addison-Wesley.

Pohl, R.W. (1975). *Elektrizitätslehre*, 21st ed. Berlin-Heidelberg: Springer-Verlag.

Reitz, J.R., and Milford, F.J. (1967). *Foundations of electromagnetic theory.* Reading, MA: Addison-Wesley.

Slater, J.C. (1950). *Microwave electronics.* New York: Van Nostrand.

Slater, J.C., and Frank, N.H. (1947). *Electromagnetism.* New York: McGraw-Hill.

Smith, W.V., and Sorokin, P.P. (1966). *The laser.* New York: McGraw-Hill.

Index